Glossary of symbols

A User's original matrix.

$\mathbf{A}^{(k)}$ Reduced matrix (of order $n{\times}n$) before step k of Gaussian elimination.

$\mathbf{A}^{[k]}$ Matrix (order $n{\times}n$) associated with k-th finite element.

$\mathbf{A}_{i\bullet}$ Row i of the matrix **A**.

$\mathbf{A}_{\bullet j}$ Column j of the matrix **A**.

B,C,... Other matrices.

b Right-hand side vector.

c Intermediate solution vector.

c_k Column count (number of entries in column k of matrix).

D Diagonal matrix.

$\mathbf{e}_i$ i-th column of **I**.

I Identity matrix.

L Lower triangular matrix, usually formed by triangular factorization (Gaussian elimination) of **A**.

L\U LU factorization of a matrix packed into a single array.

n Order of the matrix **A**.

$O(\ )$ If $f(n)/g(n){\to}k$ as $n{\to}\infty$, where k is a constant, then $f(n) = O(g(n))$.

P Permutation matrix (usually applied to matrix rows).

p_j Number of entries in column j of **U**.

Q Orthogonal matrix (usually a permutation matrix).

r Residual vector $\mathbf{r} = \mathbf{b} - \mathbf{Ax}$.

r_k Row count (number of entries in row k of matrix).

U Upper triangular matrix, usually a factor of **A**.

u Threshold for numerical pivoting.

$\mathbf{v}_i$ i-th eigenvector of **A**.

x Solution vector.

ε Relative precision.

$\kappa(\mathbf{A})$ Condition number of **A**.

λ_i i-th eigenvalue of **A**.

ρ Largest element in any reduced matrix, that is $\max\limits_{i,j,k}|a_{ij}^{(k)}|$.

τ Number of entries in the matrix **A**.

$\in$ Belongs to.

MONOGRAPHS ON NUMERICAL ANALYSIS

General Editors

G. Dahlquist   L. Fox
K. W. Morton   B. Parlett
J. Walsh

MONOGRAPHS ON NUMERICAL ANALYSIS

An introduction to numerical linear algebra   L. Fox

The algebraic eigenvalue problem   J. H. Wilkinson

The numerical treatment of integral equations   Christopher T. H. Baker

Numerical functional analysis   Colin W. Cryer

Direct methods for sparse matrices   I. S. Duff, A. M. Erisman, and J. K. Reid

Direct Methods for Sparse Matrices

I. S. Duff

Group Leader, Numerical Analysis,
Computer Science and Systems Division,
Harwell Laboratory, Oxfordshire

A. M. Erisman

Director, Engineering Technology Applications Division,
Boeing Computer Services, Seattle

J. K. Reid

Senior Research Scientist,
Computer Science and Systems Division,
Harwell Laboratory, Oxfordshire

CLARENDON PRESS · OXFORD

Oxford University Press, Walton Street, Oxford OX2 6DP

Oxford New York Toronto
Delhi Bombay Calcutta Madras Karachi
Petaling Jaya Singapore Hong Kong Tokyo
Nairobi Dar es Salaam Cape Town
Melbourne Auckland
and associated companies in
Berlin Ibadan

Oxford is a trade mark of Oxford University Press

Published in the United States
by Oxford University Press, New York

First published 1986
First published in paperback (with corrections) 1989

British Library Cataloguing in Publication Data
Duff, Iain S.
Direct methods for sparse matrices.—
(Monographs on numerical analysis)
1. Sparse matrices—Data processing
I. Title   II. Erisman, Albert Maurice
III. Reid, John K.   IV. Series
512.9′434   QA188
ISBN 0-19-853408-6
ISBN 0-19-853421-3 (pbk)

Printed in Great Britain by
St Edmundsbury Press Ltd, Bury St Edmunds, Suffolk

Preface

The subject of sparse matrices has its roots in such diverse fields as management science, power systems analysis, surveying, circuit theory, and structural analysis. Mathematical models in all of these areas give rise to very large systems of linear equations that could not be solved were it not for the fact that the matrices contain relatively few nonzeros. Only comparatively recently, in the last fifteen years or so, has it become apparent that the equations can be solved even when the pattern is irregular, and it is primarily the solution of such problems that we consider.

The subject is intensely practical and we have written this book with practicalities ever in mind. Whenever two methods are applicable we have considered their relative merits, basing our conclusions on practical experience in cases where a theoretical comparison is not possible. We hope that the reader with a specific problem may get some real help in solving it. Non-numeric computing techniques have been included, as well as frequent illustrations, in an attempt to bridge the usually wide gap between the printed page and the working computer code. Despite this practical bias, we believe that many aspects of the subject are of interest in their own right and we have aimed for the book to be suitable also as a basis for a student course, probably at M.Sc. level. Exercises have been included to illustrate and strengthen understanding of the material, as well as to extend it.

We have aimed to make modest demands on the mathematical expertise of the reader, but familiarity with elementary linear algebra is almost certainly needed. Similarly, modest computing background is expected, though familiarity with Fortran is helpful. The computational tools required to handle sparsity are explained in Chapter 2. We considered making the assumption of familiarity with the numerical analysis of full matrices but felt that this might provide a barrier for an unreasonably large number of potential readers and, in any case, have found that we could summarize the results required in this area without lengthening the book unreasonably. This summary constitutes Chapters 3 and 4. The reader with a background in computer science will find Chapter 2 straightforward, while the reader with a background in numerical analysis will find the material in Chapters 3 and 4 familiar.

Sparsity is considered in earnest in the remainder of the book. In Chapter 5, Gaussian elimination for sparse matrices is introduced, and the utility of standard software packages to realize the potential saving is emphasized. We begin our detailed discussion of sparsity in Chapter 6 by considering the preliminary step of reducing the matrix to block triangular form, which allows the problem to be solved as a sequence of smaller subproblems. An in-depth look at pivotal strategies which are designed to preserve sparsity is made in Chapter 7 and we consider the a priori ordering of sparse matrices to desirable forms in Chapter 8. The implementation of these techniques for both symmetric and unsymmetric matrices is the subject of Chapters 9 and 10. In Chapter 11 we describe related techniques for solving a sequence of problems that differ by low-rank perturbations and for solving huge problems by adding artificial perturbations. These perturbations produce easy-to-solve problems whose solutions may then be corrected to produce the required solutions. Finally, in Chapter 12, we consider other exploitations of sparsity, including using a known sparsity pattern while approximating Jacobian and Hessian matrices by finite differences, computation of selected entries of the inverse of a sparse matrix, and the re-examination of backward error analysis in the light of sparsity.

We use bold face in the body of the text when we are defining terms and we use bold italic face when we wish to give emphasis.

For the purposes of illustration, we include fragments of Fortran code conforming to the ANSI Fortran 77 standard, and some exercises ask the reader to write such code fragments. Where a more informal approach is needed, we use syntax loosely based on Algol 60, with ':=' meaning assignment.

Throughout the book we present timing data drawn from a variety of computers (including IBM 370/168, IBM 3033, IBM 3081K, CYBER 175, CRAY-1, and CRAY X-MP) to illustrate various points. This diversity of computers is due to the different papers, environments, and time periods from which our results are drawn. It does not cause a difficulty of interpretation because it is the relative performance within a particular area that is compared; absolute numbers are less meaningful.

We often refer to code by the string MAxy (for example MA37). This nomenclature is that of the Harwell Subroutine Library, where MA is the prefix of the set of routines for solving linear sets of equations. When it is important, however, we give a full reference to a publication that describes the code or its design.

Four appendices are included; they cover matrix norms, the LINPACK condition number estimate, the names of Fortran variables used in illustrations and examples, and pictures of some of the sparse matrices from various applications that we have used for test comparisons. These appendices precede solutions to selected exercises, the collected set of references, an author index, and a subject index.

We find it convenient to estimate operation counts and storage requirements by the term that dominates for large problems. We will use the symbol O for this. For instance, if a certain process needs $\frac{1}{3}n^3 - \frac{1}{6}n(n+1)$ multiplications we might write this as $\frac{1}{3}n^3 + O(n^2)$ or $O(n^3)$.

Efficient use of sparsity is a key to solving large problems in many fields. We hope that this book will supply both insight and answers for those attempting to solve these problems.

Acknowledgements

The authors wish to acknowledge the support of many individuals and institutions in the development of this book. The international co-authorship has been logistically challenging. Both AERE Harwell and Boeing Computer Services (our employers) have supported the work through several exchange visits. Other institutions have been the sites of extended visits by the authors, including the Australian National University and Argonne National Laboratory (ISD), Carnegie-Mellon University (AME), and the Technical University of Denmark (JKR).

The book was typeset at Harwell on a Linotron 202 typesetter using the TSSD (Typesetting System for Scientific Documents) package written by Mike Hopper. We wish to thank Harwell for supporting us in this way, the staff of the Harwell Reprographic Section for their rapid service, Mike Hopper for answering so many of our queries, and Rosemary Rosier for copying successive drafts for us to check.

Oxford University Press has been very supportive (and patient) over the years. We would like to thank the staff involved for the encouragement and help that they have given us, and for their rapid response to queries.

Many friends and colleagues have read and commented on chapters. We are particularly grateful to the editors, Leslie Fox and Joan Walsh, for going far beyond their expected duties in reading and commenting on the book. Others who have commented include Pat Gaffney, Ian Gladwell, Nick Gould, Nick Higham, John Lewis, and Jorge Moré.

Finally, we wish to thank our supportive families who accepted our time away from them, even when at home, during this lengthy project. Thanks to Diana, Catriona, and Hamish Duff; Nancy, Mike, Andy, and Amy Erisman; and Alison, Martin, Tom, and Pippa Reid.

Harwell, Oxon, England and   I. S. D.
Seattle, USA,   A. M. E.
May, 1986.   J. K. R.

Contents

1 Introduction

The use of graph theory to 'visualize' the relationship between sparsity patterns and Gaussian elimination is introduced. The potential of significant savings from the exploitation of sparsity is illustrated by one example. The effect of computer hardware on 'efficient computation' is discussed. Realization of sparsity means more than faster solutions; it affects the formulation of mathematical models and the feasibility of solving them.

1.1 Introduction

A matrix is sparse if many of its coefficients are zero. The interest in sparsity arises because its exploitation can lead to enormous computational savings and because many large matrix problems that occur in practice are sparse. How much of the matrix must be zero for it to be considered sparse depends on the computation to be performed, the pattern of the nonzeros, and even the architecture of the computer. Generally, we say that a matrix is **sparse** if there is an advantage in exploiting its zeros.

For practical reasons, we do not necessarily exploit all the zeros. Particularly in the intermediate calculations, it may be better to store some zeros explicitly. We therefore use the term **entry** to refer to those coefficients that are handled explicitly. All nonzeros are entries and some zero coefficients may also be entries. We label each of our test matrices with its number of nonzeros and assume that the number of entries is the same.

This book is primarily concerned with direct methods for solving sparse systems of linear equations, though other operations with sparse matrices are also discussed. The significant benefit from sparsity does not come from the cost reductions, but rather from the fact that problems that were hitherto infeasible can now be solved. They may have tens or even hundreds of thousands of equations.

Often it is possible to gain insight into sparse matrix techniques by working with the graph associated with the matrix, and this is considered in Section 1.2. There is a well-defined relationship between the pattern of the entries of a square sparse matrix and its associated graph. Furthermore, results from graph theory sometimes provide answers to questions associated with algorithms for sparse matrices. We introduce this topic here in order to be able to use it later in the book.

To illustrate the potential saving from exploiting sparsity, we consider a small example in Section 1.3. Without going into detail, which is the subject of the rest of the book, we use this example to motivate the study.

When serial computers were our only concern, there was a near-linear relationship between the number of **floating-point operations** (additions, subtractions, multiplications, and divisions) and the run time of the program. Thus a program requiring 600 000 operations was about 20 per cent more expensive than one requiring 500 000 operations unless there were unusual overheads. In such an environment, exploitation of sparsity meant reducing floating-point computation while keeping the overheads in proportion. Vector pipeline computers have introduced additional complications. It is necessary to consider the number of vector operations, since each involves a 'start-up' time equal to that of several (sometimes many) floating-point operations. Parallel architectures have added another level of difficulty for comparisons between algorithms. A brief discussion of vector and parallel architectures is contained in Section 1.4.

In Section 1.5 we discuss the formulation of mathematical models from the viewpoint of the exploitation of sparsity. If sparsity tools are available, it is useful to apply these tools in a straightforward manner within an existing formulation. However, going back to the model may make it possible to reformulate the problem so that more is achieved. This section is intended only to stimulate thinking along these lines since a full discussion is beyond the scope of this book. Finally in Section 1.6 we explain the need for a collection of sparse matrix test problems and tabulate those used in this book.

1.2 Graph theory

Matrix sparsity and graph theory are subjects that can be closely linked. The pattern of a square sparse matrix can be represented by a graph, for example, and then results from graph theory can be used to obtain sparse matrix results. George and Liu (1981), among others, do this in their book. Graph theory is also a subject in its own right, and detailed treatment is given by König (1950), Harary (1969), and Read (1972), for example.

In this book we use graph theory mainly as a tool to visualize what is happening in sparse matrix computation. As a result, we use only limited results from graph theory and make no assumption of knowledge of the subject by the reader. In this section we introduce the basic concepts. Other concepts are introduced as they are used, for example in Chapter 6.

A **directed graph** or **digraph** consists of a set of **nodes** (also called **vertices**) and directed **edges** between nodes. Any square sparse matrix pattern has an associated digraph, and any digraph has an associated square sparse matrix pattern. For a given square sparse matrix **A**, a node is associated with each row. If a_{ij} is an entry, there is an edge from node i to node j in the directed graph. This is usually written diagrammatically as a line with an arrow, as illustrated in Figure 1.2.1. For example, the line from node 1 to node 2 in the digraph corresponds to the entry a_{12} of the matrix. The more general representation of the directed graph includes self-loops on nodes corresponding to diagonal entries (see, for example, Varga 1962, p. 19), but we do not find self-loops helpful and therefore do not include them.

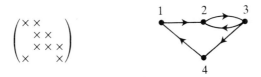

Figure 1.2.1. An unsymmetric matrix and its digraph.

For a symmetric matrix a connection from node i to node j implies that there must also be a connection from node j to node i; therefore the arrows may be dropped, and we obtain an **undirected graph** or **graph,** as illustrated in Figure 1.2.2.

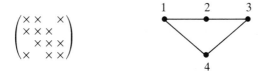

Figure 1.2.2. A symmetric matrix and its graph.

Formally, $G(\mathbf{A})$, the digraph associated with the matrix $\mathbf{A}$, is not a picture but is a set X of nodes and a set E of edges. An edge is an ordered pair of nodes (x_i, x_j) and is associated with the matrix entry a_{ij}.

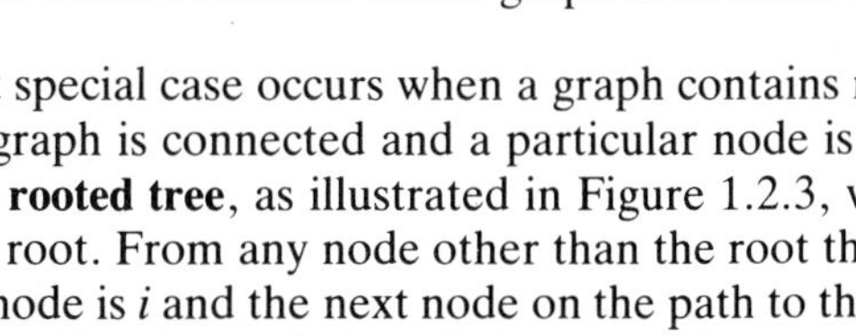

Figure 1.2.3. A matrix whose graph is a rooted tree.

An important special case occurs when a graph contains no closed paths (cycles). If the graph is connected and a particular node is labelled as the **root,** we have a **rooted tree,** as illustrated in Figure 1.2.3, where the node labelled 5 is the root. From any node other than the root there is a path to the root. If the node is i and the next node on the path to the root is j, then i is called the **son** of j and j is called the **father** of i. It is conventional to draw trees with the root at the top and with all the sons of a node at the same height. A node without a son is called a **leaf**. (These terms are standard in the literature and no sexist connotations are intended; 'parent' and 'child' would be unsuitable because a child has two parents.)

A fundamental operation used in solving equations with matrices involves adding multiples of one row, say the first, to other rows to make all entries in the first column below the diagonal equal to zero. This process is illustrated in the next section. Detailed discussion of the algorithm, which is called Gaussian elimination, is found in Chapters 3 and 4. Notice that when this process is applied to the matrix of Figure 1.2.1, adding a multiple of the first row to the fourth creates a new entry in position (4,2).

Graph theory helps in visualizing the changing pattern of entries as elimination takes place. Corresponding to the graph G, the elimination digraph G_y for node y is obtained by removing node y and adding a new edge (x,z) whenever (x,y) and (y,z) are edges of G but (x,z) is not. For example, G_1 for the digraph of Figure 1.2.1 would have the representation shown in Figure 1.2.4, with the new edge (4,2) added. Observe that this is precisely the digraph corresponding to the 3×3 submatrix that results from the elimination of the (4,1) entry by the row transformation discussed in the last paragraph. In the case of a symmetric matrix whose graph is a tree, no extra edges are introduced when a leaf node is eliminated. The corresponding elimination operations introduce no new entries into the matrix.

Figure 1.2.4. The graph G_1 for the graph of Figure 1.2.1.

This relationship between graph reduction and Gaussian elimination was first discussed by Parter (1961). It is most often used in connection with symmetric matrices since a symmetric permutation of a symmetric matrix leaves its graph unchanged except for the numbering of its nodes. For an unsymmetric matrix it is often necessary to interchange rows without interchanging the corresponding columns. This leads to a different digraph, and the correlation between the digraph and Gaussian elimination is not so apparent (Rose and Tarjan 1978). If symmetric permutations are made, the digraph remains unchanged apart from the node numbering.

1.3 Example

In the design of safety features in motor cars and aeroplanes, the dynamics of the human body in a crash environment have been studied with the aim of reducing injuries. In initial studies the simple stick figure model of a person illustrated in Figure 1.3.1 was used.

The dynamics are modelled by a set of time-dependent differential

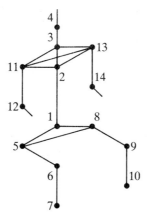

Figure 1.3.1. Stick figure modelling a person.

equations with which we are not concerned here. The body segments are connected at joints (the nodes of the graph), as illustrated in Figure 1.3.1. These segments may not move independently since the position of the end of one segment must match the position of the end of the connecting segment. This leads to 42 equations of constraint (three for each numbered joint) which must be added to the mathematical system to yield a system of algebraic and differential equations to be solved numerically. At each time step of the numerical solution (typically there will be thousands of these) a set of 42 linear algebraic equations must be solved for the reactions at the joints.

Since there are 14 numbered joints, with three constraints for each joint, the matrix representing the algebraic equations may be considered as a 14×14 block matrix with entries that are 3×3 submatrices. This is the pattern shown in Figure 1.3.2, where each $\times$ represents a full 3×3 submatrix. The pattern of this block matrix may be developed from Figure 1.3.1 by associating the given joint numbers with block numbers. For example, since joint 6 is connected by body segments to joints 5 and 7, there is a relationship between the corresponding reactions, and block row 6 has entries in block columns 5, 6, and 7.

Referring to the discussion of the last section on the relationship between sparse matrices and graph theory, note that the stick person shown in Figure 1.3.1 corresponds to the graph of the matrix in Figure 1.3.2. From the graph it is apparent that the use of node 1 in the first elimination stage introduces new edges (2,5) and (2,8). If we begin at the head (4), hands (12,14), or feet (7,10), the elimination of any of these nodes introduces no new edges.

In the remainder of this section we demonstrate that a careful utilization

	1	2	3	4	5	6	7	8	9	10	11	12	13	14
1	×	×			×			×						
2	×	×	×								×		×	
3		×	×	×							×		×	
4			×	×										
5	×					×	×		×					
6						×	×	×						
7							×	×						
8	×					×			×	×	×			
9									×	×	×			
10										×	×			
11		×	×								×	×	×	
12											×	×		
13		×	×								×		×	×
14													×	×

Figure 1.3.2. The pattern of the matrix associated with the stick man of Figure 1.3.1.

of the structure of a matrix as in Figure 1.3.2 leads to a significant saving in the computational cost of the solution of the algebraic equations. Since these equations must be solved at each time step, with the numerical values changing from step to step while the sparsity pattern remains fixed, cost savings can become significant over the whole problem. In Section 3.12 the relationship between block and single variable equations is established. Here it is sufficient to say that an efficient solution of linear equations whose coefficient matrix has the sparsity pattern of Figure 1.3.2 can readily be adapted to generate an efficient solution of the block equations. We therefore examine the solution of the 14 simultaneous equations whose sparsity pattern is shown in Figure 1.3.2.

Various methods for solving systems of n linear equations

$$\sum_{j=1}^{n} a_{ij} x_j = b_i, \quad i = 1, 2, \ldots, n, \tag{1.3.1}$$

or, in matrix notation,

$$\mathbf{A}\mathbf{x} = \mathbf{b}, \tag{1.3.2}$$

are discussed and compared in Chapters 3 and 4. For the purposes of this example, where $n = 14$, we simply solve the equations by a sequence of row operations. Multiples of equation 1 are first added to the other equations to eliminate their dependence on x_1. This leaves a revised system of equations of which only the first involves x_1. If no special account is taken of the zeros, these calculations will require 29 floating-point operations for the elimination of x_1 from each of 13 equations, making 29×13 operations in all. Then multiples of the second equation are added to the later equations to eliminate their dependence on x_2. This requires 27 operations for each of 12 equations, making 27×12 operations in total. This process is continued until the fourteenth equation depends only on x_{14} and may be solved directly. The total number of operations to perform this transformation (including operations on zeros) is

$$29 \times 13 + 27 \times 12 + 25 \times 11 + \cdots + 5 \times 1 = 1911. \qquad (1.3.3)$$

The solution of the resulting trivial fourteenth equation requires only one division. Then the thirteenth equation may be solved for x_{13}, using the computed x_{14} (requiring three operations). Similarly x_{12}, x_{11},..., x_1 are computed in turn using the previously found components of $\mathbf{x}$. Thus the calculation of the solution from the transformed equations requires

$$1 + 3 + \cdots + 27 = 196 \qquad (1.3.4)$$

operations. The process of eliminating variables from each equation in turn and then solving for the components in reverse order is called **Gaussian elimination**. Its variants are discussed further in Chapter 3.

Following the same approach, but operating only on the entries, requires far less work. For instance in the first step, multiples of equation 1 need only be added to equations 2, 5, and 8 since none of the others contain x_1. Furthermore, after a multiple has been calculated, only 4 multiplications (including one for the right-hand side) are required to multiply the first equation by it. Therefore 3 divisions, 12 multiplications, and 12 additions are needed to eliminate x_1. Note, however, that when a multiple of equation 1 is added to equation 2, new entries are created in positions 5 and 8. Figure 1.3.3 shows by black squares all the new entries (**fill-ins**) that are created in the complete process. The elimination of x_2 from each of the five other equations in which it now appears requires 1 division and 7 multiplications and causes fill-ins to equations 3, 5, 8, 11, and 13. The number of floating-point operations needed for the complete elimination is

$$2 \times (13 \times 5) + 3 \times (11 \times 4) + 4 \times (9 \times 3) + 2 \times (7 \times 2) + 2 \times (5 \times 1) = 408. \quad (1.3.5)$$

The corresponding number of operations required to solve the transformed equations is

$$7 + 2 \times 11 + 3 \times 9 + 3 \times 7 + 2 \times 5 + 2 \times 3 + 1 = 94. \qquad (1.3.6)$$

Thus, utilizing the sparsity in the solution process reduces the required number of operations from 2107 to 502, a factor of more than four. This becomes particularly significant when the saving is repeated thousands of times. To realize these savings in practice requires a specially adapted algorithm that operates only on the entries. For a large matrix it is also desirable to store only the entries. The development of data structures and special algorithms for achieving these savings is the major topic of this book.

Before leaving this example we ask another natural question. Can the equations be reordered to permit their solution with even fewer

```
        1  2  3  4  5  6  7  8  9 10 11 12 13 14
  1     ×  ×        ×        ×
  2     ×  ×  ×        ■        ■        ×     ×
  3        ×  ×  ×  ■        ■        ×     ×
  4           ×  ×  ■        ■        ■     ■
  5     ×  ■  ■  ■  ×  ×     ×        ■     ■
  6                 ×  ×  ×  ■        ■     ■
  7                    ×  ×  ■        ■     ■
  8     ×  ■  ■  ■  ×  ■  ■  ×  ×     ■     ■
  9                       ×  ×  ×  ■        ■
 10                          ×  ×  ■        ■
 11        ×  ×  ■  ■  ■  ■  ■  ■  ■  ×  ×  ×
 12                                ×  ×  ■
 13        ×  ×  ■  ■  ■  ■  ■  ■  ■  ×  ■  ×  ×
 14                                   ×  ×
```

Figure 1.3.3. The pattern of the matrix and its fill-ins.
Fill-ins are marked ■.

```
        4  7  6 10  9 12 14  5  8  1  2  3 11 13
  4     ×                             ×
  7        ×  ×
  6        ×  ×           ×
 10              ×  ×
  9              ×  ×     ×
 12                    ×              ×
 14                       ×              ×
  5     ×                 ×  ×  ×
  8           ×           ×  ×  ×
  1                       ×  ×  ×  ×
  2                       ×  ×  ×  ×  ×
  3     ×                 ×  ×  ×  ×
 11                    ×        ×  ×  ×  ×
 13                       ×     ×  ×  ×  ×
```

Figure 1.3.4. The pattern of the reordered matrix.

operations? One of the principal ordering strategies that we discuss in Chapter 7, when applied to the symmetric matrix of Figure 1.3.2, results in the reordered pattern of Figure 1.3.4. Here the original numbering is indicated by the numbering on the rows and columns. The symmetry of the matrix is preserved because rows and columns are reordered in the same way.

If we carry through the transformation process on the matrix of Figure 1.3.4, we observe that no new entries are created. Using cost formulae as before, the number of operations required is 105. Solving the transformed equations adds another 48 operations. The results of the three approaches are summarized in Table 1.3.1. The gain is very dramatic on large practical problems where the order of the matrix may be in the tens of thousands.

	Treating matrix as full	Unordered sparse	Ordered sparse
Transformation cost	1911	408	105
Solution cost	196	94	48

Table 1.3.1. Numbers of operations for the different methods.

Several comments on this reordering are appropriate. First, it produced an ordering where no new entries were generated; this *is not typical*. Second, the ordering is optimal for this problem (in the sense that another ordering could not produce fewer new entries) which also *is not typical*, because finding an optimal ordering is very expensive for a genuine problem. Third, that this ordering is based on a practical approach and is able to produce a significant saving *is typical*.

In this discussion the symmetry of the matrix has been ignored (apart from the use of a symmetric reordering). In Chapter 3 we introduce the Choleski method, which exploits the symmetry and approximately halves the work for the elimination operations on the matrix. Also we did not consider the effects of computer roundoff in this illustration. These effects are discussed in Chapter 4.

This example gives an indication that exploiting sparsity can produce dramatic computational savings and that reordering can also be significant in reducing the computation. Methods of reordering sparse equations to preserve sparsity are the major topic of Chapters 7 and 8. The order of magnitude cost reduction reflected in Table 1.3.1 for this example *is not unusual*, and indeed even greater gains can be obtained on large problems.

1.4 Advanced computer architectures

Until the early 1970s most computers were strictly serial in nature. That is, one arithmetic operation at a time ran to completion before the next commenced. Although some machines allowed operands to be prefetched (fetched before they are needed) and allowed some overlapping of instructions, the execution time of a program implementing a numerical algorithm could be well approximated by the formula

$$\text{time} = K \times (\text{number of operations}), \qquad (1.4.1)$$

where K is the average time per operation. The work in moving data between memory and functional units was either low or proportional to the computation time for the arithmetic. In recent years, however, many machines have been built for which this simple model of computation is inadequate.

These new computer architectures employ one or more degrees of parallelism. Designing computers with parallel architectures is seen as the only means of achieving significantly greater computing speeds. For example, the speed of light imposes a bound on communication time and thus limits serial computation speeds. To avoid such restrictions, it is necessary to allow the simultaneous execution of two or more instructions. This can be done in several ways. For example, arithmetic operations may be segmented into several (usually four or five) distinct phases and functional units designed so that different operands can be in different segments of the same operation at the same time. This technique, called **pipelining**, is employed by machines commonly called vector processors, since pipelining is particularly useful when performing calculations with vectors. On such machines, n operations in vector mode will be very much faster than n identical scalar operations, and so the computational model (1.4.1) must be adjusted to account for this. The model often used (see, for example, Hockney and Jesshope 1981) is that, at each clock cycle, all data in an arithmetic pipe move to the next segment so that one result is delivered every clock cycle, once the pipe is full.

The model for vector computation can thus be written

$$\text{time} = (s+n)c \text{ secs,} \qquad (1.4.2)$$

where s is the number of segments, n the length of the vector, and c the clock cycle time in seconds. The time sc before any result is produced is usually called the **start-up time**, and s is sometimes termed the $n_{\frac{1}{2}}$ value (Hockney and Jesshope 1981). This is because the rate of computation (measured in floating-point operations per second) attains half of its peak performance of $1/c$ (sometimes denoted by r_∞) at a vector length of s (see Exercise 1.1). With the $n_{\frac{1}{2}}$ and r_∞ notation, formula (1.4.2) may be written as

$$\text{time} = (n+n_{\frac{1}{2}})/r_\infty \text{ secs.} \qquad (1.4.3)$$

Although pipelining was originally employed in machines normally thought of as being serial, the idea was refined in the CRAY-1 (first delivery 1976) to permit an order of magnitude speed-up for operations on vectors whose components are evenly spaced in storage. Pipelining is also employed by the CRAY X-MP, by the Japanese machines (for example, the Fujitsu FACOM VP/400 and the Hitachi S-820), and by the CYBER 205 of Control Data. More recently (1985) many cheaper (although slightly less powerful) machines employing pipelining have been produced. These include the CONVEX C-1, the FPS 264, and the FPS 164/MAX.

Other new architectures employ different kinds of parallelism (see, for example, Dongarra and Duff 1987). Some merely duplicate the functional units allowing several results per clock cycle. An example is the Fujitsu FACOM VP/400 and, at a much higher degree of parallelism, the ICL DAP. Others allow quite independent operations to proceed simultaneously, for example the CRAY X-MP and the Denelcor HEP. Again, smaller machines employing MIMD (multiple instruction, multiple data) architecture are now available, for example ELXSI and ALLIANT and the hypercubes of NCUBE and INTEL. In such cases, the simple models (1.4.1) and (1.4.2) are inadequate, though generalizations may be possible to take account of the additional parallelism.

When discussing the merits of various algorithms throughout the remainder of this book, we will use formulae (1.4.1), (1.4.2), and (1.4.3) where appropriate and will draw attention to techniques that are especially powerful on advanced architectures.

1.5 Problem formulation

If the potential gains of sparse matrix computation indicated in this chapter are to be realized, it is necessary to consider both efficient implementation and basic problem formulation. For the dense matrix problem, the order n of the matrix controls the requirements for both the storage and solution time to solve a linear system. In fact, quite precise predictions of solution time ($O(n^3)$)) and storage ($O(n^2)$)) may be made for a properly formulated algorithm (see Section 3.11).

This type of dependence on n becomes *totally invalid* for sparse problems. This point was demonstrated for the example in Section 1.3, but it should be emphasized that it is true in general. The number of entries in the matrix, τ, is a more reliable indicator of work and storage requirements. But even using τ, the precise predictions of the dense case are not possible. We illustrate the effect on problem formulation with several examples.

A very simple demonstration that n is not the only factor in determining the cost of solving equations is given by the case where $\mathbf{A}$ is reducible, that is of the form

$$\mathbf{A} = \begin{bmatrix} \mathbf{A}_{11} & \mathbf{A}_{12} & \cdot & \cdot & \cdot & \mathbf{A}_{1k} \\ & \mathbf{A}_{22} & \cdot & \cdot & \cdot & \mathbf{A}_{2k} \\ & & \cdot & & & \cdot \\ & & & \cdot & & \cdot \\ & & & & \cdot & \cdot \\ & & & & & \mathbf{A}_{kk} \end{bmatrix}, \tag{1.5.1}$$

where each $\mathbf{A}_{ii}$ is a square submatrix. The reason for such a form being desirable may be illustrated by the case with $k=2$. If $\mathbf{x}$ and $\mathbf{b}$ are partitioned to correspond to the partitioning of $\mathbf{A}$, the system of equations (1.3.2) takes the form

$$\mathbf{A}_{11}\mathbf{x}_1 + \mathbf{A}_{12}\mathbf{x}_2 = \mathbf{b}_1 \tag{1.5.2a}$$

$$\mathbf{A}_{22}\mathbf{x}_2 = \mathbf{b}_2. \tag{1.5.2b}$$

This may be solved as two smaller systems, solving equation (1.5.2b) for $\mathbf{x}_2$, substituting $\mathbf{x}_2$ in equation (1.5.2a), and solving the equation

$$\mathbf{A}_{11}\mathbf{x}_1 = \mathbf{b}_1 - \mathbf{A}_{12}\mathbf{x}_2 \tag{1.5.3}$$

for $\mathbf{x}_1$. Observe that the entries in $\mathbf{A}_{12}$ enter the computation only in multiplying a vector, not in the factorization. Chapter 6 is devoted to permuting a given matrix to reducible form (1.5.1) if this is possible.

Another example, closer to the problem formulaticn, is the least-squares solution of the overdetermined system of m equations in n unknowns

$$\mathbf{A}\mathbf{x} = \mathbf{b}, \tag{1.5.4}$$

where $\mathbf{A}$ is $m \times n$, $m \geq n$. One method that is valid for reasonably well-conditioned problems (those for which the solution is not very sensitive to changes in problem data) is the formation and solution of the normal equations

$$\mathbf{A}^T\mathbf{A}\mathbf{x} = \mathbf{A}^T\mathbf{b}. \tag{1.5.5}$$

Exploitation of sparsity in the solution of equation (1.5.5) will often yield significant gains. A related approach is to solve the $(m+n)$ order system

$$\begin{pmatrix} \mathbf{I} & \mathbf{A} \\ \mathbf{A}^T & \mathbf{0} \end{pmatrix} \begin{pmatrix} \mathbf{r} \\ \mathbf{x} \end{pmatrix} = \begin{pmatrix} \mathbf{b} \\ \mathbf{0} \end{pmatrix}, \tag{1.5.6}$$

where $\mathbf{I}$ is the $n \times n$ identity matrix. It may be observed that the first block row of equation (1.5.6) states that the residual vector is given by the equation

$$\mathbf{r} = \mathbf{b} - \mathbf{A}\mathbf{x} \tag{1.5.7}$$

and the second block row requires that the equation

$$\mathbf{A}^T\mathbf{r} = \mathbf{0} \tag{1.5.8}$$

is satisfied. By substituting expression (1.5.7) for $\mathbf{r}$ into equation (1.5.8), we obtain equation (1.5.5). Thus, solving equation (1.5.6) is an alternative way of obtaining the solution to equation (1.5.5). If dense matrix methods are used, our measure of work clearly indicates that the solution of equation (1.5.5) is less costly than the solution of equation (1.5.6). But if $\mathbf{A}$ is sparse and sparse methods are used, the solution of equation (1.5.6) may be less costly than the formation and solution of equation (1.5.5). Additionally, formulation (1.5.6) is better behaved numerically since the solution of equation (1.5.6) is less sensitive to small changes in its data than the solution of equation (1.5.5) is to small changes in $\mathbf{A}^T \mathbf{b}$.

A third example of the influence of the problem formulation is the problem of computer-aided analysis and design of electrical networks. Standard formulations of this problem have been given in terms of loop equations, state equations, or nodal equations. While loop and state equations often give rise to matrices of smaller order, the matrices are in general dense. The nodal equations, however, are sparse. Based on more sophisticated sparse matrix methods that exploit not only the sparsity but the type of entries (for example, treating the values 1 and -1 specially and handling those that are constant from one solution to the next differently from those that vary), a new approach to electrical network analysis was developed (Hachtel, Brayton, and Gustavson 1971), called the **sparse tableau** approach. In one particular example there were 35 loop equations, 80 nodal equations, and 200 tableau equations. Without sparsity considerations, the sparse tableau would be an absurd approach. With carefully implemented sparse matrix algorithms on a serial computer, the relative solution efficiencies of these formulations are in the opposite order to the size of the matrices. The sparse tableau leads to the largest matrix but shortest solution time, the nodal representation is in between for both measures, and the loop formulation has the smallest matrix but the most computing time. This example, along with the least-squares illustration of equations (1.5.4) to (1.5.6), demonstrates that the commonly used strategy of simplifying a model by reducing the order of the matrix problem may be the wrong approach if sparsity is present.

As a further illustration of problem formulation, consider the system of ordinary differential equations of the form

$$A \frac{dx}{dt} = Bx, \quad x(0) = x_0, \tag{1.5.9}$$

where the backward Euler method is used for discretization. That is, dx/dt at $i+1$ is replaced by $(x_{i+1} - x_i)/\Delta t$, and the sequence of vectors x_i is generated starting from the given x_0. We assume that A and B are sparse and that A is nonsingular. For each i, the determination of x_{i+1} from a given x_i requires the solution of the equation

$$(A - B\Delta t)x_{i+1} = Ax_i. \tag{1.5.10}$$

An apparent simplification is to compute the matrix $C = (A - B\Delta t)^{-1}A$ and replace equation (1.5.10) by the equation

$$x_{i+1} = Cx_i, \tag{1.5.11}$$

where each new vector in the sequence may now be determined by a simple matrix by vector multiplication. The matrix $A - B\Delta t$ is straightforward to compute as a sparse matrix, but C is dense except under very special circumstances. If the sparsity of $\mathbf{A} - \mathbf{B}\Delta t$ is exploited, equation (1.5.10) may be reformulated to the sequence

$$\mathbf{y}_i = \mathbf{A}\mathbf{x}_i \qquad (1.5.12a)$$

$$(\mathbf{A} - \mathbf{B}\Delta t)\mathbf{x}_{i+1} = \mathbf{y}_i, \qquad (1.5.12b)$$

where the sparsity of $\mathbf{A}$ is used in equation (1.5.12a) and the sparsity of $\mathbf{A} - \mathbf{B}\Delta t$ is used in solving for $\mathbf{x}_{i+1}$ in equation (1.5.12b). This reformulation can lead to considerable computational savings when $\mathbf{A}$ and $\mathbf{B}$ are sparse, but would be twice as costly in the dense case.

Similar considerations apply to more sophisticated solution methods for ordinary differential equations, and many examples of a similar nature could be given from other fields. These are best given in each field by those familiar with the physical properties of their particular problems and also familiar with the gains that may be achieved through sparse matrix methods. In this very important sense, sparse matrix techniques are understood by those working at the formulation stage of mathematical models. Indeed, it is here that many of the advances in sparse matrix technology have been made over the past few years.

It is the authors' belief that most very complicated physical systems have a mathematical model with a sparse representation. In such diverse fields as structural analysis, electrical analysis, chemical engineering, atmospheric science, and operations research, the real benefits of sparsity have depended upon the formulation of the model as well as the choice of solution algorithm.

1.6   A sparse matrix test collection

Comparisons between sparse matrix strategies and computer programs are difficult because of the enormous dependence on implementation details and because the various ordering methods (introduced in Section 5.3 and discussed in detail in Chapters 7 and 8) are heuristic. This means that comparisons between them will be problem dependent. These concerns have led to the development of a set of test matrices by Duff and Reid (1979b) extended by Duff, Grimes, and Lewis (1987).

A major objective of the test collection has been to represent important features of practical problems. Sparse matrix characteristics (such as average density of entries per row, pattern of the entries, symmetry, and matrix size) can differ among matrices arising from, for example, structural analysis, circuit design, or linear programming. The test problems, though varying widely in their characteristics, have very distinctive patterns. Pictures of some of these patterns are included in Appendix D. Some of the test patterns of Duff and Reid (1979b) and Duff et al. (1987) are used later in this book. They are briefly described in Tables 1.6.1 and 1.6.2.

Order	Number of nonzeros	Description
147	2449	Matrices **A**, **B** from finite-element eigenvalue problem
147	2441	**Ax**=λ**Bx**, supplied by T. Johansson of Lunds Datacentral, Lund, Sweden.
1176	18552	Pattern of large electrical network problem, supplied by A. M. Erisman, Boeing Computer Services, Seattle.
113	655	Pattern of a matrix arising from a statistical application, supplied by W. M. Gentleman, Waterloo, Canada.
54	291	Pattern of matrix arising when solving a stiff set of biochemical ordinary differential equations, supplied by A. R. Curtis, Harwell, England.
57	281	Pattern of Jacobian matrix associated with an emitter-follower-current switch circuit (Willoughby 1971).
199	701	Pattern of a stress-analysis matrix (Willoughby 1971).
200×199	702	Above problem with added row having a nonzero in its first column.
292	2208	Patterns of normal matrices associated with
85	523	least-squares adjustment of survey data, supplied by V. Ashkenazi, Nottingham University, England.
130	1282	Jacobian matrix of a set of ordinary differential equations associated with a laser problem, supplied by A. R. Curtis, Harwell, England.
130	1296	The symmetric pattern formed from the union of the above matrix and its transpose.
363	2454	Basis matrices obtained at various stages of the
363	3157	application of the simplex method to two linear
822	4790	programming problems, supplied by M. A. Saunders,
822	4841	SOL, Stanford University.
541	4285	Four matrices having the same pattern but varying conditioning, which arose at different stages of the use of FACSIMILE (a stiff ODE package) to solve an atmospheric pollution problem involving chemical kinetics and two-dimensional transport, supplied by A. R. Curtis, Harwell, England.
331×104	662	Surveys of Scotland and Holland, supplied by
219×85	438	V. Ashkenazi, Nottingham University, England.

Table 1.6.1. Test matrices of Duff and Reid (1979b) used in examples and comparisons throughout the book.

Note that, for the symmetric matrices, we have counted the nonzeros in both the upper and lower triangular parts. For some algorithms, it is possible to store only one of the triangles.

Another group of matrices arises from solving partial differential equations on a square $N\times N$ grid, pictured for $N=5$ in Figure 1.6.1. The simplest finite-element problem has square bilinear elements, and there is

Order	Number of nonzeros	Description
503	6027	Finite-element calculations on ship structures (Everstine 1979). Symmetric patterns.
1005	8621	
265	1753	Successively finer triangulations of an L-shaped region (George and Liu 1978b). Symmetric patterns.
406	1904	
577	3889	
778	5272	
1009	6865	
1561	10681	
2233	15337	
3466	23896	
1624	6050	Two symmetric structures of power networks of Western USA and one of the entire USA (Boeing Computer Services).
1723	6511	
5300	21842	
532	3474	Unsymmetric matrices from oil reservoir modelling, produced by the PORES package, Harwell, 1983.
1224	9613	
183	1069	Unsymmetric matrix from studies of photochemical smog, produced by the FACSIMILE package, Harwell, 1983.
216	876	Unsymmetric matrices from simulation studies on computing systems, supplied by F. Cachard, Grenoble, 1981.
1107	5664	
2003	83883	Symmetric stiffness matrix from a generalized eigenproblem in fluid flow calculations (Boeing Computer Services).
225	1308	Unsymmetric matrix from hydrocarbon separation problem in chemical engineering, supplied by D. Bogle, Imperial College, London.
156	371	Unsymmetric matrix from simple chemical plant model, supplied by A. Westerberg, Carnegie Mellon University.
760	5976	Unsymmetric matrix from ozone depletion studies, produced by the FACSIMILE package, Harwell, 1983.
256	2916	Symmetric matrix from an aircraft structure problem, supplied by L. Marro, Cannes, 1981.
2529	90158	Unsymmetric matrix from economic model of Australia, supplied by K. Pearson, Melbourne, 1984.

Table 1.6.2. Test matrices of Duff *et al.* (1987) used in examples and comparisons throughout the book.

a one-to-one correspondence between nodes in the grid and variables in the system of equations. Thus the resulting symmetric matrix **A** has order $n=(N+1)^2$, and each row has entries in nine columns corresponding to a node and its eight nearest neighbours. Such a pattern can also arise from a nine-point finite-difference discretization on the same grid. Another test

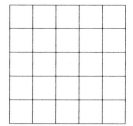

Figure 1.6.1. A 5×5 discretization.

case arises from the five-point finite-difference discretization, in which case each row has off-diagonal entries in columns corresponding to nodes connected directly to the corresponding node by a grid line. Further regular problems arise from the discretization of three-dimensional problems using 7-point and 27-point approximations. These matrices are important because they typify matrices that occur when solving partial differential equations and because very large test problems can be generated easily.

Another way of generating large sets of matrices is to use random number generators to create both the pattern and the nonzero values. Early testing of sparse matrix algorithms was done in this way. Although we use such a set occasionally, we do so sparingly because there is some debate as to how accurately random patterns reflect real-life problems.

Exercises

1.1 Show that if we use the model of computation given by equation (1.4.2), the maximum number of floating-point operations per second is $1/c$ and the vector length for half the maximum is s.

1.2 For the graph

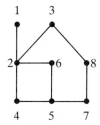

write down the corresponding sparse matrix pattern.

1.3 Find the pattern of the matrix that results from the elimination of the subdiagonal entries in columns one and two by adding multiples of rows one and two to appropriate later rows.

1.4 Can you find a reordering for the sparse matrix corresponding to the graph

for which no new entries are created when solving equations with this as coefficient matrix?

2 Sparse matrices: storage schemes and simple operations

*Some fundamental **tools** for storing and manipulating sparse vectors and matrices without undue overheads.*

2.1 Introduction

The aim of the chapter is to examine data structures suitable for holding sparse matrices and vectors. We can only do this in conjunction with a consideration of the operations we wish to perform on these matrices and vectors, so we include a discussion of the simple kinds of operations that we need.

An important distinction is between **static** structures that remain fixed and **dynamic** structures which are adjusted to accommodate fill-ins as they occur. Naturally, the overheads of adjusting a dynamic structure can be significant. Furthermore, the amount of space needed for a static structure will be known in advance, but this is not the case for a dynamic structure. Both types are widely used in the implementation of sparse Gaussian elimination. We discuss the use of them in detail in Chapters 9 and 10.

Usually what is required is a very compact representation that permits easy manipulation. There is no one best data structure; most practical computer codes use different storage patterns at different stages.

2.2 Sparse vector storage

For simplicity, we begin by considering the storage of sparse vectors. Some of our remarks generalize at once to sparse matrices whose rows (or columns) may be regarded as sets of sparse vectors. Furthermore, the Gaussian elimination process (see Chapters 3 and 4) can be viewed as a sequence of vector operations.

A sparse vector may be held in a full-length vector of storage. This is rather wasteful of storage, but is often used because of the simplicity and speed with which it may then be manipulated. For instance, the i-th component of a vector may be found directly.

To economize in storage we may hold the entries as real, integer pairs (x_i, i). In Fortran we normally use a real array (for example VAL in Figure 2.3.1) and a separate integer array (for example IND in Figure 2.3.1), each of length at least the number of entries. We call the operation of transforming from full-length to packed form a **gather** and the reverse a **scatter**. Applying an elementary Gaussian elimination operation, say adding a multiple of component i to component j, is not convenient on this packed form because we need to search for x_i and x_j. A sequence of such operations may be performed efficiently if just one full-length vector of storage is used. For each packed vector we perform the following steps:

(i) Place the nonzero entries in a full-length vector known to be set to zero.

(ii) For each operation

   (a) revise the full-length vector by applying the operation to it, and

   (b) if this changes a zero to a nonzero, then alter the integer part of the packed form to correspond.

(iii) By scanning the integers of the packed form, place the modified entries back in the packed form while resetting the full-length vector to zero.

Notice that the work performed depends only on the number of entries and not on the length of the full vector. It is very important in manipulating sparse data structures to avoid complete scans of full-length vectors.

2.3 Inner product of two packed vectors

Taking an inner product of a packed vector with a full-length vector is very simply and economically achieved by scanning the packed vector, as shown in the Fortran code of Figure 2.3.1. An inner product between two packed vectors is best achieved by first expanding one of them into a full-length vector, although if they both have their components in order then we may scan the two in phase as shown in the Fortran code of Figure 2.3.2. Note that, although the code might at first sight suggest that we do about NZX*NZY comparisons, the actual number is about NZX+NZY since the inner loop always starts from where it last finished.

```
        PROD = 0.
        DO 10 K = 1,NZ
            PROD = PROD + VAL(K)*W(IND(K))
   10 CONTINUE
```

Figure 2.3.1. Code for the inner product between packed vector (VAL, IND) with NZ entries and full-length vector W.

```
          PROD = 0.
          KY = 1
          DO 30 KX = 1, NZX
            K = KY
            DO 10 KY = K, NZY
              IF (INDY (KY) - INDX (KX) ) 10, 20, 30
   10       CONTINUE
            GO TO 40
   20       PROD = PROD + VALX (KX) *VALY (KY)
   30    CONTINUE
   40    . . . . . . . .
```

Figure 2.3.2. Code for the inner product between two packed ordered vectors (VALX, INDX) and (VALY, INDY) with NZX and NZY entries.

2.4 Adding packed vectors

Another very important operation is that of adding a multiple of one vector to another, say

$$\mathbf{x} := \mathbf{x} + \alpha \mathbf{y}. \tag{2.4.1}$$

If $\mathbf{x}$ and $\mathbf{y}$ are held in packed form with their components unordered, a satisfactory technique is again to use a full-length vector $\mathbf{w}$ known initially to be zero. We then modify $\mathbf{x}$ using $\mathbf{w}$ as a working vector but restoring it to zero as follows:

(i) Load $\mathbf{y}$ into $\mathbf{w}$.

(ii) Scan $\mathbf{x}$. For each entry x_i check w_i. If it is nonzero, modify x_i appropriately and reset w_i to zero.

(iii) Scan $\mathbf{y}$. For each entry y_i check w_i. If it is nonzero add a new component of $\mathbf{x}$ to the data structure (we have a fill-in). Reset w_i to zero.

Fortran code for these operations is shown in Figure 2.4.1. If the components are in order then an in-phase scan of the two vectors, analogous to that shown in Figure 2.3.2, may be made. Note that fill-ins in general do not occur at the end of the vector, so the order can be preserved only if a fresh vector is constructed (so that the operation is of the form $z:=x+\alpha y$). We set the writing of such code as Exercise 2.1. Perhaps surprisingly, it is more complicated than the unordered code. In fact, there appears to be little advantage in using ordered vectors except for the need of the full-length vector $\mathbf{w}$, but even this may be avoided at the expense of slightly more complicated code (see Exercise 2.2).

```
        DO 10 K = 1, NZY
          W(INDY(K)) = VALY(K)
    10 CONTINUE
        DO 20 K = 1, NZX
          I = INDX(K)
          IF(W(I).EQ.0.) GO TO 20
          VALX(K) = VALX(K) + ALPHA*W(I)
          W(I) = 0.
    20 CONTINUE
        DO 30 K = 1, NZY
          I = INDY(K)
          IF(W(I).EQ.0.) GO TO 30
          NZX = NZX + 1
          VALX(NZX) = ALPHA*W(I)
          INDX(NZX) = I
          W(I) = 0.
    30 CONTINUE
```

Figure 2.4.1. Code for the operation $\mathbf{x} := \mathbf{x} + \alpha\mathbf{y}$.

An alternative approach is as follows:

(i) Load $\mathbf{x}$ into $\mathbf{w}$.

(ii) Scan $\mathbf{y}$. For each entry y_i, check w_i. If it is nonzero, set $w_i := w_i + \alpha y_i$; otherwise set $w_i := \alpha y_i$ and add i to the data structure for $\mathbf{x}$.

(iii) Scan the revised data structure for $\mathbf{x}$. For each i, set $x_i := w_i$, $w_i := 0$.

While it is slightly more expensive than the first approach, since the revised $\mathbf{x}$ is stored first in $\mathbf{w}$ and then placed in $\mathbf{x}$, it offers one basic advantage. It permits the sequence of operations

$$\mathbf{x} := \mathbf{x} + \sum_j \alpha_j \mathbf{y}^{(j)} \tag{2.4.2}$$

to be performed with only one execution of steps (i) and (iii). Step (ii) is simply repeated with $\mathbf{y}^{(1)}$, $\mathbf{y}^{(2)}$,..., while intermediate results for $\mathbf{x}$ remain in $\mathbf{w}$. Step (ii) may be simplified if it is known in advance that the sparsity pattern of $\mathbf{x}$ contains that of $\mathbf{y}$.

We will see later (Chapter 9) that there are times during the solution of sparse equations when each of these approaches is preferable.

2.5 Use of full-sized arrays

In the last three sections we have considered the temporary use of a full-length vector of storage to aid computations with packed sparse vectors. Provided sufficient storage is available, a simpler possibility is to use full-length vectors all the time. Whether this involves more computer time will depend on the degree of sparsity, the operations to be performed and the details of the hardware. On a scalar processor, operations involving indirect addressing (as in Figure 2.3.1, for example) are typically two to four times slower than the corresponding operations using direct addressing. Therefore on scalar processors the threshold figure beyond which full-length vector storage involves less computer time is usually in the range 25 to 50 per cent.

Vector architectures such as the CRAY-1 and CYBER 205 show a much more substantial degradation for indirect addressing since it does not vectorize. However, recent advances in vector architectures (see Section 1.4) have enabled indirect addressing to vectorize on some hardware (for example, CRAY X-MP and Fujitsu FACOM VP), which leads to an asymptotic performance ratio between direct and indirect addressing that is similar to that found on scalar machines.

A further possibility is to use a full-length vector of storage spanning between the first and last entry. This allows fast execution on a vector computer and may result in a very worthwhile saving in storage (and computation) compared with holding the whole vector in full-length storage.

A sparse matrix, too, may be held in full-length storage, either as a two-dimensional array or as an array of one-dimensional arrays. Similar considerations apply as for vectors. Indeed on vector computers some further gains may be available from vectorizing over the second dimension.

2.6 Coordinate scheme for storing sparse matrices

Perhaps the most convenient way to specify a sparse matrix is as its set of entries in the form of an unordered set of triples (a_{ij}, i, j). In Fortran these are best held in one real array and two integer arrays, all of length the number of entries. The matrix of Figure 2.6.1, for example, might have the representation in Table 2.6.1, where IRN and JCN are integer arrays and VAL is a real array. In this table and in similar ones in this chapter, the row denoted 'subscripts' indicates the locations in the arrays. For example, location IRN(9) holds the value 2, JCN(9) holds the value 3, and so on. The matrix of this table is 44 per cent dense (11 entries in a 5×5 matrix) so is in the range we quoted in Section 2.5 as being borderline for the use of a full-sized array. It is interesting that if reals and integers require the same amount of storage, less total storage is required by the full-sized matrix (25 words) than by the packed form (33 words). The position is reversed for double-length reals to the ratio 25:22 and an even bigger change (to 25:16.5) takes place with half-length integers and double-length reals. The situation for complex-valued matrices is analogous to that of double-length reals.

The major difficulty with this data structure lies in the inconvenience of accessing it by rows or columns. It is perfectly suitable if we wish to use a scalar machine to multiply by a vector in full-length storage mode to give a

$$\mathbf{A} = \begin{bmatrix} 1. & 0 & 0 & -1. & 0 \\ 2. & 0 & -2. & 0 & 3. \\ 0 & -3. & 0 & 0 & 0 \\ 0 & 4. & 0 & -4. & 0 \\ 5. & 0 & -5. & 0 & 6. \end{bmatrix}$$

Figure 2.6.1. A 5×5 sparse matrix.

Subscripts	1	2	3	4	5	6	7	8	9	10	11
IRN	1	2	2	1	5	3	4	5	2	4	5
JCN	4	5	1	1	5	2	4	3	3	2	1
VAL	−1.	3.	2.	1.	6.	−3.	−4.	−5.	−2.	4.	5.

Table 2.6.1. The matrix of Figure 2.6.1 stored in the coordinate scheme.

result also in full-length storage mode. However the direct solution of a set of linear equations, for example, involves a sequence of row (or column) operations on the coefficient matrix. There are two principal storage schemes which provide ready access to this information: the collection of sparse vectors and the linked list. These we now describe.

2.7 Sparse matrix as a collection of sparse vectors

Our first alternative is to hold each row (or column) as a packed sparse vector of the kind described in Section 2.2. The components of each vector may be ordered or not. Since our conclusion in Section 2.4 was that there is little advantage in ordering them, we take them to be unordered.

For each member of the collection we will normally store a pointer to its start and the number of entries. Since we are using Fortran 77, a pointer is simply an array subscript indicating a position in an array. Thus, for example, the matrix of Figure 2.6.1 may be stored in Fortran arrays as shown in Table 2.7.1. Here LENROW(i) contains the number of entries in row i, while IROWST(i) contains the location in arrays JCN and VAL of the first entry in row i. For example, row 4 starts in position 7, so referring to position 7 in JCN and VAL we find the (4,4) entry has value −4. Since LENROW(4) = 2, the eighth position is also in row 4, specifically the (4,2) entry has value 4.

In this representation, the rows are stored in order (row 1, followed by row 2, followed by row 3, etc.). It is not really necessary to store both arrays IROWST and LENROW, since IROWST can be calculated from

Subscripts	1	2	3	4	5	6	7	8	9	10	11
LENROW	2	3	1	2	3						
IROWST	1	3	6	7	9						
JCN	4	1	5	1	3	2	4	2	3	1	5
VAL	-1.	1.	3.	2.	-2.	-3.	-4.	4.	-5.	5.	6.

Table 2.7.1. Matrix of Figure 2.6.1 stored as a collection of sparse row vectors.

LENROW or LENROW can be calculated from IROWST and a pointer to the end of the structure. If it is only necessary to access the matrix forwards or backwards, as is the case when a triangular factor is being held (see Section 9.4), then holding just LENROW is satisfactory and has the advantage that the integers are smaller and so may fit into a shorter computer word. Where the rows may need to be accessed in arbitrary sequence, we may dispense with LENROW and hold just IROWST (and a pointer to the end of the last row), but holding just LENROW would lead to a costly extra computation for finding entries in a particular row.

A basic difficulty with this structure is associated with inserting new entries. This arises when a multiple of one row is added to another, since the new row may be longer, as discussed in Section 2.4. It is usual to waste (temporarily) the space presently occupied by the row and add a fresh copy at the end of the structure. Once the rows have become disordered because of this, both the arrays LENROW and IROWST are needed (although we might replace the last column index for each row by its negation to avoid storing LENROW). After a sequence of such operations, we may not have room at the end of the structure for the new row although there is plenty of room inside the structure. Here we should 'compress' it by moving all the rows forward to become adjacent once more. It is clear that this data structure demands some 'elbow room' if an unreasonable number of compresses are not to be made.

2.8 Linked lists

Another data structure which is used widely for sparse matrices is the linked list. We introduce linked lists in this section and show how they can be used to store sparse matrices in Section 2.9. Several programming languages (for example, Pascal and Ada) have explicit pointer types, but we assume here, as in the rest of the book, that the Fortran 77 language is being used and our data structures and our examples reflect this assumption. Pointers are therefore just array subscripts.

The essence of a linked list is that there is a pointer to the first entry (header pointer) and with each entry is associated a pointer (or link) which points to the next entry or is null for the last entry. The list can be scanned by accessing the first entry through the header pointer and the other entries through the links until the null link is found. For example, we show in Table 2.8.1 a linked list for the set of column indices $(10, 3, 5, 2)$, where we have used zero for the null pointer. Scanning this linked list, we find the header points to entry 1 with the value 10, its link points to entry 2 with value 3, and its link in turn points to entry 3 with value 5. The link from entry 3 points to entry 4 whose value is 2 and whose link value is 0, indicating the end of the list.

Subscripts	1	2	3	4
Values	10	3	5	2
Links	2	3	4	0
Header	1			

Table 2.8.1. Linked list for holding $(10, 3, 5, 2)$.

When using linked lists, it is important to realize that the ordering is determined purely by the links and not by the the physical location of the entries. For example, the same ordered list $(10, 3, 5, 2)$ can be stored as in Table 2.8.2. Furthermore, the links can be adjusted so that the values are scanned in order without moving the physical locations (see Table 2.8.3).

Subscripts	1	2	3	4
Values	3	10	2	5
Links	4	1	0	3
Header	2			

Table 2.8.2. Linked list for holding $(10, 3, 5, 2)$ in a different physical order.

Subscripts	1	2	3	4
Values	3	10	2	5
Links	4	0	1	2
Header	3			

Table 2.8.3. Linked list for holding $(10, 3, 5, 2)$ in increasing order of values.

If we are storing a vector of integers in isolation it would be nonsensical to prefer a linked list, since we would have unnecessarily increased both the storage required and the complexity of accessing the entries. There are two reasons, however, why linked lists are used in sparse matrix work. The first is that entries can be added without requiring adjacent free space. For example an extra entry, say 4 between 3 and 5, could be accommodated as shown in Table 2.8.4 even if locations 5 to 8 hold data that we do not wish to disturb. Second, if we wish to remove an entry from the list, no data movement is necessary since only the links need be adjusted. The result of removing entry 3 from the list of Table 2.8.2 is shown in Table 2.8.5.

Subscripts	1	2	3	4	5	6	7	8	9
Values	3	10	2	5	*	*	*	*	4
Links	9	1	0	3	*	*	*	*	4
Header	2								

Table 2.8.4. Linked list for holding (10, 3, 4, 5, 2).

Subscripts	1	2	3	4
Values	*	10	2	5
Links	*	**4**	0	3
Header	2			

Table 2.8.5. Removal of entry 3 from the Table 2.8.2 list. The only changed link is shown in bold.

A problem with adding and deleting entries is that usually the previous entry must be identified so that its link can be reset, although additions may be made to the head of the list if the ordering is unimportant. If the list is being scanned at the time a deletion or insertion is needed, the previous entry should be available. However, if an essentially random entry is to be deleted or if insertion in order is wanted, we need to scan the list from its beginning. At the expense of more storage, this search may be avoided by using a doubly-linked list which has a second link associated with each entry and which points to the previous entry. An example is shown in Table 2.8.6. It is now straightforward to add or delete any entry (see Exercise 2.6).

A significant simplification can occur when the values are distinct integers in the range 1 to n and a full-length vector of length n is available. The location i may be associated with the value i without actually storing it. Our example (10, 3, 5, 2) is shown in Table 2.8.7 held in this form. Several such lists may be held in the same array of links in this way, provided no two lists contain a value in common. For instance, we may wish to group all the rows with the same number of entries. If rows (10, 3, 5, 2) all have one

Subscripts	1	2	3	4
Values	3	10	2	5
Forward links	4	1	0	3
Backward links	2	0	4	1
Forward header	2			
Backward header	3			

Table 2.8.6. Doubly-linked list holding (10, 3, 5, 2).

Subscripts	1	2	3	4	5	6	7	8	9	10
Links		0	5		2					3
Header	10									

Table 2.8.7. Implicit linked list holding (10, 3, 5, 2).

Subscripts	1	2	3	4	5	6	7	8	9	10
Links	9	0	5	0	2	7	8	0	4	3
Headers	10	1	6							

Table 2.8.8. Implicit linked list holding (10, 3, 5, 2), (1, 9, 4), and (6, 7, 8).

entry, rows (1, 9, 4) have two entries and rows (6, 7, 8) have three entries, the three groups may be recorded together as shown in Table 2.8.8.

We have used the value zero to indicate a null pointer, but any value that cannot be a genuine link may be used instead, so it may be used to hold other information. We illustrate this in Section 2.11.

For a more complete discussion of linked lists we refer the reader to Knuth (1969, pp. 251-257).

2.9 Sparse matrix in row-linked list

A major benefit of using linked lists for the storage of sparse matrices is that the elbow room and compressing operations associated with the data structure in Section 2.7 can be avoided entirely. To store the matrix as a collection of rows, each in a linked list, we need an array of header pointers with the i-th entry pointing to the location of the first entry for row i. We illustrate this in Table 2.9.1 for our 5×5 matrix of Figure 2.6.1. The values are physically located in the rather arbitrary order shown in Table 2.6.1, but the links have been constructed so that the rows are scanned in column order. We employ this ordering because it facilitates operations on the matrix and enables us to illustrate the flexibility of the structure better. However, it is not a requirement of the linked list scheme that the rows be accessed in order, since a variation of the technique of Section 2.4 may be used for the critical operation of adding a multiple of one row to another.

Subscripts	1	2	3	4	5	6	7	8	9	10	11
IROWST	4	3	6	10	11						
JCN	4	5	1	1	5	2	4	3	3	2	1
VAL	−1.	3.	2.	1.	6.	−3.	−4.	−5.	−2.	4.	5.
LINK	0	0	9	1	0	0	0	5	2	7	8

Table 2.9.1. The matrix of Figure 2.6.1 held as a linked list.

To illustrate the use of this structure, we work through row 4. IROWST(4) = 10, JCN(10) = 2, and VAL(10) = 4.0, so the first entry in row 4 is in the (4,2) position and has value 4.0. LINK(10) = 7, JCN(7) = 4, and VAL(7) = −4.0, so the next entry in row 4 is in the (4,4) position and has value −4.0. Since LINK(7) = 0, there are no further entries in row 4.

We can illustrate how the structure is used by supposing that a multiple of row 1 of the matrix of Figure 2.6.1 is added to row 2. A fill-in occurs in position (2,4) and this may be added to the linked list of Table 2.9.1, while preserving the order within row 2, by placing the value of the new entry in VAL(12), setting JCN(12) = 4, giving the new entry the link value that entry (2,3) used to have, and setting the link of entry (2,3) to 12. The resulting new LINK array is shown in Table 2.9.2, the changed entries being shown in bold. The IROWST array is not affected in this case.

Subscripts	1	2	3	4	5	6	7	8	9	10	11	12
LINK	0	0	9	1	0	0	0	5	**12**	7	8	**2**

Table 2.9.2. The new link array after inserting entry (2,4).

We can delete entries as described in Section 2.8 and normally find it convenient to link deleted entries together so that their storage locations are readily available for subsequent insertions.

One difficulty with linked lists is the requirement to store integers whose magnitude may be as large as the number of entries in the matrix, whereas integers stored in the collection of sparse vectors are never larger than the matrix order. This can limit the ability to use half-word storage for integers. The difficulty can be circumvented by limiting the size of the link array (and corresponding JCN and VAL arrays) to the desired bound. Additional entries can be handled in additional LINK, JCN, and VAL arrays. For each additional set of arrays required, one additional array similar to IROWST is needed to link the LINK arrays. Thus for every row that stretches across two LINK arrays, one additional indirect address calculation is required to find all of the entries in that row. Another possibility is to restrict each row to one set of LINK, JCN, and VAL arrays.

2.10 Generating collections of sparse vectors

Neither of the structures described in Sections 2.7 and 2.9 is particularly convenient for initial specification of a sparse matrix to a library subroutine. Curtis and Reid (1971b) required the matrix to be input by columns to their sparse matrix subroutines, and this proved unpopular with many users. Fortunately, we can sort from the coordinate scheme (Section 2.6) to a row-linked list or to a collection of sparse vectors (rows or columns) in $O(n) + O(\tau)$ operations, if n is the matrix order and τ is the number of entries, without using any additional workspace.

A row-linked list may very easily be constructed from an unstructured list by scanning it once, as shown in Figure 2.10.1. Essentially, we begin with a null list and add the entries in one by one without ordering them within rows. Note that we could overwrite IRN with links, thereby saving the storage occupied by the array LINK.

```
      DO 10 I = 1,N
          IROWST(I) = 0
   10 CONTINUE
      DO 20 K = 1,NZ
          I = IRN(K)
          LINK(K) = IROWST(I)
          IROWST(I) = K
   20 CONTINUE
```

Figure 2.10.1. Generating a row-linked structure.

If ordering within the rows is wanted, this can be achieved by first linking by columns and then using these links to scan the matrix in reverse column order when setting row links (Exercise 2.8). A temporary array for holding column header pointers will be needed. It is fascinating that in this way we order the τ entries in 2τ operations (of the type of the loop on label 20 in Figure 2.10.1) in view of the fact that a general-purpose sort involves $O(\tau \log_2 \tau)$ operations. This is possible because of the special nature of the data to be ordered; it consists of a set of integers in the range $(1,n)$ and we have a work vector IROWST of length n.

To transform a matrix held in the coordinate scheme, with entries (a_{ij}, i, j), to a collection of sparse row vectors in $O(n) + O(\tau)$ operations without using any vector storage in addition to that needed for the original and final forms, we may proceed as follows.

(a) Use IROWST to accumulate counts of the numbers of entries in the rows during a scan of the entries.

(b) Use the code

```
      IROWST(1) = IROWST(1) + 1
      DO 10 I = 2,N
          IROWST(I) = IROWST(I) + IROWST(I-1)
   10 CONTINUE
```

to put in IROWST pointers to just after where each row would end in a compressed collection with rows in natural order.

(c) Take the first entry, say a_{ij}; decrease IROWST(i) by one, move the entry in the location IROWST(i) into temporary storage, move a_{ij} into this position, and mark a_{ij} as now correctly ordered by negating its column index. Apply the same procedure to the displaced entry, which is now in temporary storage. Continue until an entry is placed in the first position.

(d) Scan the remaining entries looking for any not marked as in position by having a negated column index. Use each to initiate a loop similar to that in (c).

Note that each entry is moved directly to its final position and that IROWST automatically ends storing pointers to the starts of the rows. The entries are scanned three times: once in (a), once when being moved, and once when being checked in (d). The entries are not sorted into order within each row, but fortunately often this is adequate, as was illustrated in Sections 2.3 and 2.4. If an extra integer array of length n is available, it is possible to replace the scan of length τ in (d) with a scan of length n.

2.11   Access by rows and columns

We will see in Chapter 9 that, although the sparse matrix storage structures we have so far considered are sometimes perfectly suitable, there are times when access by rows and columns is needed. In both the row-linked list and the collection of sparse row vectors, the entries in a particular column cannot be discovered without a search of all or nearly all the entries. This is required, for instance, to reveal which rows are active during a single stage of Gaussian elimination.

If the matrix is held as a collection of sparse row vectors, a satisfactory solution (Gustavson 1972) is to hold also the structure of the matrix as a separate collection of sparse column vectors. In this second collection there is no need to hold the numerical values themselves, since all real arithmetic is performed on the sparse row vectors and the column collection is required only to reveal which rows are involved in the elimination steps. We show our 5×5 example stored in this format in Table 2.11.1. It differs from Table 2.7.1 only because of the addition of the integer arrays LENCOL, JCOLST, and IRN.

Similarly, for the linked list, we may add row indices and column links. This, however, means that four integers are associated with each entry. Curtis and Reid (1971a) felt that this gave unacceptable storage overheads and therefore dropped the row and column indices but made the last link of each row (or column) contain the negation of the row (or column) index instead of zero. Thus the row (or column) index of an entry could always be found by searching to the end of the row (or column). For really sparse

Subscripts	1	2	3	4	5	6	7	8	9	10	11
LENROW	2	3	1	2	3						
IROWST	1	3	6	7	9						
JCN	4	1	5	1	3	2	4	2	3	1	5
VAL	−1.	1.	3.	2.	−2.	−3.	−4.	4.	−5.	5.	6.
LENCOL	3	2	2	2	2						
JCOLST	1	4	6	8	10						
IRN	2	1	5	3	4	5	2	1	4	2	5

Table 2.11.1. Matrix of Figure 2.6.1 stored in Gustavson's format.

Subscripts	1	2	3	4	5	6	7	8	9	10	11
IROWST	4	3	6	10	11						
JCOLST	4	6	9	1	2						
VAL	−1.	3.	2.	1.	6.	−3.	−4.	−5.	−2.	4.	5.
LINKRW	−1	−2	9	1	−5	−3	−4	5	2	7	8
LINKCL	7	5	11	3	−5	10	−4	−3	8	−2	−1

Table 2.11.2. Matrix of Figure 2.6.1 stored in Curtis-Reid format.

matrices this search is not expensive. We show our 5×5 example in this format in Table 2.11.2.

Given either the collection of sparse row vectors or the row-linked list, it is extremely easy to construct these larger structures (see Exercises 2.10 and 2.11).

2.12 Matrix by vector products

Any of the storage schemes we have considered is suitable for multiplying a sparse matrix by a full-length vector to give a full-length vector result. For the operation

$$y = Ax, \qquad (2.12.1)$$

where x is sparse, access to A by columns is desirable. In fact, all we need are the columns corresponding to the nonzero components x_i, and any data structure that provides access by columns is suitable. The vector y is just a linear combination of these columns and the technique for adding sparse vectors that we described at the end of Section 2.4 is very suitable.

If column access is not available, a scan of the whole packed matrix $\mathbf{A}$ is inevitable. It is important, however, that we do not scan the packed vector $\mathbf{x}$ for each entry in $\mathbf{A}$, and this can be avoided if $\mathbf{x}$ is first expanded into a full-length vector. If access by rows is available then the result $\mathbf{y}$ can be placed at once in packed storage since y_i depends only on the i-th row of $\mathbf{A}$.

2.13 Matrix by matrix products

We now consider the matrix by matrix product

$$\mathbf{C} = \mathbf{BA}, \tag{2.13.1}$$

where $\mathbf{A}$ and $\mathbf{B}$ are conformable (though not necessarily square) sparse matrices and $\mathbf{C}$ is perhaps sparse. For full matrices, the usual way of computing $\mathbf{C}$ is to consider each entry c_{ij} as an inner product of the i-th row $\mathbf{B}_{i\bullet}$ of $\mathbf{B}$ and the j-th column $\mathbf{A}_{\bullet j}$ of $\mathbf{A}$, that is

$$c_{ij} = \sum_k b_{ik} a_{kj} = \mathbf{B}_{i\bullet} \mathbf{A}_{\bullet j}. \tag{2.13.2}$$

The trouble with this formula for a general sparse matrix is that it is very difficult to avoid performing multiplications $b_{ik} a_{kj}$ with one or other of the factors having the value zero (and an explicit test for a zero is likely to be equally expensive). For instance, if $\mathbf{A}$ is stored by columns and $\mathbf{B}$ is stored by rows, column j of $\mathbf{A}$ may be loaded into a full vector and then c_{ij} may be calculated by scanning the entries of row i of $\mathbf{B}$. This means that all entries of $\mathbf{B}$ are scanned for each column of $\mathbf{A}$. If the sparsity pattern of $\mathbf{C}$ is already known, c_{ij} need not be calculated unless it is an entry, but there are still likely to be many occasions when we multiply an entry of $\mathbf{B}$ by a zero in the full vector.

None of these unnecessary operations is performed with the **outer-product** formulation as a sum of rank-one matrices,

$$\mathbf{C} = \sum_k \mathbf{B}_{\bullet k} \mathbf{A}_{k\bullet}, \tag{2.13.3}$$

which is natural if $\mathbf{B}$ is stored by columns and $\mathbf{A}$ is stored by rows (note that $\mathbf{B}_{\bullet k}$ is a column vector and $\mathbf{A}_{k\bullet}$ is a row vector, so $\mathbf{B}_{\bullet k} \mathbf{A}_{k\bullet}$ has the shape of $\mathbf{C}$). If both matrices are stored by columns, column j of $\mathbf{C}$ may be accumulated as a linear combination of the columns of $\mathbf{B}$ by expressing (2.13.3) in the form

$$\mathbf{C}_{\bullet j} = \sum_k a_{kj} \mathbf{B}_{\bullet k}. \tag{2.13.4}$$

This is analogous to performing a sequence of matrix by vector products in the first way described in Section 2.12.

If both matrices are stored by rows, row i of $\mathbf{C}$ may be accumulated as a linear combination of the rows of $\mathbf{A}$ by the formula

$$C_{i\bullet} = \sum_k b_{ik} A_{k\bullet}, \tag{2.13.5}$$

and the required work is identical to that when $\mathbf{A}$ and $\mathbf{B}$ are both stored by columns.

An important special case is when $\mathbf{B} = \mathbf{A}^T$, which can arise in the least-squares problem, in which case $\mathbf{C}$ is the normal matrix. If $\mathbf{A}$ is stored by rows, we have the case of (2.13.3) with $\mathbf{B}_{\bullet k} = \mathbf{A}_{k\bullet}^T$ which yields

$$\mathbf{C} = \sum_k \mathbf{A}_{k\bullet}^T \mathbf{A}_{k\bullet}. \tag{2.13.6}$$

If $\mathbf{A}$ is stored by columns we have the case of (2.13.2) with $\mathbf{B}_{i\bullet} = \mathbf{A}_{\bullet i}^T$ which, in general, will not be as efficient. Indeed it may be preferable to make a copy of $\mathbf{A}$ that is stored by rows.

Another advantage of the outer-product approach is that $\mathbf{A}$ need not be stored as a whole. We may generate the rows successively (or read them from auxiliary storage) and accumulate the contribution from $\mathbf{A}_{k\bullet}$ at once, after which this data is no longer needed.

We refer the reader to Gustavson (1978) for a more detailed discussion of the topic of this section.

2.14 Permutation matrices

Permutation matrices are very special sparse matrices (identity matrices with reordered rows or columns). They are frequently used in matrix computations to represent the ordering of components in a vector or the rearrangement of rows and columns in a matrix. For this reason we want a special representation of permutation matrices which is both compact in storage and readily allows permutation operations to be performed.

If we think of the permutation as a sequence of interchanges

$$(i, p_i), \quad i = 1, 2, ..., n-1 \tag{2.14.1}$$

where $p_i \geq i$, this sequence of interchanges operating on a vector of length n can be expressed in matrix notation as

$$\mathbf{Px}, \tag{2.14.2}$$

where $\mathbf{P}$ is a permutation matrix. Note that $\mathbf{P}$ may be obtained by performing the sequence of interchanges (2.14.1) on the identity matrix. Note also that for any permutation matrix, the relation

$$\mathbf{P}\mathbf{P}^T = \mathbf{P}^T\mathbf{P} = \mathbf{I} \tag{2.14.3}$$

is true. A convenient way to store such a permutation matrix is simply as the set of integers

$$p_i, \quad i = 1, 2, ..., n-1. \tag{2.14.4}$$

In this case $\mathbf{x}$ may be permuted **in place**, that is without any additional array storage, by code such as that shown in Figure 2.14.1.

```
DO 10 I = 1,N-1
   TEMP = X(I)
   X(I) = X(ISWAP(I))
   X(ISWAP(I)) = TEMP
10 CONTINUE
```

Figure 2.14.1. Code to permute **x** in place.

In fact each interchange (i, p_i) is itself a representation of an elementary permutation matrix $\mathbf{P}_i$ (elementary because only two rows or columns of the identity matrix have been interchanged), and $\mathbf{P}$ is the product

$$\mathbf{P} = \mathbf{P}_{n-1}\mathbf{P}_{n-2} \cdots \mathbf{P}_1. \tag{2.14.5}$$

The inverse permutation

$$\mathbf{P}^T = \mathbf{P}_1^T\mathbf{P}_2^T \cdots \mathbf{P}_{n-1}^T \tag{2.14.6}$$

may therefore be represented by the same set of interchanges but now in reverse order.

An alternative representation of $\mathbf{P}$ is as the set of n integers π_i, $i = 1$, $2,\ldots, n$, which represent the positions of the components of **x** in $\mathbf{y} = \mathbf{P}\mathbf{x}$. In this case in-place permutation is not straightforward (see Exercise 2.13), but **y** can be formed from **x** as shown in Figure 2.14.2. It involves less array element accesses than the code of Figure 2.14.1, so is likely to run faster. This representation is equally convenient for the inverse permutation, again provided in-place permutation is not required.

```
DO 10 I = 1,N
   Y(IPOS(I)) = X(I)
10 CONTINUE
```

Figure 2.14.2. Code to form $\mathbf{y} = \mathbf{P}\mathbf{x}$ from **x**. The array IPOS holds the integers π_i.

Which of these representations is preferable depends on the application. Often speed is not of great importance because this is an $O(n)$ process and other transformations involving more operations will be needed too. In this case the convenience of being able to sort in place may be important. However, sometimes the average number of entries per row may be very modest (for example two or three), in which case the speed of permutations may be quite important.

To construct π_i, $i = 1, 2,\ldots, n$ from p_i, $i = 1, 2,\ldots, n-1$ is trivial. We merely have to apply the successive interchanges to the vector $(1, 2,\ldots, n)$. The reverse construction is not quite so easy, but it can be done in $O(n)$ operations if care is taken.

2.15 Clique (or finite-element) storage

Matrices that arise in finite-element calculations have the form

$$\mathbf{A} = \sum_k \mathbf{A}^{[k]}, \qquad (2.15.1)$$

where each matrix $\mathbf{A}^{[k]}$ is associated with a finite element and has nonzeros in only a few rows and columns. Therefore it is possible to hold each $\mathbf{A}^{[k]}$ as a small dense matrix together with a set $S^{[k]}$ of indices to indicate where it belongs in the overall matrix. For example, the symmetric matrix whose pattern is shown in Figure 1.3.2 consists of the sum of matrices $\mathbf{A}^{[k]}$ whose index sets $S^{[k]}$ are (1,2), (1,5,8), (2,3,11,13), (3,4), (5,6), (6,7), (8,9), (9,10), (11,12), and (13,14). Another way to store a sparse matrix $\mathbf{A}$ is therefore as the sum (2.15.1). Any sparse matrix may be written in this form by putting each entry into its own $\mathbf{A}^{[k]}$.

We will see in Section 10.9 that there can be advantages during Gaussian elimination in using this representation because elimination operations may be represented symbolically by merging index lists. Because such a merged list cannot be longer than the sum of the lengths of the lists merged, this symbolic operation can be done without any need for additional storage. In this way we can be sure not to run out of storage while finding out how much storage will be needed for the subsequent operations. Furthermore the merge of two index lists will be much faster than the corresponding elimination step since all the entries are involved in the elimination.

There can be advantages, particularly on a vector or parallel processor in retaining the form (2.15.1) for the matrix by vector multiplication

$$\mathbf{y} = \mathbf{A}\mathbf{x}, \qquad (2.15.2)$$

where $\mathbf{x}$ is a full-length vector. In this case each product

$$\mathbf{A}^{[k]}\mathbf{x} \qquad (2.15.3)$$

may be formed by

(i) gathering the required components of $\mathbf{x}$ into a full-length vector (a gather operation),

(ii) performing a full-length matrix by vector multiplication and

(iii) adding the result back into the relevant positions in the full-length vector $\mathbf{y}$ (a scatter operation).

Steps (i) and (iii) involve indirect addressing so are likely to be executed relatively slowly, but full advantage can be taken of vector or parallel processing in step (ii) and there will be a gain if there are many more operations to be performed in step (ii) than in steps (i) and (iii). Thus this technique will be successful if the number of indices in all the index sets (this is the number of operations in steps (i) and (iii)) is substantially less than the number of nonzeros in $\mathbf{A}$.

2.16   Comparisons between sparse matrix structures

We will now make some comparisons between the sparse matrix structures that we have been describing. It has already been remarked that the coordinate scheme is the most convenient for general-purpose data input but is inadequate for later use. It therefore remains necessary to consider the linked list, the collection of sparse vectors, and clique storage.

We first remark that coding for the linked list is usually simpler and shorter. Figure 2.10.1 provides an example of just how compact it can be. A further advantage is that no movement of data is needed, no elbow room is needed, nor does the collection need occasional compression. Its disadvantages are in greater storage demands (because of having to hold the links themselves) or greater demands for computer time if the row and/or column indices are overwritten by the links (because then occasional searches are needed). The disadvantage is enhanced by the fact that the links need a word length sufficient to hold the number of entries whereas the row and column indices cannot exceed the matrix order, unless the technique described at the end of Section 2.9 is used. The linked list has some run-time disadvantage because of the access time for the links themselves. Furthermore when we need a matrix entry we are likely to need the rest of its row/column too. These may be scattered around the store which can cause 'page thrashing' (an unreasonable number of page faults) on a machine with virtual storage.

We discussed some of the benefits of clique storage in Section 2.15. If the problem is from finite elements, the clique representation (2.15.1) is the natural way to hold the original matrix. However, as we remarked in Section 2.15, it is possible to 'disassemble' any symmetric matrix into a sum of the form (2.15.1). Duff and Reid (1983) investigated this further and we show some of their results in Table 2.16.1. This indicates that there is unlikely to be any gain from artificially generating the clique storage for a matrix if it is naturally in assembled form. However, when Gaussian elimination is applied to a symmetric (or symmetrically structured) matrix, cliques are automatically produced and (see Chapter 10) there are advantages in storing them in this form.

We are faced with a problem-dependent decision and no firm conclusion may be made for the best sparse matrix structure. Some computer codes make use of several structures; the reader with a particular application should consider the advantages and disadvantages described here as they relate to the application in hand.

Matrix order	147	1176	292	85
Nonzeros in matrix	2449	18 552	2208	523
Nonzeros stored	1298	9864	1250	304
Number of generated cliques	157	907	373	102
Number of entries in generated cliques	2970	11 286	2029	481

Table 2.16.1 Some statistics on artificial generation of cliques. Since each matrix is symmetric, only the nonzeros in one of its triangular halves need be stored.

Exercises

2.1 Write Fortran code, analogous to that of Figure 2.3.2, to execute the operation $z := x + \alpha y$ where x, y, and z are sparse vectors held in packed form with components in order.

2.2 Write Fortran code, analogous to that of Figure 2.4.1, for the operation $x := x + \alpha y$ where x and y are unordered packed vectors. Do not use a work vector, but assume that an integer workspace IW of length the number of entries in y is available and that a full-length integer vector ITEMP containing non-negative integers may be used temporarily as workspace, provided it is restored afterwards. [Hint: Set ITEMP(i) to a pointer to the position of y_i in the packed vector, preserving the old value of ITEMP(i) in IW.]

2.3 Write Fortran code analogous to that of Figure 2.4.1 for the operation $x := x + \alpha y$ when it is known that the sparsity pattern of y is a subset of that of x.

2.4 Design an algorithm to compress a collection of sparse vectors.

2.5 Write Fortran code to delete entry I from a linked list held in the array LINK with header IHEAD.

2.6 Write Fortran code to delete entry I from a doubly-linked list held with forward links in the array LINKFD with header IHEAD and backward links in the array LINKBK.

2.7 Suppose the array LENROW of length N has all its values lying between 1 and N. Write Fortran code to generate a doubly-linked list containing all entries with value K.

2.8 Write Fortran code to transform a matrix held in the coordinate scheme (a_{ij}, i, j) to a row-linked sparse matrix with links such that any row scan through them is in order of column numbers. [Hint: See the third paragraph of Section 2.10; replace column numbers temporarily by column links, restoring them when scanning by columns creating row links.]

2.9 Indicate how an in-place sort similar to that in Section 2.10 can be performed without scanning the matrix entries more than twice. Assume that two integer arrays of length n are available.

2.10 Write Fortran code that constructs the arrays LENCOL, JCOLST, and IRN of Gustavson's collection of sparse column vectors from a file of sparse row vectors held in LENROW, IROWST, JCN, and VAL. Do not assume that the rows are stored contiguously and ensure that the generated row numbers are in order within each column.

2.11 Write Fortran code that, given a matrix held as a row-linked list in arrays VAL, IROWST, JCN, and LINK, sets up the Curtis-Reid format, putting pointers to the starts of the columns in JCOLST and replacing column numbers in JCN by column pointers. Ensure that column links are in natural order.

2.12 Write Fortran code to perform the in-place permutation $\mathbf{P}^T\mathbf{x}$ when $\mathbf{P}$ is held as the sequence of interchanges (i,p_i), $i = 1, 2,..., n-1$.

2.13 By exploiting the fact that the permutation $(i{\rightarrow}\pi_i$, $i=1,2,...,n)$ can be expressed as a sequence of cycles $i_1 \rightarrow i_2 = \pi_{i_1}$, $i_2 \rightarrow i_3 = \pi_{i_2}$, $...$, $i_k \rightarrow i_1 = \pi_{i_k}$, write code for the in-place sort corresponding to that of Figure 2.14.1.

2.14 Write Fortran code to form $\mathbf{y} = \mathbf{P}^T\mathbf{x}$ when $\mathbf{P}$ is held as the set of integers π_i, $i = 1, 2,..., n$, which represent the positions of the components of $\mathbf{x}$ in $\mathbf{P}\mathbf{x}$.

2.15 Given the permutation $(i{\rightarrow}\pi_i$, $i=1,2,...,)$, write Fortran code that involves $O(n)$ operations to construct the same permutation as a sequence of interchanges (i,p_i), $i = 1, 2,..., n-1$.

2.16 Identify the permutation matrix $\mathbf{P}$ such that $\mathbf{PAP}^T$ is the matrix pattern in Figure 1.3.4 and $\mathbf{A}$ is the pattern of Figure 1.3.2. Write $\mathbf{P}$ in the two different compact storage schemes of Section 2.14.

2.17 Show that if $\mathbf{P}$ is a permutation matrix, $\mathbf{P}^T\mathbf{P} = \mathbf{I}$.

3 Gaussian elimination for dense matrices: the algebraic problem

> A review of the fundamental operations in the direct solution of linear equations: a simplified look at variations in computational steps without concern for rounding error caused by computer arithmetic.

3.1 Introduction

In this chapter we review the algebraic properties associated with the direct solution of the equation

$$\mathbf{A}\mathbf{x} = \mathbf{b}, \tag{3.1.1}$$

where $\mathbf{A}$ is an $n \times n$ nonsingular dense matrix, $\mathbf{x}$ is a vector of n unknowns, and $\mathbf{b}$ is a given vector of length n. The effect of numerical inaccuracies is deferred to the next chapter. Sparsity is not the subject of this chapter, though considerations of sparsity motivate the selection of material and the manner of presentation.

Names such as Gaussian elimination, the Crout algorithm, the Doolittle algorithm and $\mathbf{LU}$ factorization are associated with various direct methods for solving (3.1.1). All of these methods are algebraically equivalent, with minor differences in the sequence of computation. The different sequences may be used to advantage for sparsity work, so we discuss them in some detail.

Several books such as those of Fox (1964), Wilkinson (1965), Forsythe and Moler (1967), Stewart (1973), Strang (1980), and Golub and Van Loan (1983) give a more complete treatment of the numerical solution of dense equations. The review in this chapter and the next allows the development of the concepts needed to handle sparsity.

3.2 Solution of triangular systems

Basic to all general-purpose direct methods for solving equation (3.1.1) is the concept that triangular systems of equations are 'easy' to solve. They take the form

$$\mathbf{U}\mathbf{x} = \mathbf{c}, \tag{3.2.1}$$

where $\mathbf{U}$ is upper triangular (all entries below the main diagonal are zero), or the form

$$\mathbf{L}\mathbf{c} = \mathbf{b}, \tag{3.2.2}$$

where **L** is lower triangular (all entries above the main diagonal are zero). For example, when $n=4$ equation (3.2.1) takes the form

$$
\begin{bmatrix}
u_{11} & u_{12} & u_{13} & u_{14} \\
0 & u_{22} & u_{23} & u_{24} \\
0 & 0 & u_{33} & u_{34} \\
0 & 0 & 0 & u_{44}
\end{bmatrix}
\begin{bmatrix}
x_1 \\ x_2 \\ x_3 \\ x_4
\end{bmatrix}
=
\begin{bmatrix}
c_1 \\ c_2 \\ c_3 \\ c_4
\end{bmatrix}.
\tag{3.2.3}
$$

If the inequalities $u_{kk} \neq 0$, $k = 1, 2, 3, 4$, are true, the components of the solution can readily be obtained in the order x_4, x_3, x_2, x_1 by the steps

$$x_4 = c_4/u_{44}, \tag{3.2.4a}$$

$$x_3 = (c_3 - u_{34}x_4)/u_{33}, \tag{3.2.4b}$$

$$x_2 = (c_2 - u_{23}x_3 - u_{24}x_4)/u_{22}, \tag{3.2.4c}$$

$$x_1 = (c_1 - u_{12}x_2 - u_{13}x_3 - u_{14}x_4)/u_{11}. \tag{3.2.4d}$$

In general, systems of the form (3.2.1) may be solved by the steps

$$x_n = c_n/u_{nn} \tag{3.2.5a}$$

$$x_k = (c_k - \sum_{j=k+1}^{n} u_{kj}x_j)/u_{kk}, \quad k = n-1, n-2,..., 1, \tag{3.2.5b}$$

provided the inequalities $u_{kk} \neq 0$, $k = 1, 2,..., n$, are true. This process is known as **back-substitution**. Similarly, equation (3.2.2) may be solved forwards by the steps

$$c_1 = b_1/l_{11} \tag{3.2.6a}$$

$$c_k = (b_k - \sum_{j=1}^{k-1} l_{kj}c_j)/l_{kk}, \quad k = 2, 3,..., n \tag{3.2.6b}$$

provided the inequalities $l_{kk} \neq 0$, $k = 1, 2,..., n$, are true. This is known as **forward substitution**. The goal of our methods for solving equation (3.1.1) is to transform this equation into one of type (3.2.1) or (3.2.2), which may then be solved readily.

An upper (lower) triangular matrix is said to be **unit** upper (lower) triangular if it has ones on the main diagonal. Note that the formulae (3.2.5) and (3.2.6) simplify in this case.

An important current research area is concerned with forms of 'easy to solve' systems of equations other than triangular. The motivation for this work is for effective use of vector and parallel computers. Observe that the n steps in (3.2.5) and (3.2.6) are fundamentally serial operations, which limit the extent to which parallel architectures can be exploited. Nevertheless, most of the current work on vector and parallel computers makes use of these triangular forms and research on alternatives is too preliminary to discuss further here.

3.3 Gaussian elimination

One transformation to triangular form is Gaussian elimination, which we illustrate with the system

$$
\begin{bmatrix} a_{11} & a_{12} & a_{13} \\ a_{21} & a_{22} & a_{23} \\ a_{31} & a_{32} & a_{33} \end{bmatrix} \begin{bmatrix} x_1 \\ x_2 \\ x_3 \end{bmatrix} = \begin{bmatrix} b_1 \\ b_2 \\ b_3 \end{bmatrix}. \tag{3.3.1}
$$

Multiplying the first equation by a_{21}/a_{11} (assuming that the inequality $a_{11} \neq 0$ holds) and subtracting from the second produces the equivalent system

$$
\begin{bmatrix} a_{11} & a_{12} & a_{13} \\ 0 & a_{22}^{(2)} & a_{23}^{(2)} \\ a_{31} & a_{32} & a_{33} \end{bmatrix} \begin{bmatrix} x_1 \\ x_2 \\ x_3 \end{bmatrix} = \begin{bmatrix} b_1 \\ b_2^{(2)} \\ b_3 \end{bmatrix}, \tag{3.3.2}
$$

where

$$
a_{22}^{(2)} = a_{22} - (a_{21}/a_{11}) a_{12}, \tag{3.3.3a}
$$

$$
a_{23}^{(2)} = a_{23} - (a_{21}/a_{11}) a_{13}, \tag{3.3.3b}
$$

and

$$
b_2^{(2)} = b_2 - (a_{21}/a_{11}) b_1. \tag{3.3.3c}
$$

Similarly, multiplying the first equation by a_{31}/a_{11} and subtracting from the third produces the equivalent system

$$
\begin{bmatrix} a_{11} & a_{12} & a_{13} \\ 0 & a_{22}^{(2)} & a_{23}^{(2)} \\ 0 & a_{32}^{(2)} & a_{33}^{(2)} \end{bmatrix} \begin{bmatrix} x_1 \\ x_2 \\ x_3 \end{bmatrix} = \begin{bmatrix} b_1 \\ b_2^{(2)} \\ b_3^{(2)} \end{bmatrix}, \tag{3.3.4}
$$

where the new terms are given by the equations

$$
a_{32}^{(2)} = a_{32} - (a_{31}/a_{11}) a_{12}, \tag{3.3.5a}
$$

$$
a_{33}^{(2)} = a_{33} - (a_{31}/a_{11}) a_{13}, \tag{3.3.5b}
$$

and

$$
b_3^{(2)} = b_3 - (a_{31}/a_{11}) b_1. \tag{3.3.5c}
$$

Finally multiplying the new second row by $a_{32}^{(2)}/a_{22}^{(2)}$ (assuming that the inequality $a_{22}^{(2)} \neq 0$ holds) and subtracting from the new third row produces the system

$$\begin{bmatrix} a_{11} & a_{12} & a_{13} \\ 0 & a_{22}^{(2)} & a_{23}^{(2)} \\ 0 & 0 & a_{33}^{(3)} \end{bmatrix} \begin{bmatrix} x_1 \\ x_2 \\ x_3 \end{bmatrix} = \begin{bmatrix} b_1 \\ b_2^{(2)} \\ b_3^{(3)} \end{bmatrix}, \qquad (3.3.6)$$

where the new terms are given by the equations

$$a_{33}^{(3)} = a_{33}^{(2)} - (a_{32}^{(2)}/a_{22}^{(2)})\, a_{23}^{(2)} \qquad (3.3.7a)$$

and

$$b_3^{(3)} = b_3^{(2)} - (a_{32}^{(2)}/a_{22}^{(2)})\, b_2^{(2)}. \qquad (3.3.7b)$$

Notice that equation (3.3.6) has the upper triangular form (3.2.1) with the correspondences

$$\mathbf{U} = \begin{bmatrix} a_{11} & a_{12} & a_{13} \\ 0 & a_{22}^{(2)} & a_{23}^{(2)} \\ 0 & 0 & a_{33}^{(3)} \end{bmatrix} \quad \text{and} \quad \mathbf{c} = \begin{bmatrix} b_1 \\ b_2^{(2)} \\ b_3^{(3)} \end{bmatrix}. \qquad (3.3.8)$$

This process may be performed in general by creating zeros in the first column, then the second, and so forth. For $k = 1, 2,..., n-1$ we use the formulae

$$a_{ij}^{(k+1)} = a_{ij}^{(k)} - (a_{ik}^{(k)}/a_{kk}^{(k)})\, a_{kj}^{(k)}, \quad i, j > k \qquad (3.3.9a)$$

and

$$b_i^{(k+1)} = b_i^{(k)} - (a_{ik}^{(k)}/a_{kk}^{(k)})\, b_k^{(k)}, \quad i > k, \qquad (3.3.9b)$$

where $a_{ij}^{(1)} = a_{ij}$, $i, j = 1, 2,..., n$. The only assumption required is that the inequalities $a_{kk}^{(k)} \neq 0$, $k = 1, 2,..., n$, hold. These entries are called **pivots** in Gaussian elimination. It is convenient to use the notation

$$\mathbf{A}^{(k)}\mathbf{x} = \mathbf{b}^{(k)} \qquad (3.3.10)$$

for the system obtained after $(k-1)$ stages, $k = 1, 2,..., n$, with $\mathbf{A}^{(1)} = \mathbf{A}$ and $\mathbf{b}^{(1)} = \mathbf{b}$. The final matrix $\mathbf{A}^{(n)}$ is upper triangular (see equation (3.3.6), for example).

3.4 Required row interchanges

The Gaussian elimination process described above breaks down when the relation $a_{kk}^{(k)} = 0$ holds, illustrated by the case

$$\begin{pmatrix} 0 & 1 \\ 2 & 3 \end{pmatrix} \begin{pmatrix} x_1 \\ x_2 \end{pmatrix} = \begin{pmatrix} 4 \\ 5 \end{pmatrix}. \tag{3.4.1}$$

Exchanging the equations to give

$$\begin{pmatrix} 2 & 3 \\ 0 & 1 \end{pmatrix} \begin{pmatrix} x_1 \\ x_2 \end{pmatrix} = \begin{pmatrix} 5 \\ 4 \end{pmatrix} \tag{3.4.2}$$

completely avoids the difficulty. This simple observation is the basis for the solution to the problem for a matrix of any order. We illustrate by working with $n = 5$. The extension to the general case is obvious.

Suppose we have proceeded through two stages on a system of order 5 and the system has the form

$$\begin{bmatrix} a_{11}^{(1)} & a_{12}^{(1)} & a_{13}^{(1)} & a_{14}^{(1)} & a_{15}^{(1)} \\ & a_{22}^{(2)} & a_{23}^{(2)} & a_{24}^{(2)} & a_{25}^{(2)} \\ & & 0 & a_{34}^{(3)} & a_{35}^{(3)} \\ & & a_{43}^{(3)} & a_{44}^{(3)} & a_{45}^{(3)} \\ & & a_{53}^{(3)} & a_{54}^{(3)} & a_{55}^{(3)} \end{bmatrix} \begin{bmatrix} x_1 \\ x_2 \\ x_3 \\ x_4 \\ x_5 \end{bmatrix} = \begin{bmatrix} b_1^{(1)} \\ b_2^{(2)} \\ b_3^{(3)} \\ b_4^{(3)} \\ b_5^{(3)} \end{bmatrix}. \tag{3.4.3}$$

If one of the inequalities $a_{43}^{(3)} \neq 0$, $a_{53}^{(3)} \neq 0$ holds, we interchange the third row with either the fourth or fifth row and proceed. Interchanging to obtain a nonzero pivot is called **pivoting**. The only way this process can break down is if all the equalities

$$a_{33}^{(3)} = a_{43}^{(3)} = a_{53}^{(3)} = 0 \tag{3.4.4}$$

hold. In this case, equations 3, 4, and 5 of system (3.4.3) do not involve x_1, x_2, or x_3. If they have a solution, x_3 can be given an arbitrary value and x_1 and x_2 can be determined by back-substitution in equations 1 and 2 of system (3.4.3), as in steps (3.2.5). In fact, the matrix is **singular**, that is $\det \mathbf{A} = 0$, since expanding by minors produces the equation

$$\det \mathbf{A}^{(3)} = a_{11}^{(1)} a_{22}^{(2)} \det \begin{vmatrix} 0 & a_{34}^{(3)} & a_{35}^{(3)} \\ 0 & a_{44}^{(3)} & a_{45}^{(3)} \\ 0 & a_{54}^{(3)} & a_{55}^{(3)} \end{vmatrix} = 0 \tag{3.4.5}$$

and previous transformations

$$\mathbf{A} = \mathbf{A}^{(1)} \rightarrow \mathbf{A}^{(2)} \rightarrow \mathbf{A}^{(3)} \tag{3.4.6}$$

can at most have changed the sign of the determinant. This indeterminacy of a variable, in this case x_3, is a typical facet of singularity.

Extending to the general case, we see that as long as $\mathbf{A}$ is nonsingular, the equations may always be reordered through interchanging rows so that $a_{kk}^{(k)}$ is nonzero.

3.5 Relationship with LU factorization

We now examine the computation from another point of view. Assume for the moment that all the relations $a_{kk}^{(k)} \neq 0$ hold and define l_{ik} for $i > k$ by the equation

$$l_{ik} = a_{ik}^{(k)}/a_{kk}^{(k)}. \qquad (3.5.1)$$

Referring to equations (3.3.9), we see that this number is precisely what is used to multiply the k-th (pivot) row and subsequently subtract from the i-th row in building the new i-th row. Hence l_{ik} is called a **multiplier**.

Now let $\mathbf{L}^{(k)}$ be the unit lower triangular matrix

$$\mathbf{L}^{(k)} = \begin{bmatrix} 1 & & & & & \\ & \cdot & & & & \\ & & \cdot & & & \\ & & & 1 & & \\ & & & -l_{k+1,k} & \cdot & \\ & & & \cdot & & \cdot \\ & & & -l_{n,k} & & 1 \end{bmatrix}, \qquad (3.5.2)$$

which differs from the identity matrix $\mathbf{I}$ only in the k-th column below the main diagonal, where the negatives of the multipliers l_{ik} appear. Matrices of the form (3.5.2) are sometimes called **elementary lower triangular matrices**. This permits the relations (3.3.9a) to be expressed in matrix notation as the relation

$$\mathbf{A}^{(k+1)} = \mathbf{L}^{(k)} \mathbf{A}^{(k)}. \qquad (3.5.3)$$

Using this relation for all values of k gives the equation

$$\mathbf{U} = \mathbf{A}^{(n)} = \mathbf{L}^{(n-1)} \mathbf{L}^{(n-2)} \ldots \mathbf{L}^{(1)} \mathbf{A}. \qquad (3.5.4)$$

Now the inverse of $\mathbf{L}^{(k)}$ is given by the equation

$$\left(\mathbf{L}^{(k)}\right)^{-1} = \begin{bmatrix} 1 & & & & & \\ & \cdot & & & & \\ & & \cdot & & & \\ & & & 1 & & \\ & & & l_{k+1,k} & \cdot & \\ & & & \cdot & & \cdot \\ & & & \cdot & & & \cdot \\ & & & l_{n,k} & & & 1 \end{bmatrix}, \tag{3.5.5}$$

as may readily be verified by directly multiplying the matrix (3.5.2) by the matrix (3.5.5). Multiplying equation (3.5.4) successively by $\left(\mathbf{L}^{(n-1)}\right)^{-1},\dots,$ $\left(\mathbf{L}^{(1)}\right)^{-1}$ gives the equation

$$\mathbf{A} = \left(\mathbf{L}^{(1)}\right)^{-1}\left(\mathbf{L}^{(2)}\right)^{-1}\dots\left(\mathbf{L}^{(n-1)}\right)^{-1}\mathbf{U}. \tag{3.5.6}$$

By direct multiplication using (3.5.5), we find the equation

$$\left(\mathbf{L}^{(1)}\right)^{-1}\left(\mathbf{L}^{(2)}\right)^{-1}\dots\left(\mathbf{L}^{(n-1)}\right)^{-1} = \begin{bmatrix} 1 & & & & & \\ l_{21} & \cdot & & & & \\ l_{31} & & \cdot & & & \\ \cdot & & & 1 & & \\ \cdot & & & l_{k+1,k} & \cdot & \\ \cdot & & & \cdot & & \cdot \\ l_{n1} & & l_{n,k} & & & 1 \end{bmatrix} = \mathbf{L}. \tag{3.5.7}$$

We therefore see that equation (3.5.6) can be written as the triangular factorization

$$\mathbf{A} = \mathbf{LU}, \tag{3.5.8}$$

where $\mathbf{U} = \mathbf{A}^{(n)}$ and $\mathbf{L}$ is the unit lower triangular matrix of multipliers shown in equation (3.5.7). We illustrate with an example of order 3 in Figure 3.5.1.

Observe that in doing this computation on a computer, we may use a single two-dimensional array if we overwrite $\mathbf{A}^{(1)}$ by $\mathbf{A}^{(2)}$, $\mathbf{A}^{(3)}$, etc. Furthermore, each multiplier l_{ij} may overwrite the zero it creates. Thus the array finally contains both $\mathbf{L}$ and $\mathbf{U}$ in packed form, excluding the unit diagonal of $\mathbf{L}$. This is often called the L\U array.

If any row interchanges are needed during elimination, they are applied to the whole L\U array and the result is exactly as if they had been applied to $\mathbf{A}$ and the elimination had proceeded without interchanges (Exercise 3.1). Such row interchanges may be represented in matrix notation as the premultiplication of $\mathbf{A}$ by the permutation matrix $\mathbf{P}$ which is obtained from $\mathbf{I}$ by applying the same sequence of row interchanges (see Section 2.14).

$$\mathbf{A} = \mathbf{A}^{(1)} = \begin{pmatrix} 1 & 2 & 1 \\ 1 & 1 & 1 \\ -1 & 0 & 1 \end{pmatrix}$$

$$\mathbf{L}^{(1)} = \begin{pmatrix} 1 & 0 & 0 \\ -1 & 1 & 0 \\ 1 & 0 & 1 \end{pmatrix} \qquad \mathbf{L}^{(1)}\mathbf{A}^{(1)} = \mathbf{A}^{(2)} = \begin{pmatrix} 1 & 2 & 1 \\ 0 & -1 & 0 \\ 0 & 2 & 2 \end{pmatrix}$$

$$\mathbf{L}^{(2)} = \begin{pmatrix} 1 & 0 & 0 \\ 0 & 1 & 0 \\ 0 & 2 & 1 \end{pmatrix} \qquad \mathbf{L}^{(2)}\mathbf{A}^{(2)} = \mathbf{A}^{(3)} = \mathbf{U} = \begin{pmatrix} 1 & 2 & 1 \\ 0 & -1 & 0 \\ 0 & 0 & 2 \end{pmatrix}$$

$$\mathbf{L} = (\mathbf{L}^{(1)})^{-1}(\mathbf{L}^{(2)})^{-1} = \begin{pmatrix} 1 & 0 & 0 \\ 1 & 1 & 0 \\ -1 & 0 & 1 \end{pmatrix}\begin{pmatrix} 1 & 0 & 0 \\ 0 & 1 & 0 \\ 0 & -2 & 1 \end{pmatrix} = \begin{pmatrix} 1 & 0 & 0 \\ 1 & 1 & 0 \\ -1 & -2 & 1 \end{pmatrix}$$

$$\mathbf{LU} = \begin{pmatrix} 1 & 0 & 0 \\ 1 & 1 & 0 \\ -1 & -2 & 1 \end{pmatrix}\begin{pmatrix} 1 & 2 & 1 \\ 0 & -1 & 0 \\ 0 & 0 & 2 \end{pmatrix} = \begin{pmatrix} 1 & 2 & 1 \\ 1 & 1 & 1 \\ -1 & 0 & 1 \end{pmatrix} = \mathbf{A}$$

Figure 3.5.1. An example of the equivalence of Gaussian elimination and triangular factorization.

Hence we will have computed the triangular factorization of $\mathbf{PA}$ (that is $\mathbf{PA}$ = $\mathbf{LU}$). To simplify notation, we ignore the effect of interchanges in Sections 3.6 to 3.12.

In this development of the $\mathbf{LU}$ factorization of the matrix $\mathbf{A}$ and its equivalence to Gaussian elimination, we have not commented on the effect of the $\mathbf{L}^{(k)}$ matrices on $\mathbf{b}$. To see this, we substitute $\mathbf{LU}$ for $\mathbf{A}$ in equation (3.1.1) to obtain the equation

$$\mathbf{LUx} = \mathbf{b}. \tag{3.5.9}$$

If $\mathbf{c}$ is given by the equation

$$\mathbf{c} = \mathbf{Ux}, \tag{3.5.10}$$

we can compute it from the equation

$$\mathbf{Lc} = \mathbf{b} \tag{3.5.11}$$

by forward substitution, see equations (3.2.6). It may readily be verified that the relations

$$c_i = b_i^{(i)}, i = 1, 2,..., n \tag{3.5.12}$$

(see equation (3.3.10)) hold. Now $\mathbf{x}$ may be computed using back-substitution on system (3.5.10).

Notice that the effect of row and column permutations can be incorporated easily into this solution scheme. If $\mathbf{P}$ and $\mathbf{Q}$ are permutation matrices such that the factorization can be written

$$\mathbf{PAQ} = \mathbf{LU}, \tag{3.5.13}$$

then the solution of equation (3.1.1) is given by the solution $\mathbf{x}$ to the system

$$\mathbf{PAQQ}^T\mathbf{x} = \mathbf{Pb} \tag{3.5.14}$$

so that the steps (3.5.10) and (3.5.11) are replaced by the steps

$$\mathbf{y} = \mathbf{Pb}, \tag{3.5.15a}$$

$$\mathbf{Lz} = \mathbf{y}, \tag{3.5.15b}$$

$$\mathbf{Uw} = \mathbf{z}, \tag{3.5.15c}$$

and

$$\mathbf{x} = \mathbf{Qw}. \tag{3.5.15d}$$

3.6 Equivalent factorization methods, including Doolittle and Crout

The **LU** factorization of **A** may be derived in a way quite different from the approach of the previous section. This alternative derivation provides useful insight for different computational sequences.

Formally, using the rules of matrix multiplication in the equation $\mathbf{A} = \mathbf{LU}$, where **L** is unit lower triangular and **U** is upper triangular, we have the relations

$$a_{ij} = \sum_{p=1}^{\min(i,j)} l_{ip} u_{pj}, \quad i, j = 1, 2,..., n. \tag{3.6.1}$$

Rearranging terms gives the relations

$$l_{ij} = \left(a_{ij} - \sum_{p=1}^{j-1} l_{ip} u_{pj}\right)/u_{jj}, \quad i > j \tag{3.6.2}$$

and

$$u_{ij} = \left(a_{ij} - \sum_{p=1}^{i-1} l_{ip} u_{pj}\right), \quad i \leqslant j, \tag{3.6.3}$$

where a sum $\sum_{k=l}^{m}$ with $m < l$ is taken as zero (for example equation (3.6.3) reduces to the form $u_{1j} = a_{1j}$ when $i = 1$).

We may use (3.6.2) and (3.6.3) to compute the entries of **L** and **U**, provided an order is chosen such that in each case those quantities required on the right-hand side are already computed.

One convenient computational sequence involves row-wise order in the **L\U** array; that is at the k-th stage compute in order $l_{k1}, l_{k2}, ..., l_{k,k-1}, u_{kk}, ..., u_{kn}$. This is known as the **Doolittle** algorithm. To see the relationship with Gaussian elimination, we note that the partial sums

$$a_{ij} - \sum_{p=1}^{k-1} l_{ip} u_{pj} \qquad (3.6.4)$$

involved in (3.6.2) and (3.6.3) are just the intermediate numbers $a_{ij}^{(k)}$ of Gaussian elimination. Thus partial sums are stored and then retrieved in Gaussian elimination, but the entire sum is accumulated and stored once with the Doolittle algorithm. Sometimes additional accuracy is obtained by accumulating this sum in greater precision than that used to store $\mathbf{A}$, $\mathbf{L}$, and $\mathbf{U}$, but this is likely to be worthwhile only for computer hardware where the cost of greater precision arithmetic is not much more than that of standard precision. The Doolittle algorithm may also be arranged by columns, where u_{1k}, u_{2k}, ..., u_{kk}, $l_{k+1,k}$, ..., l_{nk} are computed at stage k.

By revising these operations only slightly, we generate the **Crout algorithm**. Here it is usually assumed that $\mathbf{U}$ rather than $\mathbf{L}$ has ones on the diagonal. This results in modifying equations (3.6.2) and (3.6.3) to the formulae

$$l_{ij} = a_{ij} - \sum_{p=1}^{j-1} l_{ip} u_{pj}, \quad i \geq j \qquad (3.6.5)$$

and

$$u_{ij} = (a_{ij} - \sum_{p=1}^{i-1} l_{ip} u_{pj})/l_{ii}, \quad i < j. \qquad (3.6.6)$$

The sequence of computation usually associated with the Crout algorithm is the first column of $\mathbf{L}$, then the first row of $\mathbf{U}$, then the second column of $\mathbf{L}$, then the second row of $\mathbf{U}$, etc. For a 4×4 example, entries in the $\mathbf{L}\backslash\mathbf{U}$ array are computed in the order shown in Figure 3.6.1. Note that this now contains entries of $\mathbf{L}$ rather than $\mathbf{U}$ on its diagonal.

$$\begin{bmatrix} 1 & 5 & 6 & 7 \\ 2 & 8 & 11 & 12 \\ 3 & 9 & 13 & 15 \\ 4 & 10 & 14 & 16 \end{bmatrix}$$

Figure 3.6.1. The order of computation using the Crout algorithm.

This may be compared with Gaussian elimination by writing equation (3.3.9a) in the form

$$a_{ij}^{(k+1)} = a_{ij}^{(k)} - a_{ik}^{(k)} (a_{kj}^{(k)}/a_{kk}^{(k)}) \qquad (3.6.7)$$

and making the associations

$$l_{ik} = a_{ik}^{(k)} \text{ and } u_{kj} = a_{kj}^{(k)}/a_{kk}^{(k)}, \qquad (3.6.8)$$

that is dividing the pivotal row by the pivot $a_{kk}^{(k)}$ rather than dividing the pivotal column by it. As with the Doolittle algorithm, sums are accumulated in the Crout algorithm rather than calculated one stage at a time.

More generally, we could develop the factorization

$$A = LDU, \tag{3.6.9}$$

where **D** is a diagonal matrix and both **L** and **U** have ones on their main diagonals. Then the Crout factorization corresponds to the association (**LD**)**U**, while the Doolittle factorization corresponds to **L**(**DU**).

3.7 Computational sequences

We have seen that in exact arithmetic Gaussian elimination and the various factorization methods lead to the same result, though the sequence of computations may be quite different. The examination of the different sequences is of more than academic interest, since it is of fundamental importance in the efficient implementation of sparse matrix factorization.

The key advantage of the factorization methods of the previous section is the ability to compute each entry in **L** and **U** at one time, accumulating the sum. The repeated storage and retrieval of partial sums is avoided and the accumulation may be done very efficiently.

Gaussian elimination offers an advantage which becomes apparent when numerical considerations are made. This advantage is that Gaussian elimination makes the entire matrix $\mathbf{A}^{(k)}$ available at each stage. Numerically this is useful because the size of the entries of $\mathbf{A}^{(k)}$ has an impact of the success of the computation. In the sparse case we shall see that the matrices $\mathbf{A}^{(k)}$ have a changing sparsity pattern which we also need to know.

Another area of distinction in the algorithms which will be useful in the sparse case is the different sets of 'active' entries. Figures 3.7.1. to 3.7.3 show what is stored at the beginning of the third major processing step on a 5×5 matrix. Quantities to be altered during the step are marked * and quantities to be used without alteration are marked †. Figure 3.7.1. shows the standard Gaussian elimination sequence, as already described. In **LU** factorization, we observed that the only constraint on the order of computation is that the entries of **L** and **U** on the right-hand side of (3.6.2) and (3.6.3) must be already computed so that the expression may be evaluated. The orderings that correspond to the Doolittle and Crout algorithms of the previous section are illustrated in Figures 3.7.2 and 3.7.3, respectively.

Another important variation, suggested by Epton (private communication 1979) and Dongarra, Gustavson, and Karp (1984), involves rewriting the Crout algorithm as matrix by vector operations rather than simply as

$$
\begin{matrix}
u_{11} & u_{12} & u_{13} & u_{14} & u_{15} \\
l_{21} & u_{22} & u_{23} & u_{24} & u_{25} \\
l_{31} & l_{32} & u_{33}^{\dagger} & u_{34}^{\dagger} & u_{35}^{\dagger} \\
l_{41} & l_{42} & a_{43}^{(3)*} & a_{44}^{(3)*} & a_{45}^{(3)*} \\
l_{51} & l_{52} & a_{53}^{(3)*} & a_{54}^{(3)*} & a_{55}^{(3)*}
\end{matrix}
$$

Figure 3.7.1. Computational sequence for Gaussian elimination. † Entries used. * Entries changed.

$$
\begin{matrix}
u_{11}^{\dagger} & u_{12}^{\dagger} & u_{13}^{\dagger} & u_{14}^{\dagger} & u_{15}^{\dagger} \\
l_{21}^{\dagger} & u_{22}^{\dagger} & u_{23}^{\dagger} & u_{24}^{\dagger} & u_{25}^{\dagger} \\
a_{31}^{*} & a_{32}^{*} & a_{33}^{*} & a_{34}^{*} & a_{35}^{*} \\
a_{41} & a_{42} & a_{43} & a_{44} & a_{45} \\
a_{51} & a_{52} & a_{53} & a_{54} & a_{55}
\end{matrix}
$$

Figure 3.7.2. Computational sequence for row Doolittle LU decomposition. † Entries used. * Entries changed.

$$
\begin{matrix}
l_{11} & u_{12} & u_{13}^{\dagger} & u_{14}^{\dagger} & u_{15}^{\dagger} \\
l_{21} & l_{22} & u_{23}^{\dagger} & u_{24}^{\dagger} & u_{25}^{\dagger} \\
l_{31}^{\dagger} & l_{32}^{\dagger} & a_{33}^{*} & a_{34}^{*} & a_{35}^{*} \\
l_{41}^{\dagger} & l_{42}^{\dagger} & a_{43}^{*} & a_{44} & a_{45} \\
l_{51}^{\dagger} & l_{52}^{\dagger} & a_{53}^{*} & a_{54} & a_{55}
\end{matrix}
$$

Figure 3.7.3. Computational sequence for Crout LU decomposition. † Entries used. * Entries changed.

vector operations. Referring to Figure 3.7.3 and using equations (3.6.5) and (3.6.6), we see that stage 3 of the Crout algorithm applied to a matrix of order 5 can be written in the form

$$
\begin{pmatrix} l_{33} \\ l_{43} \\ l_{53} \end{pmatrix} = \begin{pmatrix} a_{33} \\ a_{43} \\ a_{53} \end{pmatrix} - \begin{pmatrix} l_{31} & l_{32} \\ l_{41} & l_{42} \\ l_{51} & l_{52} \end{pmatrix} \begin{pmatrix} u_{13} \\ u_{23} \end{pmatrix}
\tag{3.7.1a}
$$

and

$$
\begin{pmatrix} u_{34} \\ u_{35} \end{pmatrix} = \frac{1}{l_{33}} \left(\begin{pmatrix} a_{34} \\ a_{35} \end{pmatrix} - \begin{pmatrix} u_{14} & u_{24} \\ u_{15} & u_{25} \end{pmatrix} \begin{pmatrix} l_{31} \\ l_{32} \end{pmatrix} \right).
\tag{3.7.1b}
$$

In general at the k-th stage in an $n \times n$ matrix, the computation involves the multiplication of vectors of length $(k-1)$ by matrices of dimension

$(n-k)\times(k-1)$ and $(n-k+1)\times(k-1)$. If special coding is used for the inner computation, for example by calling a member of a library of computational kernels, the greater computation gives more scope for taking advantage of particular hardware.

Until we get to the discussion of sparse solution methods, it is difficult to compare the utility of these different sequences. Both Figures 3.7.1 and 3.7.2 have all entries active at some stage of the computation (the first stage for Figure 3.7.1 and the last stage for Figure 3.7.2), so the Crout ordering illustrated in Figure 3.7.3 appears to have a decided advantage. On a vector or parallel computer, there is advantage in using the matrix by vector version. A disadvantage of the Crout sequence is that it involves calculating $\mathbf{L}$ by columns and $\mathbf{U}$ by rows, and then uses columns of $\mathbf{U}$ and rows of $\mathbf{L}$ in subsequent computations; this complicates the data structure.

The sequence suggested by Figure 3.7.2 (or its column-oriented counterpart) can be applied to a variation of Gaussian elimination (where $a_{ij}^{(k)}$ is explicitly computed and stored for each k) rather than $\mathbf{LU}$ factorization (where the entries in $\mathbf{L}$ and $\mathbf{U}$ are accumulated). Referring to Figure 3.7.2, we may compute $a_{32}^{(2)}$, $a_{33}^{(2)}$, $a_{34}^{(2)}$, and $a_{35}^{(2)}$, using the multiplier a_{31}/a_{11}. Then $a_{33}^{(3)}$, $a_{34}^{(3)}$, and $a_{35}^{(3)}$ can be computed using the multiplier $a_{32}^{(2)}/a_{22}^{(2)}$. While this apparently combines the disadvantages of Gaussian elimination (no accumulated sums), the disadvantages of $\mathbf{LU}$ factorization (no updated submatrix), and the disadvantage of this computational sequence (all entries active at the final stage), it has the advantage that all entries are accessed sequentially within their rows. For some sparsity applications this advantage outweighs all the disadvantages. This approach is called **row Gaussian elimination** by Tinney and Walker (1967). We could equally well take take this approach by columns instead of by rows.

3.8 Symmetric matrices

An important special case of these algorithms occurs when $\mathbf{A}$ is symmetric (that is $a_{ij} = a_{ji}$ for all i, j). This is not simply a mathematical phenomenon since matrices in structural analysis, some areas of power systems analysis, optimization, and many other application areas have this property. In this case, assuming that pivots are chosen from the diagonal, the active parts of $\mathbf{A}^{(k)}$, $k = 1, 2,..., n$ are symmetric, that is the relations

$$a_{ij}^{(k)} = a_{ji}^{(k)}, \quad i \geqslant k, \ j \geqslant k \qquad (3.8.1)$$

hold. This is easily deduced from the fundamental Gaussian elimination equation (3.3.9a). Equation (3.8.1) is true for $k = 1$ and equation (3.3.9a) tells us that it is true for $k+1$ if it is true for k. It follows that only about half of the arithmetic operations need be performed in the symmetric case.

The relation (3.8.1) also allows the immediate deduction that in the **LDU** factorization of **A** the relation

$$l_{ij} = u_{ji}, \quad i > j \tag{3.8.2}$$

holds, that is $\mathbf{L} = \mathbf{U}^T$. The factorization is usually written $\mathbf{LDL}^T$ in this case and of course **U** is not stored.

If **A** is also positive definite (that is if $\mathbf{x}^T \mathbf{A} \mathbf{x} > 0$ for all nonzero vectors **x**), then it can be shown that each d_{ii} must be positive (see Exercise 3.12). In this case, the factorization can be written

$$\mathbf{A} = (\mathbf{LD}^{\frac{1}{2}})(\mathbf{D}^{\frac{1}{2}}\mathbf{L}^T) = \bar{\mathbf{L}}\,\bar{\mathbf{L}}^T. \tag{3.8.3}$$

This is usually called the **Choleski** factorization. Note that in practice many symmetric matrices are positive definite because an energy minimizing principle is associated with the underlying mathematical model.

As in the unsymmetric case, the process breaks down if any $a_{kk}^{(k)}$ is zero, though this cannot happen in the positive-definite case (see Exercise 3.13). Unfortunately, we destroy symmetry if we use row interchanges to avoid this difficulty. If another diagonal entry $a_{jj}^{(k)}$, $j > k$ is nonzero, the symmetric permutation that interchanges rows j and k and columns j and k will bring the nonzero $a_{jj}^{(k)}$ to the required position. There results a factorization

$$\mathbf{PAP}^T = \mathbf{LDL}^T, \tag{3.8.4}$$

where **P** is the permutation matrix that represents all the interchanges.

Unfortunately even this process can break down without **A** being singular, as the example

$$\mathbf{A} = \begin{pmatrix} 0 & 1 \\ 1 & 0 \end{pmatrix} \tag{3.8.5}$$

shows. We return to this in the next chapter (Section 4.7).

As in the unsymmetric case, there is some freedom in the choice of computational sequence. Those corresponding to storage of the lower triangular part of the matrix are shown in Figures 3.8.1 to 3.8.3 and correspond to those in Figures 3.7.1 to 3.7.3.

$$
\begin{array}{l}
d_{11} \\
l_{21} \ d_{22} \\
l_{31} \ l_{32} \ d_{33}^{\dagger} \\
l_{41} \ l_{42} \ a_{43}^{(3)*} \ a_{44}^{(3)*} \\
l_{51} \ l_{52} \ a_{53}^{(3)*} \ a_{54}^{(3)*} \ a_{55}^{(3)*}
\end{array}
$$

Figure 3.8.1. Computational sequence for symmetric Gaussian elimination. † Entries used. * Entries changed.

$$
\begin{array}{lllll}
d_{11}^{\dagger} & & & & \\
l_{21}^{\dagger} & d_{22}^{\dagger} & & & \\
a_{31}^{*} & a_{32}^{*} & a_{33}^{*} & & \\
a_{41} & a_{42} & a_{43} & a_{44} & \\
a_{51} & a_{52} & a_{53} & a_{54} & a_{55}
\end{array}
$$

Figure 3.8.2. Computational sequence for row Doolittle $\mathbf{LDL}^T$ decomposition. † Entries used. * Entries changed.

$$
\begin{array}{lllll}
d_{11}^{\dagger} & & & & \\
l_{21} & d_{22}^{\dagger} & & & \\
l_{31}^{\dagger} & l_{32}^{\dagger} & a_{33}^{*} & & \\
l_{41}^{\dagger} & l_{42}^{\dagger} & a_{43}^{*} & a_{44} & \\
l_{51}^{\dagger} & l_{52}^{\dagger} & a_{53}^{*} & a_{54} & a_{55}
\end{array}
$$

Figure 3.8.3. Computational sequence for Crout $\mathbf{LDL}^T$ decomposition. † Entries used. * Entries changed.

3.9 Gauss-Jordan elimination

The back-substitution phase of Gaussian elimination may be avoided by adding row operations that make entries above the diagonal zero. The operations (3.3.9) are replaced by the operations

$$a_{ij}^{(k+1)} = a_{ij}^{(k)} - (a_{ik}^{(k)}/a_{kk}^{(k)})\,a_{kj}^{(k)}, \quad i \neq k, j > k, \tag{3.9.1a}$$

$$b_i^{(k+1)} = b_i^{(k)} - (a_{ik}^{(k)}/a_{kk}^{(k)})\,b_k^{(k)}, \quad i \neq k, \tag{3.9.1b}$$

and

$$b_k^{(k+1)} = b_k^{(k)}, \tag{3.9.1c}$$

and the system is reduced to the diagonal form

$$a_{kk}^{(k)} x_k = b_k^{(n)}, \quad k = 1, 2, \ldots, n, \tag{3.9.2}$$

which is trivial to solve.

This procedure seems very attractive, but unfortunately involves more work in the full case (see Section 3.11). Unless great care is taken, as for the Hellerman-Rarick algorithm of Chapter 8, it can be particularly uneconomical in the sparse case because of the likelihood that many extra nonzeros are created by the additional row operations (see, for example, Brayton, Gustavson, and Willoughby 1970).

3.10 Multiple right-hand sides and inverses

So far we have considered the case with a single right-hand side **b**, but it is trivial to extend the algorithms to the case with m right-hand sides. In the forward-substitution phase of Gaussian elimination, equation (3.3.9b) extends to

$$b_{ij}^{(k+1)} = b_{ij}^{(k)} - (a_{ik}^{(k)}/a_{kk}^{(k)})\,b_{kj}^{(k)}, \quad i > k, \ j = 1, 2, ..., m \qquad (3.10.1)$$

and in the back-substitution phase equations (3.2.5) extend to

$$x_{nj} = c_{nj}/u_{nn}, \quad j = 1, 2, ..., m, \qquad (3.10.2a)$$

$$x_{kj} = (c_{kj} - \sum_{p=k+1}^{n} u_{kp} x_{pj})/u_{kk}, \quad k = n-1, n-2, ..., 1, \ j = 1, 2, ..., m. \qquad (3.10.2b)$$

An important special case is where the inverse $\mathbf{A}^{-1}$ is required, since this may be obtained by solving

$$\mathbf{AX} = \mathbf{I} \qquad (3.10.3)$$

by taking columns of **I** as successive right-hand side vectors. Note that in this case a worthwhile computational saving is available in equation (3.10.1) because the elimination operations always leave a right-hand side that is a lower triangular matrix (that is $b_{ij}^{(i)} = 0$ for $i < j$).

If a sequence of problems with the same matrix but different right-hand sides **b** is to be solved, it is tempting to calculate the inverse and use it to form the product

$$\mathbf{x} = \mathbf{A}^{-1}\mathbf{b}, \qquad (3.10.4)$$

but this is not a sufficient reason for calculating the inverse. This is because in the dense case the number of floating-point operations needed to form the product is approximately that of solving

$$\mathbf{LUx} = \mathbf{b} \qquad (3.10.5)$$

given the factorization $\mathbf{A} = \mathbf{LU}$ (see next section). The form (3.10.4) may be advantageous on parallel architectures, however. In the sparse case the inverse $\mathbf{A}^{-1}$ is usually dense (see Section 12.6), whereas the factors **L** and **U** are usually sparse, so using relation (3.10.4) may be very much more expensive (perhaps by factors of hundreds or thousands) than using relation (3.10.5).

3.11 Computational cost

One commonly used measure of computational cost is the number of floating-point operations and we use it here because for the algorithms we analyse it represents the relative performances adequately. Notice that we adhere to our definition of floating-point operation in Section 1.1, although some authors have in the past counted the number of multiply-add pairs.

Since all of the factorization methods presented in this chapter except for Gauss-Jordan elimination perform the same operations, we examine the costs of Gaussian elimination and Gauss-Jordan elimination. We continue to examine the algebraic problem and assume that the matrix has been reordered, if necessary, so that each inequality $a_{kk}^{(k)} \neq 0$ holds.

Let r_k be the number of entries to the right of the main diagonal in row k of $\mathbf{A}^{(k)}$. Let c_k be the number of entries below the main diagonal in column k of $\mathbf{A}^{(k)}$. For a dense matrix, these have values $r_1 = n-1$, $c_1 = n-1$, $r_2 = n-2$,..., $r_n = 0$. Computing $\mathbf{A}^{(k+1)}$ from $\mathbf{A}^{(k)}$ involves computing c_k multipliers, and then performing $c_k r_k$ multiply-add pairs. The total cost of Gaussian elimination, excluding the work on the right-hand side, is therefore given by the formula

$$2\sum_{k=1}^{n-1} c_k r_k + \sum_{k=1}^{n-1} c_k. \qquad (3.11.1)$$

Straightforward computation shows that in the dense case this has the value

$$\tfrac{2}{3}n^3 - \tfrac{1}{2}n^2 - \tfrac{1}{6}n. \qquad (3.11.2)$$

As n gets large the leading term dominates the cost, and formula (3.11.2) is usually written in the form

$$\tfrac{2}{3}n^3 + O(n^2). \qquad (3.11.3)$$

When $\mathbf{A}$ is symmetric and can be factorized symmetrically, only the upper (or lower) triangle of the matrices $\mathbf{A}^{(k)}$ need be calculated, which yields the formula

$$\tfrac{1}{3}n^3 + O(n^2). \qquad (3.11.4)$$

Referring to equations (3.2.5b) we see that the back-substitution operation involves one multiply-add pair for each off-diagonal entry of $\mathbf{U}$. Hence the back-substitution cost, including the divide for each u_{kk}, is given by the simple formula

$$n + 2\sum_{i=1}^{n-1} r_i. \qquad (3.11.5)$$

Similarly, the forward substitution cost is given by the formula

$$2 \sum_{i=1}^{n-1} c_i. \tag{3.11.6}$$

For dense matrices, formula (3.11.5) reduces to the expression

$$n^2, \tag{3.11.7}$$

and formula (3.11.6) becomes the expression

$$n^2 - n. \tag{3.11.8}$$

Similar calculations for the cost of Gauss-Jordan elimination (Section 3.9) shows this to be $\frac{1}{3}n^3 + O(n^2)$ greater than Gaussian elimination. Since solving an upper triangular system costs $n^2 + O(n)$, this cost penalty is not recovered.

To compute $\mathbf{A}^{-1}$ efficiently requires solving the systems of equations

$$\mathbf{AX} = \mathbf{I} \tag{3.11.9}$$

(see Section 3.10), taking care to exploit the fact that the right-hand side is lower triangular throughout forward elimination. The overall cost is $2n^3 + O(n^2)$, approximately three times the cost of factorization itself (see Exercise 3.9). Note that using it to compute the solution of $\mathbf{Ax} = \mathbf{b}$ as $\mathbf{A}^{-1}\mathbf{b}$ and computing $\mathbf{x}$ from the factorization both require $2n^2 + O(n)$ operations. In the sparse case, using the inverse may be much more expensive. Note that a quite different way of computing certain entries of the inverse of a sparse matrix is discussed in Section 12.7.

Using these cost figures we note that, on a scalar machine, it is better not to use $\mathbf{A}^{-1}$ explicitly even when computing the matrix transformation

$$\mathbf{Q} = \mathbf{A}^{-1}\mathbf{B}\,\mathbf{A}. \tag{3.11.10}$$

The cost of multiplying two $n \times n$ dense matrices in the straightforward way is $2n^3$ (see Exercise 3.7). Thus actually using $\mathbf{A}^{-1}$ to compute $\mathbf{Q}$ requires

$$
\left.
\begin{aligned}
&\text{1. Compute } \mathbf{A}^{-1}: \quad 2n^3 \\
&\text{2. Form } \mathbf{A}^{-1}\mathbf{B}: \quad\ \ 2n^3 \\
&\text{3. Form } (\mathbf{A}^{-1}\mathbf{B})\mathbf{A}: 2n^3
\end{aligned}
\right\} \ \text{Total } 6n^3. \tag{3.11.11}
$$

Alternatively, we may use the following steps

$$
\left.
\begin{aligned}
&\text{1. Factor } \mathbf{A} \text{ as } \mathbf{LU}: \frac{2}{3}n^3 \\
&\text{2. Solve } \mathbf{AX} = \mathbf{B}: \quad 2n^3 \\
&\text{3. Form } \mathbf{XA}: \qquad\ \ 2n^3
\end{aligned}
\right\} \ \text{Total } 4\tfrac{2}{3}n^3. \tag{3.11.12}
$$

Again we emphasize that in the sparse case the advantage of avoiding the computation of $\mathbf{A}^{-1}$ is much more marked.

Unfortunately, simply counting floating-point operations is not adequate when vector and parallel computers are used. Two algorithms with identical operation counts can have quite different performances on such computers. In this case we really need to add estimates for vector start-up times.

There is scope for exploiting parallelism in the use of A^{-1}, since the computation of each component of $A^{-1}x$ can be regarded as an independent calculation that involves a row of A^{-1}. It may therefore be worthwhile to calculate A^{-1} explicitly, although parallelism can also be used in the solution of $AX = B$.

There are algorithms which reduce the number of operations for $n \times n$ matrix multiplications below $O(n^3)$. They are based on an algorithm of Strassen (1969) and the currently asymptotically fastest variant can multiply two $n \times n$ matrices in $O(n^{2.496})$ operations (Pan 1984). These algorithms are only applicable to full systems and the multiplicative constants require n to be very large for noticeable benefits over the $O(n^3)$ procedure. They are hence only of peripheral interest in sparse matrix work. For this reason, we always take the cost of multiplying two $n \times n$ matrices to be $2n^3$ operations.

3.12 Block factorization

The ideas discussed so far in this chapter may be extended to block matrices, the blocks being treated similarly to the single entries of the previous sections. It will suffice for our purposes to consider the 2×2 case

$$A = \begin{pmatrix} A_{11} & A_{12} \\ A_{21} & A_{22} \end{pmatrix}, \tag{3.12.1}$$

where both A_{11} and A_{22} are square submatrices. We will refer to the submatrices on the diagonal as the **diagonal blocks**; note that in general they are not diagonal matrices. The **LU** factorization of A may be written in the form

$$A = \begin{pmatrix} A_{11} & A_{12} \\ A_{21} & A_{22} \end{pmatrix} = \begin{pmatrix} L_{11} & \\ L_{21} & L_{22} \end{pmatrix} \begin{pmatrix} U_{11} & U_{12} \\ & U_{22} \end{pmatrix}, \tag{3.12.2}$$

where L_{11} and L_{22} are lower triangular submatrices and U_{11} and U_{22} are upper triangular submatrices. This is just a partitioning of the standard **LU** factorization. By equating corresponding blocks we find the relations

$$A_{11} = L_{11} U_{11}, \tag{3.12.3a}$$

$$L_{11} U_{12} = A_{12}, \tag{3.12.3b}$$

$$L_{21} U_{11} = A_{21}, \tag{3.12.3c}$$

and

$$\dot{\mathbf{A}}_{22} = \mathbf{A}_{22} - \mathbf{L}_{21}\mathbf{U}_{12} = \mathbf{L}_{22}\mathbf{U}_{22}. \qquad (3.12.3d)$$

Therefore form (3.12.2) may be constructed by triangular factorization of $\mathbf{A}_{11}$, formation of the columns of $\mathbf{U}_{12}$ and the rows of $\mathbf{L}_{21}$ by forward substitution, and finally the triangular factorization of $\mathbf{A}_{22} - \mathbf{L}_{21}\mathbf{U}_{12}$.

For each of the triangular factorizations we can use the techniques of this chapter including the use of row and/or column interchanges where appropriate. For simplicity of notation, we regard any such permutation as being absorbed into the overall permutations needed for the block form.

Just as for **LU** factorization, where we can choose either **L** or **U** to be unit triangular, so here we may choose one or other of the block factors to be block unit triangular. This gives the alternatives

$$\mathbf{A} = \begin{pmatrix} \mathbf{A}_{11} & \mathbf{A}_{12} \\ \mathbf{A}_{21} & \mathbf{A}_{22} \end{pmatrix} = \begin{pmatrix} \mathbf{A}_{11} & \\ \mathbf{A}_{21} & \bar{\mathbf{A}}_{22} \end{pmatrix} \begin{pmatrix} \mathbf{I} & \bar{\mathbf{U}}_{12} \\ & \mathbf{I} \end{pmatrix}, \qquad (3.12.4)$$

and

$$\mathbf{A} = \begin{pmatrix} \mathbf{A}_{11} & \mathbf{A}_{12} \\ \mathbf{A}_{21} & \mathbf{A}_{22} \end{pmatrix} = \begin{pmatrix} \mathbf{I} & \\ \ddot{\mathbf{L}}_{21} & \mathbf{I} \end{pmatrix} \begin{pmatrix} \mathbf{A}_{11} & \mathbf{A}_{12} \\ & \ddot{\mathbf{A}}_{22} \end{pmatrix}. \qquad (3.12.5)$$

Forms (3.12.4) and (3.12.5) are alternative block triangular factorizations. By equating blocks we find the relations

$$\mathbf{A}_{11}\bar{\mathbf{U}}_{12} = \mathbf{A}_{12}, \qquad (3.12.6a)$$

$$\bar{\mathbf{A}}_{22} = \mathbf{A}_{22} - \mathbf{A}_{21}\bar{\mathbf{U}}_{12}, \qquad (3.12.6b)$$

and

$$\ddot{\mathbf{L}}_{21}\mathbf{A}_{11} = \mathbf{A}_{21}, \qquad (3.12.7a)$$

$$\ddot{\mathbf{A}}_{22} = \mathbf{A}_{22} - \ddot{\mathbf{L}}_{21}\mathbf{A}_{12}. \qquad (3.12.7b)$$

Constructing these forms requires the solution of the sets of equations with matrix $\mathbf{A}_{11}$ in (3.12.6a) and (3.12.7a) and the matrix operations in (3.12.6b) and (3.12.7b). The solution might be obtained by triangular factorization of $\mathbf{A}_{11}$, but any other convenient method may be used including partitioning (that is, using the concept recursively). Indeed the partition may be chosen specially so that calculating these solutions is easy even though $\mathbf{A}_{11}$ has large order (see Section 11.2, for example).

The matrices $\dot{\mathbf{A}}_{22}$ of (3.12.3d), $\bar{\mathbf{A}}_{22}$ of (3.12.6b), and $\ddot{\mathbf{A}}_{22}$ of (3.12.7b) are identical (apart from rounding errors) since in each case we are merely performing Gaussian elimination on the block $\mathbf{A}_{11}$ (see Exercise 3.14). The (2,2) block is known as the **Schur complement**.

Given a block factorization of the matrix **A**, the forward substitution and back-substitution can proceed using the partitioned form of the equations. We illustrate this process by developing the computational steps for the second block factorization (3.12.4). The development for the other two forms is left to Exercise 3.15.

Given $\mathbf{A} = \bar{\mathbf{L}}\bar{\mathbf{U}}$, we solve $\mathbf{Ax} = \mathbf{b}$ by organizing $\bar{\mathbf{L}}(\bar{\mathbf{U}}\mathbf{x}) = \mathbf{b}$ as

$$\bar{\mathbf{L}}\mathbf{y} = \mathbf{b} \qquad (3.12.8a)$$

and

$$\bar{\mathbf{U}}\mathbf{x} = \mathbf{y}. \qquad (3.12.8b)$$

Using the partition (3.12.4), equation (3.12.8a) becomes

$$\begin{pmatrix} \mathbf{A}_{11} & \\ \mathbf{A}_{21} & \bar{\mathbf{A}}_{22} \end{pmatrix} \begin{pmatrix} \mathbf{y}_1 \\ \mathbf{y}_2 \end{pmatrix} = \begin{pmatrix} \mathbf{b}_1 \\ \mathbf{b}_2 \end{pmatrix}, \qquad (3.12.9)$$

which dictates the steps

$$\mathbf{A}_{11}\mathbf{y}_1 = \mathbf{b}_1 \qquad (3.12.10a)$$

and

$$\bar{\mathbf{A}}_{22}\mathbf{y}_2 = \mathbf{b}_2 - \mathbf{A}_{21}\mathbf{y}_1. \qquad (3.12.10b)$$

Then, corresponding to (3.12.8b), we have

$$\begin{pmatrix} \mathbf{I} & \bar{\mathbf{U}}_{12} \\ & \mathbf{I} \end{pmatrix} \begin{pmatrix} \mathbf{x}_1 \\ \mathbf{x}_2 \end{pmatrix} = \begin{pmatrix} \mathbf{y}_1 \\ \mathbf{y}_2 \end{pmatrix}, \qquad (3.12.11)$$

which gives the steps

$$\mathbf{x}_2 = \mathbf{y}_2 \qquad (3.12.12a)$$

and

$$\mathbf{x}_1 = \mathbf{y}_1 - \bar{\mathbf{U}}_{12}\mathbf{x}_2. \qquad (3.12.12b)$$

An interesting variation, known as **implicit block factorization**, results if the factorizations $\mathbf{A}_{11} = \mathbf{L}_{11}\mathbf{U}_{11}$ and $\bar{\mathbf{A}}_{22} = \mathbf{L}_{22}\mathbf{U}_{22}$ are stored, but $\bar{\mathbf{U}}_{12}$ is not. When $\bar{\mathbf{U}}_{12}$ is needed as a multiplier, as in equation (3.12.12b), $\mathbf{A}_{11}^{-1}\mathbf{A}_{12}$ is used instead. This has little merit in the full case, but in the sparse case it is extremely likely that $\bar{\mathbf{U}}_{12}$ has many more entries than $\mathbf{A}_{12}$ so less storage will be needed and sometimes less computation also. There is a corresponding implicit version of the factorization (3.12.5) in which $\ddot{\mathbf{L}}_{21}$ is not stored.

Note that just as for ordinary factorization when we need to avoid zero pivots, so here we must avoid $\mathbf{A}_{11}$ being a singular block pivot. The fact that $\mathbf{A}$ is nonsingular does not guarantee the nonsingularity of $\mathbf{A}_{11}$ as the example

$$\mathbf{A} = \begin{pmatrix} 1 & 1 & 1 \\ 1 & 1 & -1 \\ 0 & 1 & 1 \end{pmatrix} \qquad (3.12.13)$$

illustrates. We discuss such issues in Section 4.4.

Exercises

3.1 With row permutations at each step of the elimination process, equation (3.5.4) takes the form

$$U = L^{(n-1)}P^{(n-1)}L^{(n-2)}P^{(n-2)}\ldots L^{(1)}P^{(1)}A, \qquad (3X.1.1)$$

where $P^{(k)}$ represents an interchange between row k and a later row and $L^{(k)}$ has the form shown in equation (3.5.2). Show that this can also be expressed in the form

$$U = \bar{L}^{(n-1)}\bar{L}^{(n-2)}\ldots \bar{L}^{(1)}P^{(n-1)}P^{(n-2)}\ldots P^{(1)}A, \qquad (3X.1.2)$$

where $\bar{L}^{(k)}$ is the matrix

$$\bar{L}^{(k)} = P^{(n-1)}P^{(n-2)}\ldots P^{(k+1)}L^{(k)}P^{(k+1)}\ldots P^{(n-2)}P^{(n-1)}. \qquad (3X.1.3)$$

Interpret equation (3X.1.2) as Gaussian elimination applied to the permuted matrix $P^{(n-1)}\ldots P^{(1)}A$.

3.2 For $A = \begin{pmatrix} 1 & 2 & 3 \\ 4 & 5 & 6 \\ 7 & 8 & 10 \end{pmatrix}$, find L and U by Gaussian elimination.

3.3 For the matrix A of Exercise 3.2, find L and U using the Crout algorithm.

3.4 Compute $L^{(1)}$ and $(L^{(1)})^{-1}$ for the matrix A of Exercise 3.2.

3.5 Compute the inverse of the matrix $\begin{pmatrix} 1 & 0 & 0 & 0 \\ 2 & 1 & 0 & 0 \\ 3 & 0 & 1 & 0 \\ 4 & 0 & 0 & 1 \end{pmatrix}$.

3.6 Find the product $\begin{pmatrix} 1 & 0 & 0 & 0 \\ 2 & 1 & 0 & 0 \\ 3 & 0 & 1 & 0 \\ 4 & 0 & 0 & 1 \end{pmatrix} \begin{pmatrix} 1 & 0 & 0 & 0 \\ 0 & 1 & 0 & 0 \\ 0 & 5 & 1 & 0 \\ 0 & 6 & 0 & 1 \end{pmatrix}$.

3.7 Determine the cost of multiplying two general $n \times n$ matrices in the standard way.

3.8 What is the computational cost of forward substitution given in equations (3.2.6) for the special case where $b = (0\ 0\ \cdots\ 0\ 1)^T$?

3.9 Show that, given the factorization $A = LU$, the number of operations needed to solve $AX = I$ is $\frac{4}{3}n^3 + O(n^2)$. (Hint: Use Exercise 3.8.)

3.10 Determine the formulae for computing L, D, and U where $A = LDU$, L is unit lower triangular, U is unit upper triangular, and D is diagonal.

3.11 Given $A = LDU$ as in Exercise 3.10, what are the steps needed to solve $Ax = b$?

3.12 Prove that if A is symmetric and positive definite, then for $A = LDL^T$ defined in Section 3.8, each d_{ii} must be positive.

3.13 Prove that if A is symmetric and positive definite then no interchanges are required in computing $A = LDL^T$ and the algorithm cannot fail, that is the inequalities $a_{kk}^{(k)} \neq 0$, $k = 1, 2,\ldots, n$, hold. (Hint: Use Exercise 3.12.)

3.14 Show that the matrices $\dot{\mathbf{A}}_{22}$, $\bar{\mathbf{A}}_{22}$, and $\ddot{\mathbf{A}}_{22}$ from the factorizations (3.12.2), (13.12.4), and (3.12.5) are identical.

3.15 Show how one can use the block factorizations (3.12.2) and (3.12.5) to solve a partitioned system of equations.

3.16 If $\mathbf{A} = \begin{pmatrix} \mathbf{A}_{11} & \mathbf{A}_{12} \\ \mathbf{A}_{21} & \mathbf{A}_{22} \end{pmatrix}$, where $\mathbf{A}_{11}$ is a lower triangular $k \times k$ matrix, $\mathbf{A}_{22}$ is 1×1, and $\mathbf{A}_{12}$ and $\mathbf{A}_{21}$ are both dense, compare the computational costs of using factorization (3.12.4) and its implicit variant.

4 Gaussian elimination for dense matrices: numerical considerations

A review of the impact of inexact computation (caused by data uncertainty and computer rounding error) on the Gaussian elimination process: controlling and assessing errors caused by the algorithm and identifying errors inherent in the problem itself (ill-conditioning).

4.1 Introduction

Neither problem data nor computer arithmetic is exact. Thus, when the algorithms of the previous chapter are implemented on a computer, the solutions will not be exact. As we prepare to solve very large systems of equations, perhaps with thousands of unknowns, it is natural to be concerned about the effect of these inaccuracies on the final result.

There are two ways to study the effect of arithmetic errors on the solution of sets of equations. The most obvious way, accumulating their effect through every stage of the algorithm, is extremely tedious and usually produces very pessimistic results unless full account is taken of the correlations between errors. There is a less direct way to assess this error which has become standard practice for numerical linear algebra computations. It involves answering two questions:

(1) Is the computed solution $\tilde{x}$ the exact solution of a 'nearby' problem?

(2) If small changes are made to the given problem, are changes in the exact solution also small?

A positive answer to the first question means that we have been able to control the computing error. That is, the error in the computed solution is no greater than would result from making small perturbations to the original problem and then solving the perturbed problem exactly. In this case we say that the algorithm is **stable**. We seek always to achieve this.

If the answer to the second question is positive, that is if small changes in the data produce small changes in the solution, we say that the problem is **well-conditioned**. Conversely, when small changes in the data produce large changes in the solution, we say that the problem is **ill-conditioned**. Note that this is a property of the problem and has nothing to do with the method used to solve it. Care is always needed in the ill-conditioned case. Data errors and errors introduced by the algorithm may cause large changes to the solution. Sometimes (see for example the simple case discussed in Section 4.8) the ill-conditioning may be so severe that only a few figures, or even none, are computed with certainty.

This discussion has been deliberately vague, without defining the precise meanings of large and small or giving rigorous bounds. These issues are addressed and illustrated in the rest of this chapter.

4.2 Computer arithmetic error

Different computers represent numbers to different numbers of digits (typically the equivalent of between six and sixteen decimals), so representation errors occur as soon as numbers are entered into the computer. For instance, $\frac{1}{3}$ in its decimal representation is not exact with any finite number of digits. Since the internal representation of real numbers is usually floating-point with a radix that is a power of 2 (for example 2, 8, 16), even $\frac{1}{5}$ is unlikely to be represented exactly. So for virtually all problems, even those with no inexact data, as soon as $\mathbf{A}$ and $\mathbf{b}$ are entered into the computer they are replaced by $\mathbf{A} + \mathbf{\Delta A}$ and $\mathbf{b} + \mathbf{\Delta b}$, where $\mathbf{\Delta A}$ and $\mathbf{\Delta b}$ are errors.

Once the arithmetic begins with these numbers, further error is introduced. A fundamental problem is that the associativity property of addition does not hold; for example the relation

$$a \mp (b \mp c) = (a \mp b) \mp c, \qquad (4.2.1)$$

where $\mp$ represents computer addition, is usually untrue.

For purposes of illustration in this chapter, we use a hypothetical computer with a 3-decimal floating-point representation of the form $\pm 0.d_1 d_2 d_3 \times 10^i$ where i is an integer, d_1, d_2, d_3 are decimal digits and $d_1 \neq 0$ unless $d_1 = d_2 = d_3 = 0$. For convenience, we will write such numbers in fixed-point form (for example 1.75, 0.00171, 14400). In this case note that the relations

$$2420 \mp 1.58 = 2420, \qquad (4.2.2)$$

$$-2420 \mp (2420 \mp 1.58) = 0, \qquad (4.2.3)$$

and

$$(-2420 \mp 2420) \mp 1.58 = 1.58 \qquad (4.2.4)$$

are all true.

This kind of error, without proper control, can have a devastating effect on the solution process. On the other hand, with the care discussed in the next five sections ensuring algorithm stability, very large systems of equations can be solved with little loss of accuracy from this source.

We find it convenient to refer to the **relative precision** ε of the computer. This is the smallest number such that if $\dagger$ is any of the arithmetic operators $+, -, \times, /,$ and $\tilde{\dagger}$ is the corresponding computer operation, then $a \, \tilde{\dagger} \, b$ has the value $a(1+\varepsilon_a) \, \dagger \, b(1+\varepsilon_b)$, where ε_a and ε_b are perturbations satisfying the inequalities $|\varepsilon_a| < \varepsilon$, $|\varepsilon_b| < \varepsilon$. In our hypothetical 3-digit computer we assume that the arithmetic operations are performed as accurately as possible and hence ε has the value 0.0005.

4.3 Algorithm instability

Using our hypothetical 3-digit computer, we illustrate the effects of computer errors on the algorithms of Chapter 3 with the matrix

$$\mathbf{A} = \begin{pmatrix} 0.001 & 2.42 \\ 1.00 & 1.58 \end{pmatrix}. \qquad (4.3.1)$$

Since the relation $a_{11} \neq 0$ holds, the factorization proceeds without interchanges and we compute

$$a_{22}^{(2)} = -2420 \mp 1.58 = -2420. \qquad (4.3.2)$$

The effect of this error is seen, for example, in solving the equation

$$\mathbf{Ax} = \begin{pmatrix} 0.001 & 2.42 \\ 1.00 & 1.58 \end{pmatrix} \begin{pmatrix} x_1 \\ x_2 \end{pmatrix} = \begin{pmatrix} 5.20 \\ 4.57 \end{pmatrix}. \qquad (4.3.3)$$

The triangular system which results from applying Gaussian elimination is

$$\begin{pmatrix} 0.001 & 2.42 \\ 0 & -2420 \end{pmatrix} \begin{pmatrix} x_1 \\ x_2 \end{pmatrix} = \begin{pmatrix} 5.20 \\ -5200 \end{pmatrix}. \qquad (4.3.4)$$

The computed solution is therefore

$$\tilde{\mathbf{x}} = \begin{pmatrix} \tilde{x}_1 \\ \tilde{x}_2 \end{pmatrix} = \begin{pmatrix} 0.00 \\ 2.15 \end{pmatrix}, \qquad (4.3.5)$$

while the 3-place approximation to the true solution is

$$\mathbf{x} = \begin{pmatrix} x_1 \\ x_2 \end{pmatrix} \simeq \begin{pmatrix} 1.18 \\ 2.15 \end{pmatrix}. \qquad (4.3.6)$$

In this example, the second equation in the transformed system (4.3.4) is nearly proportional to the first. Since x_1 was not well determined from the first equation in (4.3.3), and information about x_1 in the second equation (4.3.3) has been destroyed, x_1 can no longer be determined accurately. This is manifest in the computed LU factorization of $\mathbf{A}$. Instead of the exact LU factorization of Chapter 3 we have computed an approximate factorization $\mathbf{A} \simeq \tilde{\mathbf{L}}\tilde{\mathbf{U}}$, given by

$$\begin{pmatrix} 0.001 & 2.42 \\ 1.00 & 1.58 \end{pmatrix} \simeq \begin{pmatrix} 1 & \\ 1000 & 1 \end{pmatrix} \begin{pmatrix} 0.001 & 2.42 \\ & -2420 \end{pmatrix}, \qquad (4.3.7)$$

and then tried to solve the equation (4.3.3) with **A** replaced by

$$\tilde{\mathbf{L}}\tilde{\mathbf{U}} = \mathbf{A} + \mathbf{H}, \qquad (4.3.8)$$

where **H** is the factorization-error matrix

$$\mathbf{H} = \tilde{\mathbf{L}}\tilde{\mathbf{U}} - \mathbf{A}. \qquad (4.3.9)$$

This must be small for a stable algorithm (the term 'stable' was defined in Section 4.1), and for the current example we find

$$\mathbf{H} = \begin{pmatrix} 0.0 & 0.0 \\ 0.0 & -1.58 \end{pmatrix}, \qquad (4.3.10)$$

which is not small relative to **A**, so we conclude that our algorithm is unstable.

For simplicity we have used a 2×2 illustration, which always leads to one component being calculated accurately. This can happen also in a larger problem if only the last elimination step is unstable. For a 3×3 problem which has inaccuracies in all components because the first step is unstable, see Exercise 4.1. In general we wish to calculate the whole vector **x** accurately, and having just one component accurate is insufficient.

Another way to show that the algorithm is unstable is to examine the residual

$$\mathbf{r} = \mathbf{b} - \mathbf{A}\tilde{\mathbf{x}}, \qquad (4.3.11)$$

and find that $\|\mathbf{r}\|$ is not small compared with $\|\mathbf{b}\|$ or compared with $\|\mathbf{A}\|\,\|\tilde{\mathbf{x}}\|$ (see Appendix A for definitions of norms). In our example the residual is

$$\mathbf{r} = \begin{pmatrix} -0.003 \\ 1.17 \end{pmatrix}, \qquad (4.3.12)$$

whose norm is not small compared with $\|\mathbf{b}\|$.

Note that the damage that led to the inaccurate factorization $\tilde{\mathbf{L}}\tilde{\mathbf{U}}$ (see equations (4.3.9) and (4.3.10)) was done within the computation (4.3.2), which treated 1.58 as zero. In fact it would have treated any number in the interval $(-5, 5)$ as zero. The reason that it did so was the large growth in size that took place in forming $a_{22}^{(2)}$ from a_{22}.

Wilkinson's (1965) backward error analysis, as extended by Reid (1971b), shows that Gaussian elimination (or **LU** factorization) is a stable process provided such growth does not take place. Reid's results, when modified for our wider bounds on roundoff (see the bottom of page 66), yield the inequality

$$|h_{ij}| \leqslant 5.01\,\varepsilon\,n\,\max_k |a_{ij}^{(k)}| \qquad (4.3.13)$$

for coefficients of the factorization-error matrix $\mathbf{H} = \tilde{\mathbf{L}}\tilde{\mathbf{U}} - \mathbf{A}$, where ε is the relative precision and n is the matrix order.

Unfortunately it is not practical to control the sizes of all the coefficients $a_{ij}^{(k)}$ and, in any case, some growth in the size of a coefficient that begins by being very small may be perfectly acceptable. It is normal practice, therefore, to control the largest coefficient and weaken inequality (4.3.13) to the bound

$$|h_{ij}| \leqslant 5.01 \ \varepsilon \ n \ \rho, \qquad (4.3.14)$$

where ρ is given by the equation

$$\rho = \max_{i,j,k} |a_{ij}^{(k)}|. \qquad (4.3.15)$$

Note, however, the crucial influence of scaling. For example, if the entries in one row are much bigger than those in the others, we will tolerate large growth in these other rows, which may be disastrous to the accuracy of the solution. We return to the consideration of scaling in Section 4.12, assuming for the present that our problem is well-scaled. Fortunately most problems occur naturally this way. Poor scaling is often associated with such features as mixed units (for example distances measured in millimetres and kilometres).

We use matrix and vector norms extensively in this chapter, again on the assumption that our problems are well-scaled. Since norms take little or no account of small coefficients, results based on their use may be of little value for badly-scaled problems, but we lack any alternative theoretical approach.

4.4 Controlling algorithm stability

The usual way to control growth (that is to control $\max |a_{ij}^{(k)}|$) is to require that the inequality $|l_{ij}| \leqslant 1$ should hold for the coefficients of the matrix **L**. This is readily achieved by reordering the rows of the matrix so that the new pivot $a_{kk}^{(k)}$ satisfies the inequalities

$$|a_{kk}^{(k)}| \geqslant |a_{ik}^{(k)}|, \ i > k. \qquad (4.4.1)$$

The k-th column must be scanned below the main diagonal to determine its entry of largest magnitude. The row containing this entry is exchanged with the k-th row and so becomes the new k-th row. This strategy is called **partial pivoting**.

Applying this strategy to example (4.3.1), we would interchange rows one and two of **A** before beginning the reduction and would compute the factors

$$\begin{pmatrix} 1.00 & 1.58 \\ 0.001 & 2.42 \end{pmatrix} = \begin{pmatrix} 1 & \\ 0.001 & 1 \end{pmatrix} \begin{pmatrix} 1.00 & 1.58 \\ & 2.42 \end{pmatrix}. \qquad (4.4.2)$$

Using these factors, we compute

$$\tilde{\mathbf{x}} = \begin{pmatrix} 1.17 \\ 2.15 \end{pmatrix}, \tag{4.4.3}$$

which is almost correct to three decimal places ($x_1 = 1.176$).

Note that this strategy for computation subject to rounding errors is similar to the strategy used for treating a zero pivot in exact arithmetic. In the exact case we were concerned only to avoid a zero pivot; in the presence of roundoff, small pivots must also be avoided. Note also that a zero pivot gives the ultimate disaster of infinite growth and total loss of accuracy.

In practice, Gaussian elimination with partial pivoting is considered to be a stable algorithm. This is based on experience rather than rigorous analysis, since the best a priori bound that can be given for a dense matrix is

$$\rho \leqslant 2^{n-1} \max_{i,j} |a_{ij}|. \tag{4.4.4}$$

This bound is easy to establish (see Exercise 4.2). If row and column interchanges are performed at each stage to ensure that the inequalities

$$|a_{kk}^{(k)}| \geqslant |a_{ij}^{(k)}|, \ i \geqslant k, \ j \geqslant k \tag{4.4.5}$$

all hold, then the stronger bound

$$\rho \leqslant f(n) \max |a_{ij}|, \ f(n) = \sqrt{n \left(2^1 3^{1/2} 4^{1/3} \ldots n^{1/(n-1)}\right)} \tag{4.4.6}$$

has been obtained by Wilkinson (1961). This strategy is known as **full** or **complete** pivoting. Note that $f(n)$ is much smaller than 2^{n-1} for large n (for example $f(100) \approx 3570$). It has been speculated that $f(n)$ may be replaced by n, but as far as we know this has not been established.

The bound (4.4.4) for partial pivoting is achieved for a carefully contrived matrix, as shown by Wilkinson (1965, p. 212), but in practice the algorithm has proved to be very satisfactory. Full pivoting is not used much in practice because it involves $\frac{1}{3}n^3 + O(n^2)$ comparisons for a full matrix. For a further discussion of this topic see Stewart (1973, pp. 148-158) or Golub and Van Loan (1983, pp. 64-69).

The difficulty with partial pivoting (or with full pivoting) is that requiring the satisfaction of inequalities (4.4.1) (or inequalities (4.4.5)) is too restrictive when dealing with sparse systems. The alternative strategy of threshold pivoting requires the satisfaction of the inequalities

$$|a_{kk}^{(k)}| \geqslant u \, |a_{ik}^{(k)}|, \ i > k, \tag{4.4.7}$$

where u is a suitable value in the range $0 < u \leqslant 1$. If u has the value 1, we have the partial pivoting condition. If u is less than 1, we may allow several pivot candidates, each of which may introduce limited growth. The bound (4.4.4) is replaced by

$$\rho \le (1+u^{-1})^{n-1} \max_{i,j} |a_{ij}|. \tag{4.4.8}$$

In exchange for a slight increase in potential growth, we have additional freedom which may be used to preserve sparsity. Threshold pivoting offers little benefit for the case where **A** is dense and can be held in main memory, but is very helpful in the sparse matrix case. We return to this topic in Section 5.4. We show there that the bounds (4.4.4) and (4.4.8) are very pessimistic in the sparse case.

An alternative approach to Gaussian elimination, which avoids much of the numerical difficulty discussed in this and the previous section, is that of orthogonal factorization. That is, the matrix **A** is factorized as

$$\mathbf{A} = \mathbf{QU}, \tag{4.4.9}$$

where **Q** is orthogonal and **U** is upper triangular, so that the solution to the equation

$$\mathbf{Ax} = \mathbf{b} \tag{4.4.10}$$

is easily effected through premultiplication of **b** by $\mathbf{Q}^T$ and back-substitution through **U**. Methods using the factorization (4.4.9) are sometimes used because of their good numerical properties (Wilkinson 1965) but are at least twice as costly in arithmetic as Gaussian elimination. We discuss their use on sparse systems in Section 7.12.

For the block factorization of Section 3.12, there is no really satisfactory method of guaranteeing the numerical stability. Two potential numerical difficulties must be considered. First, with the partitioning

$$\mathbf{A} = \begin{pmatrix} \mathbf{A}_{11} & \mathbf{A}_{12} \\ \mathbf{A}_{21} & \mathbf{A}_{22} \end{pmatrix}, \tag{4.4.11}$$

we must verify the assumption that $\mathbf{A}_{11}$ is nonsingular since this may not be true even though **A** is nonsingular. The example (3.12.13) illustrates this point. Second, the stability of the matrix factorization cannot be assured by simply controlling the stability of the factorizations of $\mathbf{A}_{11}$ and $\bar{\mathbf{A}}_{22} = \mathbf{A}_{22} - \mathbf{A}_{21}\mathbf{A}_{11}^{-1}\mathbf{A}_{12}$. The reason for this is that the factorization may be unstable in the computation of $\bar{\mathbf{A}}_{22}$ due to either ill-conditioning or loss of information when forming $\bar{\mathbf{A}}_{22}$ because of poor scaling of $\mathbf{A}_{11}$. This is illustrated by

$$\mathbf{A} = \begin{pmatrix} 0.000130 & 1.27189 \\ \hline 3.21673 & 2.12571 \end{pmatrix}, \tag{4.4.12}$$

partitioned as indicated.

Since the factorization of a 1×1 matrix is necessarily stable, and a 1×1 matrix cannot be ill-conditioned, the factorization of **A** does not have difficulty from these sources. Rather, forming $\bar{\mathbf{A}}_{22}$ involves the computation (in 6-digit arithmetic)

$$\bar{A}_{22} = 2.12571 - 3.21673 \times \left(\frac{1}{0.000130}\right) \times 1.27189$$

$$= 2.12571 - 31471.7 \qquad (4.4.13)$$

$$= -31469.6$$

which is unstable because growth destroyed significant information.

Stability will be controlled in the partitioned algorithm if it is controlled in factorizing A_{11} and $\bar{A}_{22}$ and if there is no significant growth relative to $\|A\|$ when computing $\bar{A}_{22} = A_{22} - A_{21} A_{11}^{-1} A_{12}$.

This stability problem cannot occur for the symmetric positive-definite case as long as A_{11} is a principal submatrix of A because of the equivalence with ordinary factorization and its proven stability.

4.5 Monitoring the stability

Because of the limited usefulness of the a priori bounds (4.4.4) and (4.4.8) and because sometimes the pivot order is predetermined (see Section 5.3), it is often advisable to monitor the stability of the factorization or assess it a posteriori. If we are not reassured, further work will be necessary to get a solution with as much accuracy as the data warrants. This might involve iterative refinement (Section 4.11), choosing the pivotal sequence afresh, or even working with greater precision.

Two forms of stability monitoring are suggested by the discussion of Section 4.3. After the factorization has been found we may compute $H = \tilde{L}\tilde{U} - A$, or after the solution has been found we may compute $r = b - A\tilde{x}$.

Computing r is easier, and in practice this measure for the stability of the solution is often used. Indeed, if $\|r\|$ is small compared with $\|b\|$, we have obviously solved a nearby problem since $A\tilde{x} = b - r$; if $\|r\|$ is small compared with $\|A\| \|\tilde{x}\|$, then $\tilde{x}$ is the exact solution of the equation $(A+H)\tilde{x} = b$ where $\|H\|$ is small compared with $\|A\|$ (see Exercise 4.3). Thus, if $\|r\|$ is small compared with $\|b\|$, $\|A\tilde{x}\|$, or $\|A\|\|\tilde{x}\|$, we have done a good job in solving the equation.

For the converse, it is important to compare $\|r\|$ with $\|A\| \|\tilde{x}\|$ rather than with $\|b\|$ since $\|r\|$ may be large compared with $\|b\|$ in spite of an accurately computed approximate solution. The set of equations

$$\begin{pmatrix} 0.287 & 0.512 \\ 0.181 & 0.322 \end{pmatrix} \begin{pmatrix} x_1 \\ x_2 \end{pmatrix} = \begin{pmatrix} -0.232 \\ 0.358 \end{pmatrix} \qquad (4.5.1)$$

has as exact solution

$$\begin{pmatrix} x_1 \\ x_2 \end{pmatrix} = \begin{pmatrix} 1000 \\ -561 \end{pmatrix}, \tag{4.5.2}$$

yet the approximate solution

$$\begin{pmatrix} \tilde{x}_1 \\ \tilde{x}_2 \end{pmatrix} = \begin{pmatrix} 1000 \\ -560 \end{pmatrix}, \tag{4.5.3}$$

which would be considered very accurate on a 3-decimal computer, has residual

$$\mathbf{r} = \begin{pmatrix} -0.512 \\ -0.322 \end{pmatrix} \tag{4.5.4}$$

and certainly $\|\mathbf{r}\|$ is large relative to $\|\mathbf{b}\|$ (though not with respect to $\|\mathbf{A}\| \, \|\tilde{\mathbf{x}}\|$). This is an example of ill-conditioning, a subject to be explored later in this chapter.

Moreover, $\mathbf{r}$ tells us nothing about the behaviour for other vectors $\mathbf{b}$. For example, the linear system with matrix (4.3.1) and right-hand side vector

$$\mathbf{b} = \begin{pmatrix} 0.001 \\ 1.00 \end{pmatrix} \tag{4.5.5}$$

has exact solution

$$\mathbf{x} = \begin{pmatrix} 1.00 \\ 0.00 \end{pmatrix}, \tag{4.5.6}$$

and this would be produced exactly by our 3-digit computer with the algorithm of Section 4.3. This does not imply that the factorization is stable, but rather that the instability does not affect the solution for this particular $\mathbf{b}$. We may not assume that the same will be true for other vectors $\mathbf{b}$, and indeed we know that the algorithm of Section 4.3 is unstable.

For this reason, stability monitoring is best done by checking the size of $\|\mathbf{H}\| = \|\tilde{\mathbf{L}}\tilde{\mathbf{U}} - \mathbf{A}\|$. To be sure that this is adequate, we must also consider the effects of forward substitution and back-substitution. If $\mathbf{T}$ is a triangular matrix, the computed solution $\tilde{\mathbf{x}}$ of the equation $\mathbf{T}\mathbf{x} = \mathbf{b}$ satisfies the equation

$$(\mathbf{T}+\mathbf{E})\,\tilde{\mathbf{x}} = \mathbf{b}, \tag{4.5.7}$$

where the coefficients of the error matrix $\mathbf{E}$ (see Stewart 1973, p. 150) satisfy the inequality

$$|e_{ij}| \leq (n+1)\,\psi\,\varepsilon\,|t_{ij}|; \tag{4.5.8}$$

here ψ is a numerical constant (like the constant 5.01 in equation (4.3.13)) and ε is the relative precision. Thus forward substitution and back-substitution are fundamentally stable. Therefore our computed solution $\tilde{\mathbf{x}}$ to $\mathbf{A}\mathbf{x} = \mathbf{b}$ satisfies the equation

$$(\tilde{\mathbf{L}} + \delta\tilde{\mathbf{L}})(\tilde{\mathbf{U}} + \delta\tilde{\mathbf{U}})\,\tilde{\mathbf{x}} = \mathbf{b}, \tag{4.5.9}$$

which may be rewritten in the forms

$$(\tilde{\mathbf{L}}\,\tilde{\mathbf{U}} + \delta\tilde{\mathbf{L}}\,\tilde{\mathbf{U}} + \tilde{\mathbf{L}}\,\delta\tilde{\mathbf{U}} + \delta\tilde{\mathbf{L}}\,\delta\tilde{\mathbf{U}})\,\tilde{\mathbf{x}} = \mathbf{b} \tag{4.5.10}$$

and

$$(\mathbf{A}+\delta\mathbf{A})\,\tilde{\mathbf{x}} = \mathbf{b}, \; \delta\mathbf{A} = (\tilde{\mathbf{L}}\tilde{\mathbf{U}}-\mathbf{A}) + \delta\tilde{\mathbf{L}}\,\tilde{\mathbf{U}} + \tilde{\mathbf{L}}\,\delta\tilde{\mathbf{U}} + \delta\tilde{\mathbf{L}}\,\delta\tilde{\mathbf{U}}. \tag{4.5.11}$$

It is possible to perform the analysis to include a perturbation $\delta\mathbf{b}$ with a different perturbation to $\mathbf{A}$, but (4.5.11) is somewhat easier for practical purposes. Since inequality (4.5.8) shows that the inequalities

$$\|\delta\tilde{\mathbf{L}}\| \le (n+1)\,\psi\,\varepsilon\,\|\tilde{\mathbf{L}}\| \quad \text{and} \quad \|\delta\tilde{\mathbf{U}}\| \le (n+1)\,\psi\,\varepsilon\,\|\tilde{\mathbf{U}}\| \tag{4.5.12}$$

hold, we may concentrate on looking at $\mathbf{H} = \tilde{\mathbf{L}}\tilde{\mathbf{U}} - \mathbf{A}$ (or its norm) and $\|\tilde{\mathbf{L}}\|\,\|\tilde{\mathbf{U}}\|$.

Computing $\mathbf{H}$ and $\|\tilde{\mathbf{L}}\|\,\|\tilde{\mathbf{U}}\|$ is often impractical. Even computing $\rho = \max\limits_{i,j,k} |a_{ij}^{(k)}|$ is expensive since it involves testing $|a_{ij}^{(k)}|$ over all i, j, k, which is an $O(n^3)$ process for dense matrices. We seek instead to bound ρ. A simple bound has been provided by Erisman and Reid (1974), based on the relation

$$a_{ij}^{(k)} = a_{ij} - \sum_{p=1}^{k-1} l_{ip} u_{pj}, \; i,j \geqslant k \tag{4.5.13}$$

(see equations (3.6.5) and (3.6.8)). The bound

$$|a_{ij}^{(k)}| \le |a_{ij}| + \|(l_{i1}, l_{i2}, \dots, l_{i,k-1})\|_p \; \|(u_{1j}, u_{2j}, \dots, u_{k-1,j})\|_q, \tag{4.5.14}$$

where $1/p + 1/q = 1$, follows by the application of Holder's inequality to relation (4.5.13), and the weaker bound

$$|a_{ij}^{(k)}| \le |a_{ij}| + \|(l_{i1}, \dots, l_{i,i-1})\|_p \; \|(u_{1j}, \dots, u_{j-1,j})\|_q \tag{4.5.15}$$

is a simple deduction. The $2n$ numbers in inequality (4.5.15) are convenient to calculate if $p = 1, 2,$ or ∞, and notice that each entry in $\mathbf{L}$ and $\mathbf{U}$ needs to be used only once, so the cost is proportional to the number of entries in $\mathbf{L}$ and $\mathbf{U}$ and not to the number of operations required to calculate $\mathbf{L}$ and $\mathbf{U}$. The growth parameter $\rho = \max\limits_{i,j,k} |a_{ij}^{(k)}|$ may be bounded by calculating expression (4.5.15) for each pair (i,j) for which $a_{ij}^{(k)}$ can ever be nonzero (this set is usually known in sparse matrix work), or the bound may be further weakened to

$$\rho \le \max\limits_{i,j} |a_{ij}| + \max\limits_{i} \|(l_{i1}, \dots, l_{i,i-1})\|_p \; \max\limits_{j} \|(u_{1j}, \dots, u_{j-1,j})\|_q. \tag{4.5.16}$$

With the factorization $\mathbf{A} = \mathbf{L}\mathbf{D}\mathbf{L}^T$ in the symmetric case, if we take $p = q = 2$, inequality (4.5.16) is replaced by

$$\rho \leqslant \max_{i,j} |a_{ij}| + \max_i \sum_{k=1}^{i-1} |d_k l_{ik}^2|. \tag{4.5.17}$$

The bound (4.5.17) in the symmetric case is usually a good estimate, but the bound (4.5.16) in the unsymmetric case can be very pessimistic since it may happen that large components of the rows of $\mathbf{L}$ are in different positions from large components of the columns of $\mathbf{U}$ (see Exercise 4.4).

A slightly improved version of the Erisman–Reid bound has been suggested by Barlow (1986), based on the relation

$$a_{ij}^{(k)} = \sum_{p=k}^{\min(i,j)} l_{ip} u_{pj}, \quad i, j \geqslant k. \tag{4.5.18}$$

For $k = 1$ this is just an expansion of the relationship

$$\mathbf{A} = \mathbf{A}^{(1)} = \mathbf{L}\mathbf{U}, \tag{4.5.19}$$

as in equation (3.6.1). For $k>1$, it expresses the fact that Gaussian elimination on the submatrix of $\mathbf{A}^{(k)}$ consisting of rows and columns k to n would involve exactly the same operations as are performed in steps k onwards of Gaussian elimination on $\mathbf{A}$, and therefore would produce a triangular factorization consisting of rows and columns k to n of $\mathbf{L}$ and $\mathbf{U}$. The bounds

$$|a_{ij}^{(k)}| \leqslant \|(l_{ik},\ldots,l_{ii})\|_p \ \|(u_{kj},\ldots,u_{jj})\|_q, \tag{4.5.20}$$

$$|a_{ij}^{(k)}| \leqslant \|(l_{i1},\ldots,l_{ii})\|_p \ \|(u_{1j},\ldots,u_{jj})\|_q, \tag{4.5.21}$$

and

$$\rho \leqslant \max_i \|(l_{i1},\ldots,l_{ii})\|_p \ \max_j \|(u_{1j},\ldots,u_{jj})\|_q. \tag{4.5.22}$$

are analogous to the bounds (4.5.14) to (4.5.16) and are obtained similarly.

In the symmetric case, the bound

$$\rho \leqslant \max_i \sum_{k=1}^{i} |d_k l_{ik}^2| \tag{4.5.23}$$

corresponds to the bound (4.5.17).

If partial pivoting is used, the inequality

$$\max_i \|(l_{i1},\ldots,l_{ii})\|_1 \leqslant n \tag{4.5.24}$$

holds, so with $p = 1$ and $q = \infty$ the bound (4.5.22) reduces to the inequality

$$\rho \leqslant n \max_{i,j} |u_{ij}|, \tag{4.5.25}$$

which is very close to the result of Businger (1971).

4.6 Special stability considerations

Sometimes particular properties of the matrix (either mathematical or related to the physical origin of the model) will either assure the stability of the factorization or give reason to simplify the pivoting strategies. Some of these are discussed in this section.

A real symmetric matrix is said to be **positive definite** if for any nonzero vector $\mathbf{x}$ the inequality

$$\mathbf{x}^T \mathbf{A} \mathbf{x} > 0 \tag{4.6.1}$$

holds. Equivalently, $\mathbf{A}$ is positive definite if all its eigenvalues are positive (see Exercise 4.5). Such matrices arise frequently in practice because in many applications $\mathbf{x}^T \mathbf{A} \mathbf{x}$ corresponds to the energy, which is a fundamentally positive quantity. For this case, Wilkinson (1961) has shown that no growth takes place in Gaussian elimination with diagonal pivots in the sense that the inequality

$$\rho = \max_{i,j,k} |a_{ij}^{(k)}| \leq \max_{i,j} |a_{ij}| \tag{4.6.2}$$

is true. Hence there is no need for stability monitoring in this case.

As was briefly stated in Section 3.8, symmetry can be preserved in this factorization, thereby saving both storage and time. An algorithm often used is the Choleski factorization

$$\mathbf{A} = \mathbf{L}\mathbf{L}^T, \tag{4.6.3}$$

see equation (3.8.3), or the variant

$$\mathbf{A} = \mathbf{L}\mathbf{D}\mathbf{L}^T, \tag{4.6.4}$$

which avoids the calculation of square roots.

In another class of matrices which arise in applications, $\mathbf{A}$ is **diagonally dominant**, that is its coefficients satisfy the inequalities

$$|a_{kk}| \geq \sum_{i \neq k} |a_{ik}|, \ k = 1, 2, \ldots, n \tag{4.6.5}$$

or the inequalities

$$|a_{kk}| \geq \sum_{j \neq k} |a_{kj}|, \ k = 1, 2, \ldots, n. \tag{4.6.6}$$

Here it may be shown (see Exercise 4.6) that the inequality

$$\rho = \max_{i,j,k} |a_{ij}^{(k)}| \leq 2 \max_{i,j} |a_{ij}| \tag{4.6.7}$$

is true, so that the factorization is always stable with diagonal pivots.

Another class to consider is the set of indefinite symmetric matrices. Here choosing pivots from the diagonal may be unstable or impossible, as is illustrated by the case

$$\mathbf{A} = \begin{pmatrix} 0 & 1 & 1 \\ 1 & 0 & 1 \\ 1 & 1 & 0 \end{pmatrix}. \tag{4.6.8}$$

Standard partial pivoting destroys the symmetry and hence we are faced with the storage and computing demands of an unsymmetric matrix. A stable way of preserving symmetry is discussed in the next section.

Sometimes the relationship between the physical model and the matrix suggests a pivoting strategy that may not be justifiable mathematically. An example of this is the complex symmetric matrix arising in frequency domain analysis of electrical networks, which is discussed by Erisman and Spies (1972). Since this matrix is neither positive definite nor diagonally dominant, a possible strategy would be to ignore the symmetry and use partial pivoting to assure numerical stability.

In this case, with diagonal pivoting, the trailing $(n-k+1) \times (n-k+1)$ submatrix of the reduced matrix $\mathbf{A}^{(k)}$ of Gaussian elimination can be associated with a model of another physical system. While cases can be constructed where the reduced order problem is unstable and will give rise to a zero diagonal pivot, such cases were assumed to be rare. For these reasons, Erisman and Spies (1972) ignored the potential dangers of instability and simply selected diagonal pivots to preserve sparsity and symmetry. To assess the validity of the answers, however, they accompanied this with stability monitoring using inequality (4.5.17).

Over a long history of the use of this approach the only failures indicated by the stability monitor were deliberately contrived (examples showing that this factorization is sometimes unstable) and data input errors. Had failures occurred more often, a more conservative pivoting strategy would have been adopted.

4.7 Solving indefinite symmetric systems

When the matrix $\mathbf{A}$ is symmetric but indefinite choosing pivots from the diagonal may be unstable or impossible, as illustrated by example (4.6.8). Yet ignoring symmetry leads to twice the cost in storage and computation. Fortunately, an algorithm has been developed (Bunch and Parlett 1971, Bunch 1974) which preserves symmetry and maintains stability for this case.

The idea is simply to extend the notion of a pivot to 2×2 blocks as well as single entries. Thus if $\mathbf{A}$ has the form

$$\mathbf{A} = \begin{pmatrix} \mathbf{P} & \mathbf{C}^T \\ \mathbf{C} & \mathbf{B} \end{pmatrix}, \tag{4.7.1}$$

where $\mathbf{P}$ and $\mathbf{B}$ are symmetric, the first step of the $\mathbf{L}\,\mathbf{D}\,\mathbf{L}^T$ factorization may be expressed in the form

$$A = \begin{pmatrix} I & 0 \\ CP^{-1} & I \end{pmatrix}\begin{pmatrix} P & 0 \\ 0 & B-CP^{-1}C^T \end{pmatrix}\begin{pmatrix} I & P^{-1}C^T \\ 0 & I \end{pmatrix}. \tag{4.7.2}$$

In the usual case P has order 1, but in the extended case P may have order 1 or 2. Note that $B - CP^{-1}C^T$ is symmetric. Bunch, Kaufman, and Parlett (1976) search at most two columns of the lower triangular part of $A^{(k)}$ and perform at most one symmetric interchange to ensure that either $a_{kk}^{(k)}$ is

suitable as a 1×1 pivot or $\begin{pmatrix} a_{kk}^{(k)} & a_{k,k+1}^{(k)} \\ a_{k+1,k}^{(k)} & a_{k+1,k+1}^{(k)} \end{pmatrix}$ is suitable as a 2×2 pivot.

They report timings comparable with those of the Choleski algorithm, and their stability bound is

$$\rho \leqslant 2.57^{n-1} \max |a_{ij}|. \tag{4.7.3}$$

Barwell and George (1976) also report that using 2×2 pivots need involve little extra expense.

4.8 Ill-conditioning: introduction

We now assume that we have been successful in computing a vector $\tilde{x}$ satisfying the equation

$$(A + H)\tilde{x} = b, \tag{4.8.1}$$

where $\|H\|$ is small, that is $\tilde{x}$ has been computed stably. As yet we have no assurance that $\tilde{x}$ is accurate. The simple example

$$A = \begin{pmatrix} 0.287 & 0.512 \\ 0.181 & 0.322 \end{pmatrix}, \quad b = \begin{pmatrix} 0.799147 \\ 0.502841 \end{pmatrix} \tag{4.8.2}$$

bears this out, if we continue to use 3-decimal arithmetic. The exact solution is

$$x = \begin{pmatrix} 0.501 \\ 1.28 \end{pmatrix}. \tag{4.8.3}$$

Representation error requires rounding b on input to

$$\tilde{b} = \begin{pmatrix} 0.799 \\ 0.503 \end{pmatrix}. \tag{4.8.4}$$

The computed 3-digit LU factors of A are

$$\begin{pmatrix} 1 & \\ 0.631 & 1 \end{pmatrix}\begin{pmatrix} 0.287 & 0.512 \\ & -0.001 \end{pmatrix}. \tag{4.8.5}$$

The perturbed system $A\tilde{x} = \tilde{b}$ has

$$\tilde{\mathbf{x}} = \begin{pmatrix} 1.00 \\ 1.00 \end{pmatrix} \qquad (4.8.6)$$

as its exact solution and this is the solution which is obtained using the factors (4.8.5). The quantities $\|\mathbf{r}\| = \|\mathbf{b} - \mathbf{A}\tilde{\mathbf{x}}\|$ and $\|\tilde{\mathbf{r}}\| = \|\tilde{\mathbf{b}} - \mathbf{A}\tilde{\mathbf{x}}\|$ are both relatively small; in fact, when $\mathbf{r}$ and $\tilde{\mathbf{r}}$ are computed in 3-decimal arithmetic, both are zero. The cause of the relatively large value of $\|\mathbf{x} - \tilde{\mathbf{x}}\|$ is thus in the data rather than in the algorithm. In this case we are faced with an ***ill-conditioned problem : small changes in the data can cause large changes in the solution.***

The next few sections are devoted to understanding and working with ill-conditioned problems.

4.9   Ill-conditioning: theoretical discussion

A problem is ill-conditioned if small changes in the data can produce large changes in the solution. In the next paragraph we show that this would not be possible if the matrix $\mathbf{A}$ scaled the norms of all vectors about equally, that is if $\|\mathbf{A}\mathbf{x}\|$ had about the same size for all vectors $\mathbf{x}$ with $\|\mathbf{x}\| = 1$. Thus we can characterize the ill-conditioning of $\mathbf{A}$ in terms of the variation in $\|\mathbf{A}\mathbf{x}\|$.

Suppose that the matrix $\mathbf{A}$ is such that there are two vectors $\mathbf{v}$ and $\mathbf{w}$ satisfying the relations

$$\|\mathbf{v}\| = \|\mathbf{w}\| \qquad (4.9.1)$$

and

$$\|\mathbf{A}\mathbf{v}\| \gg \|\mathbf{A}\mathbf{w}\|. \qquad (4.9.2)$$

If $\mathbf{b}$ has the value $\mathbf{A}\mathbf{v}$, the equation

$$\mathbf{A}\mathbf{x} = \mathbf{b} \qquad (4.9.3)$$

has solution $\mathbf{x} = \mathbf{v}$. If $\mathbf{b}$ is changed by the vector $\mathbf{A}\mathbf{w}$, which is relatively small in view of inequality (4.9.2), the solution is changed by $\mathbf{w}$, which is not small in view of equality (4.9.1). Thus the problem is ill-conditioned.

Clearly the amount of ill-conditioning is dependent on how large the ratio of the two sides of relation (4.9.2) can be. The ratio can be written in the form

$$\frac{\|\mathbf{A}\mathbf{v}\|}{\|\mathbf{A}\mathbf{w}\|} \frac{\|\mathbf{w}\|}{\|\mathbf{v}\|} = \frac{\|\mathbf{A}\mathbf{v}\|}{\|\mathbf{v}\|} \frac{\|\mathbf{A}^{-1}\mathbf{y}\|}{\|\mathbf{y}\|}, \qquad (4.9.4)$$

if $\mathbf{y}$ is the vector $\mathbf{A}\mathbf{w}$. The first term has maximum value $\|\mathbf{A}\|$ (see Appendix A for a discussion of norms) and the second term has maximum value $\|\mathbf{A}^{-1}\|$, so the ratio $\|\mathbf{A}\mathbf{v}\| / \|\mathbf{A}\mathbf{w}\|$ can be as large as $\|\mathbf{A}\| \, \|\mathbf{A}^{-1}\|$. Because of its relationship to the condition of $\mathbf{A}$, the quantity

$$\kappa(\mathbf{A}) = \|\mathbf{A}\| \, \|\mathbf{A}^{-1}\| \qquad\qquad (4.9.5)$$

is known as the **condition number** for $\mathbf{A}$. In the extreme case $\mathbf{v}$ is the vector that maximizes $\|\mathbf{Av}\|/\|\mathbf{v}\|$ and $\mathbf{w}$ is the vector $\mathbf{A}^{-1}\mathbf{y}$ where $\mathbf{y}$ maximizes $\|\mathbf{A}^{-1}\mathbf{y}\|/\|\mathbf{y}\|$. If the norm is the two-norm (see Appendix A), $\|\mathbf{Av}\|_2/\|\mathbf{v}\|_2$ and $\|\mathbf{Aw}\|_2/\|\mathbf{w}\|_2$ are the largest and smallest singular values of $\mathbf{A}$, and $\mathbf{v}$ and $\mathbf{w}$ are the corresponding singular vectors.

For the matrix $\mathbf{A}$ defined in (4.8.2) it is interesting to observe the exact relationships

$$\mathbf{A}\begin{pmatrix}1.0\\1.0\end{pmatrix} = \begin{pmatrix}0.799\\0.503\end{pmatrix}, \quad\text{and}\quad \mathbf{A}\begin{pmatrix}1.0\\-0.561\end{pmatrix} = \begin{pmatrix}-0.000232\\0.000358\end{pmatrix}, \qquad (4.9.6)$$

which shows the wide variation of $\|\mathbf{Ax}\|_\infty$ for vectors such that $\|\mathbf{x}\|_\infty = 1.0$. This wide variation explains the reason for the large difference $\mathbf{x}-\tilde{\mathbf{x}}$.

For the general problem

$$\mathbf{Ax} = \mathbf{b}, \qquad\qquad (4.9.7)$$

if $\mathbf{b}$ is perturbed by $\boldsymbol{\delta}\mathbf{b}$ and the corresponding perturbation of $\mathbf{x}$ is $\boldsymbol{\delta}\mathbf{x}$ so that the equation

$$\mathbf{A}(\mathbf{x}+\boldsymbol{\delta}\mathbf{x}) - \mathbf{b}+\boldsymbol{\delta}\mathbf{b} \qquad\qquad (4.9.8)$$

is satisfied, then by subtraction followed by multiplication by $\mathbf{A}^{-1}$ we find the relation

$$\boldsymbol{\delta}\mathbf{x} = \mathbf{A}^{-1}\boldsymbol{\delta}\mathbf{b}. \qquad\qquad (4.9.9)$$

By taking norms in both equations (4.9.7) and (4.9.9), we find the inequalities

$$\|\mathbf{A}\| \, \|\mathbf{x}\| \geqslant \|\mathbf{b}\| \qquad\qquad (4.9.10)$$

and

$$\|\boldsymbol{\delta}\mathbf{x}\| \leqslant \|\mathbf{A}^{-1}\| \, \|\boldsymbol{\delta}\mathbf{b}\|, \qquad\qquad (4.9.11)$$

from which follows the inequality

$$\frac{\|\boldsymbol{\delta}\mathbf{x}\|}{\|\mathbf{x}\|} \leqslant \|\mathbf{A}\| \, \|\mathbf{A}^{-1}\| \, \frac{\|\boldsymbol{\delta}\mathbf{b}\|}{\|\mathbf{b}\|}, \qquad\qquad (4.9.12)$$

which again illustrates the role of the condition number (4.9.5).

For simplicity of exposition, we have so far in this section confined attention to perturbations of the vector $\mathbf{b}$. In this case the bound may be far from sharp. For example, if $\mathbf{x}$ is the vector $\mathbf{w}$ of (4.9.1) and (4.9.2), $\|\mathbf{A}\| \, \|\mathbf{x}\|$ will grossly overestimate $\|\mathbf{b}\|$ in (4.9.10), which in turn leads to the inequality (4.9.12) grossly overestimating $\|\boldsymbol{\delta}\mathbf{x}\|/\|\mathbf{x}\|$. To see that the condition number realistically estimates the conditioning we must consider the effect of perturbations to $\mathbf{A}$. Let the perturbed system be

$$(\mathbf{A}+\boldsymbol{\delta}\mathbf{A})(\mathbf{x}+\boldsymbol{\delta}\mathbf{x}) = \mathbf{b}. \qquad\qquad (4.9.13)$$

Subtracting equation (4.9.7) and rearranging gives the equation

$$\mathbf{A}\boldsymbol{\delta}\mathbf{x} = -\boldsymbol{\delta}\mathbf{A}(\mathbf{x}+\boldsymbol{\delta}\mathbf{x}). \tag{4.9.14}$$

If we now multiply by $\mathbf{A}^{-1}$ and take norms, we find the inequality

$$\|\boldsymbol{\delta}\mathbf{x}\| \le \|\mathbf{A}^{-1}\| \, \|\boldsymbol{\delta}\mathbf{A}\| \, \|\mathbf{x}+\boldsymbol{\delta}\mathbf{x}\|, \tag{4.9.15}$$

which can be rearranged to

$$\frac{\|\boldsymbol{\delta}\mathbf{x}\|}{\|\mathbf{x}+\boldsymbol{\delta}\mathbf{x}\|} \le \|\mathbf{A}^{-1}\| \, \|\mathbf{A}\| \, \frac{\|\boldsymbol{\delta}\mathbf{A}\|}{\|\mathbf{A}\|}. \tag{4.9.16}$$

Comparing this with inequality (4.9.12) we see that the condition number $\kappa(\mathbf{A}) = \|\mathbf{A}\| \, \|\mathbf{A}^{-1}\|$ plays a similar role in relating the relative error in the solution with the relative error in the data. Now, however, we are considering a whole class of perturbations $\boldsymbol{\delta}\mathbf{A}$ and would expect almost all to be such that the inequality (4.9.15) is reasonably sharp, which implies that inequality (4.9.16) can be expected to be sharp. This should be contrasted with the possible lack of sharpness of inequality (4.9.10) for particular vectors $\mathbf{b}$.

Referring to bounds (4.9.12) and (4.9.16), we might be tempted to regard the condition number as a bound on the ratio between the relative solution error and the relative data error. Many have viewed it as such, but what we have shown indicates another way of looking at it. The given problem data $(\mathbf{A}, \mathbf{b})$ on a finite word-length computer represents a family of problems $(\mathbf{A}+\boldsymbol{\Delta}\mathbf{A}, \mathbf{b}+\boldsymbol{\Delta}\mathbf{b})$ where $\boldsymbol{\Delta}\mathbf{A}$ and $\boldsymbol{\Delta}\mathbf{b}$ correspond to the range of problems represented by the data. The condition number and the precision together tell us about the range of admissible solutions to this family of problems.

Since the condition number is critical to assessing the validity of the solution, it is important to be able to calculate at least some reasonable approximation to it. In the next section we discuss and compare various schemes to do this.

4.10 Ill-conditioning: automatic detection

We now ask the question: in the course of solving $\mathbf{A}\mathbf{x}=\mathbf{b}$ can ill-conditioning of $\mathbf{A}$ be detected readily? We explore seven possibilities, totally rejecting the first three on the grounds of cost or inadequacy and turning our attention to less costly ad hoc schemes. The objective is to estimate the amount of ill-conditioning in the problem inexpensively (relative to the cost of solution of the problem itself). In general we are interested only in the order of magnitude of the condition number.

(1) Compute the range of the singular values of A

Computing the singular values is very expensive and provides only the two-norm condition number, so we do not pursue it further here.

(2) Compute A^{-1} and evaluate $\|A\| \, \|A^{-1}\|$

For a dense matrix, the additional cost for computing A^{-1} is double that of the factorization itself. For a sparse problem, the factor can be much greater, although it can sometimes be reduced by computing only part of the inverse (see Section 12.7). The direct evaluation of $\|A\| \, \|A^{-1}\|$ is therefore expensive.

(3) Compute the determinant or eigenvalues of A

For the factorization $A = LU$ with L unit lower triangular, the determinant is given by the expression

$$\det(A) = \prod_{i=1}^{n} u_{ii}, \qquad (4.10.1)$$

which is very inexpensive to compute. Since a zero determinant implies that the matrix is singular and ill-conditioning is approximate singularity, the use of the determinant to assess conditioning has often been proposed. However the determinant may be relatively large when the matrix is very ill-conditioned and it may be relatively small when the matrix is well-conditioned. Hence this measure can never be used with confidence.

To illustrate this point, recall that for a real symmetric matrix the singular values are equal to the absolute values of the eigenvalues and hence the eigenvalues determine the two-norm condition number. Further, the determinant of a matrix is equal to the product of its eigenvalues. Thus for a real symmetric matrix A_1 of order 100 with all eigenvalues equal to 0.1,

$$\det A_1 = 0.1^{100} = 10^{-100} \text{ and } \kappa_2(A_1) = 1, \qquad (4.10.2)$$

while for another real symmetric matrix A_2 of order 100 with 99 eigenvalues equal to 1.1 and one eigenvalue equal to $(1.1)^{-99}$,

$$\det A_2 = 1 \text{ and } \kappa_2(A_2) = \frac{1.1}{1.1^{-99}} = 1.1^{100} \simeq 13781. \qquad (4.10.3)$$

Because of the relationship between eigenvalues and singular values in the real symmetric case, it is sometimes thought that the eigenvalues may provide a good estimate of the condition number for unsymmetric matrices as well. Generally, the eigenvalues are far too expensive to compute compared with other measures and are also an unreliable indicator for ill-conditioning. In the case of

$$A = \begin{pmatrix} 1 & 10^8 \\ & 1 \end{pmatrix}, \tag{4.10.4}$$

the eigenvalues are available but are not useful. Here the eigenvalues are both 1.0 yet the value of $\kappa(A)$ is about 10^{16}. Note that the determinant in this case has value one.

The eigenvalues do help, however, in that ill-conditioning is indicated if one is very small in magnitude compared with the largest in magnitude.

We seek an indicator that is both more reliable and less expensive to compute.

(4) Look for a small pivot

Perturbing A by subtracting $a_{ik}^{(k)}$ from a_{ik} for all i in the range $k \leqslant i \leqslant n$ leads to a zero pivot for the perturbed problem. Thus if $a_{ik}^{(k)}$, $i = k, k+1, \ldots, n$ are all small, a nearby matrix is singular and A is certainly ill-conditioned. In fact the relation

$$\|A^{-1}\|_p \geqslant \|a_k^{(k)}\|_p^{-1}, \tag{4.10.5}$$

where $a_k^{(k)}$ is the vector $(0,0,\ldots,0,a_{kk}^{(k)},\ldots,a_{nk}^{(k)})^T$, is true (see Exercise 4.8). Therefore if row interchanges are in use, the pivots should certainly be watched and a small pivot is a warning that demands attention. However the matrix may be ill-conditioned without any small pivots appearing. For example, the matrix

$$U = \begin{bmatrix} 1 & -1 & -1 & \cdot & \cdot & \cdot & -1 \\ & 1 & -1 & \cdot & \cdot & \cdot & -1 \\ & & 1 & \cdot & \cdot & \cdot & -1 \\ & & & \cdot & & & \cdot \\ & & & & \cdot & & \cdot \\ & & & & & \cdot & \cdot \\ & & & & & & 1 \end{bmatrix} \tag{4.10.6}$$

of order n has one-norm condition number $\kappa_1(A) = n\, 2^{n-1}$. Notice again that all the eigenvalues in (4.10.6) are equal to unity, with no very small eigenvalue, and yet the matrix is quite ill-conditioned for large n.

(5) Solve a problem with a known solution and check the result

For a given matrix A it is a simple matter to construct a right-hand side c with known solution vector v (simply by computing $c = Av$). The solution with right-hand side c may be computed alongside b at the cost of another forward substitution and another back-substitution. An indication of the error in $\tilde{x}$ may be based on the error in the computed solution $\tilde{v}$. The vector

$\mathbf{v}$ is often chosen as $(1,1,...,1)^T$ which means that the computation of $\mathbf{c}$ consists of finding the row sums of $\mathbf{A}$, but this may lead to atypically small errors. A better choice is a vector of generated random numbers with values between $\frac{1}{2}$ and 1.

This seems to be more effective than (4), but is likely to be less effective than (6) or (7). We know of no detailed comparisons.

(6) The LINPACK estimate

Cline, Moler, Stewart, and Wilkinson (1979) suggested estimating $\|\mathbf{A}^{-1}\|_1$ by calculating the vector $\mathbf{x}$ from the equation

$$\mathbf{A}^T\mathbf{x} = \mathbf{b} \tag{4.10.7}$$

for a specially constructed vector $\mathbf{b}$, then solving the equation

$$\mathbf{A}\mathbf{y} = \mathbf{x} \tag{4.10.8}$$

and using $\|\mathbf{y}\|_1/\|\mathbf{x}\|_1$ as an estimate for $\|\mathbf{A}^{-1}\|_1$. The vector $\mathbf{b}$ has components ±1 and signs chosen during the course of solution of (4.10.7) with the aim of making $\|\mathbf{x}\|_1$ large. The choice of signs at most doubles the work in the forward substitution phase of solving (4.10.7) so the overall cost is equivalent to solving 2 to $2\frac{1}{2}$ sets of equations using the already computed factors of $\mathbf{A}$. Details are given in Appendix B, which also contains a justification of the procedure in terms of the singular value decomposition of $\mathbf{A}$.

Cline *et al.* (1979) report that good estimates were obtained from their runs on test matrices, although Cline and Rew (1983) have shown that it is possible to construct examples for which the estimate is poor. O'Leary (1980) showed experimentally that a worthwhile improvement, particularly for small n, is available by using

$$\max \left(\frac{\|\mathbf{y}\|_1}{\|\mathbf{x}\|_1} , \|\mathbf{x}\|_\infty \right) . \tag{4.10.9}$$

She recommended applying the procedure to $\mathbf{A}^T$ if an estimate of $\|\mathbf{A}^{-1}\|_\infty$ is wanted, rather than changing the norm. Cline, Conn, and Van Loan (1982) discuss 'look behind' as well as 'look ahead' algorithms for obtaining an estimate of $\kappa_2(\mathbf{A})$ and show by experiment that they give good estimates. None of these papers make a theoretical analysis of the quality of their estimates.

(7) Use of iterative refinement

Iterative refinement allows the improvement of the computed solution $\tilde{\mathbf{x}}$ as well as an estimation of the conditioning. As a quantitative estimate of the condition number, iterative refinement has not been competitive with the

LINPACK estimate, so we do not discuss this further. But as a way of improving the solution and evaluating solution sensitivity to data perturbation, iterative refinement has proved valuable. This is discussed in the next section.

4.11 Iterative refinement

We now discuss a strategy designed to improve the accuracy of a computed solution $\tilde{\mathbf{x}}$, but which offers as a by-product a measure of conditioning similar to the seventh strategy of the previous section.

Because we are defining an iterative process, we change notation slightly and let $\mathbf{x}^{(1)}$ be the computed solution to $\mathbf{Ax} = \mathbf{b}$. Then let the residual vector be

$$\mathbf{r}^{(1)} = \mathbf{b} - \mathbf{Ax}^{(1)}. \tag{4.11.1}$$

Formally we observe the relations

$$\mathbf{A}^{-1}\mathbf{r}^{(1)} = \mathbf{A}^{-1}\mathbf{b} - \mathbf{x}^{(1)} = \mathbf{x} - \mathbf{x}^{(1)}. \tag{4.11.2}$$

Hence we may find a correction $\mathbf{\Delta x}^{(1)}$ to $\mathbf{x}^{(1)}$ by solving the equation

$$\mathbf{A}\,\mathbf{\Delta x}^{(1)} = \mathbf{r}^{(1)}. \tag{4.11.3}$$

In practice, of course, we cannot solve this exactly, but we may construct a new approximate solution by the equation

$$\mathbf{x}^{(2)} = \mathbf{x}^{(1)} + \mathbf{\Delta x}^{(1)}, \tag{4.11.4}$$

and in general define an iterative process, with steps

$$\mathbf{r}^{(k)} = \mathbf{b} - \mathbf{Ax}^{(k)}, \tag{4.11.5a}$$

$$\mathbf{A}\,\mathbf{\Delta x}^{(k)} = \mathbf{r}^{(k)}, \tag{4.11.5b}$$

and

$$\mathbf{x}^{(k+1)} = \mathbf{x}^{(k)} + \mathbf{\Delta x}^{(k)}, \tag{4.11.5c}$$

for $k = 1, 2, \ldots$.

If the residuals $\mathbf{r}^{(k)}$ are computed using additional precision, the process usually converges to full precision.

As well as improving the solution, iterative refinement can be used to indicate the likely error in $\tilde{\mathbf{x}}$. Indeed this is perhaps its most important role.

Erisman and Reid (1974) suggested applying random artificial perturbations to the entries of $\mathbf{A}$ and $\mathbf{b}$ when calculating the residuals, in order that the consequent changes $\mathbf{\Delta x}^{(k)}$ indicate the uncertainties in the solution. Exact zeros (usually caused by the lack of certain physical connections) are not perturbed. In this way, data uncertainties in the values of the entries can be taken into account. Often iteration to full working precision is not justified.

Note that we will have replaced $\mathbf{A}$ by $\tilde{\mathbf{L}}\tilde{\mathbf{U}}$ and the zeros in fill-in positions will have been replaced by numbers of size about $\varepsilon \max_k |a_{ij}^{(k)}|$, which is a qualitative change, but we can correct the solution with iterative refinement. This qualitative change is discussed further in Section 12.8.

4.12 Scaling

We have assumed so far that the given matrix $\mathbf{A}$ and the solution $\mathbf{x}$ are well-scaled. However, norms and condition numbers are affected by scaling. For instance the bound (4.9.12) may be very poor because $\kappa(\mathbf{A})$ is unnecessarily large and it will give very poor relative bounds for any components of $\mathbf{x}$ that are much smaller than the rest (for instance if $\mathbf{x} = \begin{pmatrix} 1.27 \times 10^6 \\ 3.26 \end{pmatrix}$, the bound $\|\boldsymbol{\delta}\mathbf{x}\|_\infty \leqslant 6.4$ is only useful for x_1).

To focus our discussion of matrix scaling we consider the matrix

$$\mathbf{A} = \begin{pmatrix} 1.00 & 2420 \\ 1.00 & 1.58 \end{pmatrix}. \tag{4.12.1}$$

The partial pivoting strategy discussed at the beginning of Section 4.4 apparently justifies the use of the (1,1) pivot, and the resulting factorization

$$\tilde{\mathbf{L}}\tilde{\mathbf{U}} = \begin{pmatrix} 1 & \\ 1.00 & 1 \end{pmatrix} \begin{pmatrix} 1.00 & 2420 \\ & -2420 \end{pmatrix} \tag{4.12.2}$$

does not have any growth, if growth is measured relative to $\max |a_{ij}|$. Since this matrix factorization is closely related to the unstable factorization in Section 4.3, we see that the presence of the large entry in (4.12.1) may cause failure in both the pivot selection and growth assessment strategies.

The matrix $\mathbf{A}$ in (4.12.1) is *poorly scaled* because of the wide range of entries which are, by default, presumed to all have the same relative accuracy. It is surprisingly difficult to deal in any automatic way with poorly-scaled problems because, judging only from the given data, it is generally not possible to determine what is significant. Forsythe and Moler (1967, pp. 37–46) discuss this in detail. One possibility is **equilibration**, where diagonal matrices $\mathbf{D}_1$ and $\mathbf{D}_2$ are selected so that

$$\mathbf{D}_1 \mathbf{A} \mathbf{D}_2 \tag{4.12.3}$$

has the largest entry in each row and column of the same magnitude (see Bauer 1963, for example).

Unfortunately, this choice allows $\mathbf{D}_1$ and $\mathbf{D}_2$ to vary widely. In our example, choosing

$$\mathbf{D}_1 = \begin{pmatrix} 10^{-3} & \\ & 1 \end{pmatrix} \quad \text{and} \quad \mathbf{D}_2 = \mathbf{I} \tag{4.12.4}$$

produces the scaled matrix

$$\mathbf{D}_1 \mathbf{A} \mathbf{D}_2 = \begin{pmatrix} 0.00100 & 2.42 \\ 1.00 & 1.58 \end{pmatrix}, \tag{4.12.5}$$

while choosing

$$\mathbf{D}_3 = \mathbf{I} \quad \text{and} \quad \mathbf{D}_4 = \begin{pmatrix} 1 & \\ & 10^{-3} \end{pmatrix} \tag{4.12.6}$$

produces the scaled matrix

$$\mathbf{D}_3 \mathbf{A} \mathbf{D}_4 = \begin{pmatrix} 1.00 & 2.42 \\ 1.00 & 0.00158 \end{pmatrix}. \tag{4.12.7}$$

Both are well-conditioned equilibrated matrices but they lead to significance being attached to different entries and to different pivotal choices. While careful computation with 2×2 problems will still allow acceptable results, in larger cases the different scalings can result in very different solutions.

If the unscaled matrix (4.12.1) had been caused by a badly-scaled first equation, then scaling (4.12.5) would be the proper choice. If it had been caused by choosing units for variable x_2 that are 10^3 times too large then scaling (4.12.7) would be the proper choice.

Note that when computing a stable solution to a system of equations with scaled matrix $\mathbf{D}_1 \mathbf{A} \mathbf{D}_2$ we solve the equation

$$(\mathbf{D}_1 \mathbf{A} \mathbf{D}_2)(\mathbf{D}_2^{-1} \mathbf{x}) = \mathbf{D}_1 \mathbf{b}. \tag{4.12.8}$$

The computed solution $(\mathbf{D}_2^{-1} \mathbf{x})$ is the solution to a problem 'near'

$$(\mathbf{D}_1 \mathbf{A} \mathbf{D}_2)\mathbf{y} = \mathbf{D}_1 \mathbf{b}. \tag{4.12.9}$$

This $\mathbf{x}$ may not be the solution to a problem 'near'

$$(\mathbf{D}_3 \mathbf{A} \mathbf{D}_4)(\mathbf{D}_4^{-1} \mathbf{x}) = \mathbf{D}_3 \mathbf{b}. \tag{4.12.10}$$

for some other scaling (see Exercises 4.16 to 4.19).

For these reasons, using consistent units in data and variables and making consistent modelling assumptions are usually the best way to achieve a well-scaled problem. Automatic scaling methods should be considered only when these modelling considerations are not feasible.

A practical automatic scaling algorithm has been used by Curtis and Reid (1972) following a suggestion of Hamming (1971), which is to choose scaling matrices $\mathbf{D}_1 = \mathrm{diag}(\beta^{-\rho_i})$ and $\mathbf{D}_2 = \mathrm{diag}(\beta^{-\gamma_i})$, where β is the machine radix and ρ_i and γ_i, $i=1,2,...,n$ are chosen to minimize the expression

$$\sum_{a_{ij} \neq 0} (\log_\beta |a_{ij}| - \rho_i - \gamma_j)^2 . \tag{4.12.11}$$

The scaled matrix $\mathbf{D}_1 \mathbf{A} \mathbf{D}_2$ then has the logarithms of its nonzeros as small as possible in the least-squares sense, which corresponds to the coefficients of the scaled matrix having modulus near to one. The minimization of expression (4.12.11) is a linear least-squares problem which Curtis and Reid were able to solve very effectively by the method of conjugate gradients. The major cost, particularly in the full case, lies in the computation of the logarithms.

4.13 Prevention of ill-conditioning

Unlike stability, which we can control through the solution process, ill-conditioning is fundamentally associated with the data and is independent of the algorithm. Once $\mathbf{A}$ is given, $\kappa(\mathbf{A})$ is fixed and solutions will exhibit sensitivity to perturbations in the data consistent with $\kappa(\mathbf{A})$ regardless of computer roundoff error.

   The only effective means of preventing or controlling ill-conditioning is in the choice of $\mathbf{A}$ itself. We should ask the question 'does the physical process being modelled display the same sensitivity to perturbation as the model?'. If not, we should try to identify the source of ill-conditioning and change the model to avoid it. We give two examples.

   In a linear least-squares problem we solve, in the sense of minimizing $\|\mathbf{b} - \mathbf{A}\mathbf{x}\|_2^2$, the equation

$$\mathbf{A}\mathbf{x} = \mathbf{b}, \tag{4.13.1}$$

where $\mathbf{A}$ has more rows than columns. It can be shown that the two-norm conditioning relation (see (4.9.12))

$$\frac{\|\delta\mathbf{x}\|_2}{\|\mathbf{x}\|_2} \leq \kappa_2(\mathbf{A})\frac{\|\delta\mathbf{b}\|_2}{\|\mathbf{b}\|_2} \tag{4.13.2}$$

holds if $\mathbf{b}$ is perturbed by $\delta\mathbf{b}$ and $\delta\mathbf{x}$ is the corresponding perturbation to the least-squares solution, where $\kappa_2(\mathbf{A})$ is the ratio of largest to smallest singular value of $\mathbf{A}$ (the singular values are the square roots of the eigenvalues of $\mathbf{A}^T\mathbf{A}$). The solution of equation (4.13.1) may be found from the equation

$$\mathbf{A}^T\mathbf{A}\,\mathbf{x} = \mathbf{A}^T\mathbf{b} \tag{4.13.3}$$

where now $\mathbf{A}^T\mathbf{A}$ is square, symmetric, and positive definite. Unfortunately, the relation

$$\kappa_2(\mathbf{A}^T\mathbf{A}) = (\kappa_2(\mathbf{A}))^2 \tag{4.13.4}$$

holds between condition numbers (see Exercise 4.20). Thus if $\mathbf{A}$ is ill-conditioned, $A^T A$ is more so. A more direct approach to solving equation (4.13.1) is needed. This is possible (see for example, Stewart 1973, Chapter 5), but further discussion of the least-squares problem is beyond the scope of this book.

In many network problems the objective is to compute flow differences between nodes (for example voltage differences in an electrical network). Without a reference node the basic model is singular since individual flows (or voltages) are meaningless. Once a reference node is selected the flows (or voltages) are really differences with respect to the reference node, and there is no singularity problem. However, if the reference node connects only weakly to the rest of the system, the corresponding flows will be poorly determined with respect to the reference node and the resulting model may be ill-conditioned.

A proper understanding of ill-conditioning and how it may hurt the modelling process should help the modeller to identify and eliminate it whenever possible.

4.14 Overall error bound

For simplicity of exposition in Section 4.9 we considered the effects of perturbations to A and b separately, but of course in practice we will have solved a problem of the form

$$(A+\delta A)\,(x+\delta x) = b+\delta b, \tag{4.14.1}$$

with perturbations to both A and b. The perturbations can be taken to include both data uncertainty and algorithmic error. If A is reasonably well-scaled and the algorithm is stable, we can expect the algorithmic contribution to be small compared with $\|A\|$ and $\|b\|$. By subtracting the given equation from (4.14.1) we find

$$A\,\delta x = \delta b - \delta A\,x - \delta A\,\delta x, \tag{4.14.2}$$

and multiplying by A^{-1} and taking norms yields the inequality

$$\|\delta x\| \leqslant \|A^{-1}\|\big(\|\delta b\| + \|\delta A\|\,\|x\| + \|\delta A\|\,\|\delta x\|\big), \tag{4.14.3}$$

which may be rearranged to the form

$$\|\delta x\|\big(1 - \|A^{-1}\|\,\|\delta A\|\big) \leqslant \|A^{-1}\|\big(\|\delta b\| + \|\delta A\|\,\|x\|\big). \tag{4.14.4}$$

Provided the inequality

$$\|A^{-1}\|\,\|\delta A\| < 1 \tag{4.14.5}$$

holds, we deduce the relation

$$\|\delta x\| \leqslant \frac{\|A^{-1}\|}{1 - \|A^{-1}\|\,\|\delta A\|}\big(\|\delta b\| + \|\delta A\|\,\|x\|\big) \tag{4.14.6}$$

for the absolute error. For the relative error we divide this by $\|\mathbf{x}\|$ and use the inequality $\|\mathbf{b}\| \leq \|\mathbf{A}\| \, \|\mathbf{x}\|$ to give the inequality

$$\frac{\|\delta\mathbf{x}\|}{\|\mathbf{x}\|} \leq \frac{\kappa(\mathbf{A})}{1 - \kappa(\mathbf{A})\dfrac{\|\delta\mathbf{A}\|}{\|\mathbf{A}\|}} \left(\frac{\|\delta\mathbf{b}\|}{\|\mathbf{b}\|} + \frac{\|\delta\mathbf{A}\|}{\|\mathbf{A}\|} \right), \tag{4.14.7}$$

where $\kappa(\mathbf{A}) = \|\mathbf{A}\| \, \|\mathbf{A}^{-1}\|$ is the condition number. This final inequality shows the relationship between the relative solution error and the relative error in $\mathbf{A}$ and $\mathbf{b}$, together with the key role played by the condition number.

Observe that inequality (4.14.7) substantiates the remarks made in Section 4.1. If we have a stable algorithm this assures that a neighbouring problem has been solved, that is that $\left(\dfrac{\|\delta\mathbf{b}\|}{\|\mathbf{b}\|} + \dfrac{\|\delta\mathbf{A}\|}{\|\mathbf{A}\|} \right)$ is small. This only assures an accurate solution if neighbouring problems have neighbouring solutions, that is if $\kappa(\mathbf{A})$ is small. Finally, with algorithm instability and ill-conditioning, there is a cumulative effect.

Exercises

4.1 The 3-digit approximation to the solution of the sets of equations

$$
\begin{aligned}
-0.0101\,x_1 + 1.40\,x_2 - 0.946\,x_3 &= 1.50 \\
4.96\,x_1 + 0.689\,x_2 + 0.330\,x_3 &= 0.496 \\
-1.50\,x_1 - 2.68\,x_2 + 0.100\,x_3 &= -0.290
\end{aligned}
$$

is

$$
\begin{aligned}
x_1 &= 0.225, \\
x_2 &= -0.0815, \\
x_3 &= -1.71.
\end{aligned}
$$

Without pivoting and using 3-digit arithmetic (assume that each operation produces a 3-digit result and is performed with sufficient guard digits for the only error to arise from the final rounding to 3 significant figures), compute the solution to this system of equations. Note that no component of the solution is correct. What is the indication that something has gone wrong?

4.2 Show that the inequality (4.4.4) holds with partial pivoting in use.

4.3 (i) For $\mathbf{r} = \mathbf{b} - \mathbf{A}\tilde{\mathbf{x}}$, show that $\tilde{\mathbf{x}}$ satisfies the equation $(\mathbf{A}+\mathbf{H})\tilde{\mathbf{x}} = \mathbf{b}$, where
$$\mathbf{H} = \frac{\mathbf{r}\tilde{\mathbf{x}}^T}{\|\tilde{\mathbf{x}}\|_2^2}.$$

 (ii) If $\|\mathbf{r}\|_2 = \alpha\|\mathbf{A}\|_2\|\tilde{\mathbf{x}}\|_2$, show that $\|\mathbf{H}\|_2 \leq \alpha\|\mathbf{A}\|_2$.

 (iii) Hence show that $\tilde{\mathbf{x}}$ is the solution of a nearby set of equations if and only if $\|\mathbf{r}\|_2$ is small compared with $\|\mathbf{A}\|_2\|\tilde{\mathbf{x}}\|_2$.

(iv) Show also that the fact that $\|\mathbf{r}\|_2$ is small compared with $\|A\tilde{\mathbf{x}}\|_2$ or $\|\mathbf{b}\|_2$ is a sufficient condition.

4.4 Factorize the matrix $\begin{pmatrix} 1.134 & 2.183 \\ 92.69 & 95.18 \end{pmatrix}$ using 4-decimal arithmetic without interchanges and find $\rho = \max_{i,j,k} |a_{ij}^{(k)}|$. Compare this with the Erisman-Reid bound (4.5.16), taking $p=1$ and $q=\infty$.

4.5 Show that a symmetric matrix is positive definite if and only if all its eigenvalues are positive.

4.6 Show that, if the matrix $\mathbf{A}$ is diagonally dominant by rows (inequality (4.6.6)), then so is $\mathbf{A}^{(2)}$ and the inequalities $\sum_{j=2}^{n} |a_{ij}^{(2)}| \leq \sum_{j=1}^{n} |a_{ij}|$ hold. Deduce that the numbers encountered in Gaussian elimination satisfy the inequalities $|a_{ij}^{(k)}| \leq 2 \max_{i,j} |a_{ij}|$. Show that the same result is true if $\mathbf{A}$ is diagonally dominant by columns (inequality (4.6.5)).

4.7 Show that if $\mathbf{A}$ is symmetric, with eigenvalues λ_i, $\|\mathbf{A}\|_2 = \max_i |\lambda_i|$.

4.8 Show that the inequality $\|A^{-1}\|_p \geq \|\mathbf{a}_k^{(k)}\|_p^{-1}$ is true if $\mathbf{a}_k^{(k)}$ is the vector $(0,0,\ldots,0,a_{kk}^{(k)},\ldots,a_{nk}^{(k)})^T$ of subdiagonal coefficients encountered after applying $k-1$ steps of Gaussian elimination to $\mathbf{A}$. Hint: Consider the singular matrix $(\mathbf{A} - \mathbf{a}_k^{(k)}\mathbf{e}_k^T)$, where $\mathbf{e}_k$ is column k of $\mathbf{I}$.

4.9 The matrix (4.3.1) has inverse

$$\mathbf{A}^{-1} \simeq \begin{pmatrix} -0.6533 & 1.0007 \\ 0.4135 & -0.0004 \end{pmatrix}.$$

Show that $\kappa_\infty(\mathbf{A}) \simeq 4.26$ and that if $\mathbf{D}_1 = \begin{pmatrix} 10^3 \\ & 1 \end{pmatrix}$, $\kappa_\infty(\mathbf{D}_1\mathbf{A}) \simeq 2424$.

4.10 Compute the solution of the equation

$$\begin{pmatrix} 0.287 & 0.512 \\ 0.181 & 0.322 \end{pmatrix} \begin{pmatrix} x_1 \\ x_2 \end{pmatrix} = \begin{pmatrix} -2.32 \\ 3.58 \end{pmatrix}.$$

Hint: Use (4.9.6).

4.11 Perturb the factor $\mathbf{U}$ in equation (4.8.5) by changing -0.001 to -0.0006 and use the revised factorization to solve the problem in Exercise 4.10. Hence demonstrate that the problem in Exercise 4.10 is ill-conditioned.

4.12 From the solution obtained in Exercise 4.11, compute the associated residual for the problem of Exercise 4.10. Is it small relative to $\mathbf{b}$? Explain.

4.13 The ill-conditioning in the matrix of (4.8.2) is demonstrated by a small pivot in the factorization (4.8.5). Thus method (4) of Section 4.10 for estimating conditioning works for this case. What is the key difference between this example and the one given in (4.10.6)?

4.14 Use trial vectors

$$\mathbf{v} = \begin{pmatrix} 1 \\ 1 \end{pmatrix} \quad \text{and} \quad \mathbf{v} = \begin{pmatrix} 0.723 \\ 0.656 \end{pmatrix}$$

for method (5) of Section 4.10 to demonstrate ill-conditioning in the matrix in (4.8.2). Which is more effective and why?

4.15 Compute the LINPACK estimate (see Section 4.10) for the condition number of the matrix (4.8.2).

4.16 For the badly-scaled matrix $\begin{pmatrix} 1.00 & 2420 \\ 1.00 & 1.58 \end{pmatrix}$, used for illustration in Section 4.11, consider the solution of equations with two different right-hand sides $\begin{pmatrix} 3570 \\ 1.47 \end{pmatrix}$, $\begin{pmatrix} 2.15 \\ 3.59 \end{pmatrix}$. Using the row scaling of equation (4.12.4) produces the scaled problems

$$\begin{pmatrix} 0.001 & 2.42 \\ 1.00 & 1.58 \end{pmatrix} \mathbf{x,y} = \begin{pmatrix} 3.57 \\ 1.47 \end{pmatrix}, \begin{pmatrix} 0.00215 \\ 3.59 \end{pmatrix}.$$

Evaluate $\|\mathbf{r}\|_\infty$ for the candidate solutions $\tilde{\mathbf{x}} = \begin{pmatrix} -0.863 \\ 1.48 \end{pmatrix}$ and $\tilde{\mathbf{y}} = \begin{pmatrix} 3.59 \\ -0.001 \end{pmatrix}$. Could these candidate solutions be considered solutions to problems 'near' the scaled one?

4.17 Using column scaling, equation (4.12.6), on the problems of Exercise 4.16, produces the scaled problems

$$\begin{pmatrix} 1.00 & 2.42 \\ 1.00 & 0.00158 \end{pmatrix} \mathbf{x,y} = \begin{pmatrix} 3570 \\ 1.47 \end{pmatrix}, \begin{pmatrix} 2.15 \\ 3.59 \end{pmatrix}.$$

Evaluate $\|\mathbf{r}\|_\infty$ for the candidate solutions $\tilde{\mathbf{x}} = \begin{pmatrix} -10.0 \\ 1480 \end{pmatrix}$ and $\tilde{\mathbf{y}} = \begin{pmatrix} 3.59 \\ -0.595 \end{pmatrix}$. Could these candidate solutions be considered solutions to problems 'near' the scaled one?

4.18 Rescale the solutions of Exercise 4.16 to give candidate solutions to the scaled problems of Exercise 4.17 and evaluate $\|\mathbf{r}\|_\infty$ for them. Are they solutions to nearby problems?

4.19 Rescale the solutions of Exercise 4.17 to give candidate solutions to the original problems of Exercise 4.16 and evaluate $\|\mathbf{r}\|_\infty$ for them. Are they solutions to nearby problems?

4.20 Show that the 2–norm condition numbers of $\mathbf{A}$ and $\mathbf{A}^T\mathbf{A}$ are related by the equation

$$\kappa_2(\mathbf{A}^T\mathbf{A}) = (\kappa_2(\mathbf{A}))^2.$$

5 Gaussian elimination for sparse matrices: an introduction

This chapter provides an introduction to the use of Gaussian elimination for solving sparse sets of linear equations by looking at features in two widely-used general-purpose codes. The three principal phases of most codes, ANALYSE, FACTORIZE, and SOLVE, are explained and illustrated. We stress the importance of acceptable overheads and of numerical stability, but are not concerned with algorithmic or implementation details which are the subjects of later chapters.

5.1 Introduction

With the background of the previous chapters we can now begin the discussion of the central topic of this book: the effective solution of large sparse linear systems (with perhaps thousands of equations) by direct methods. This subject is divided into three major topics: adapting the numerical methods from the dense case to the sparse case; ordering the equations to maintain sparsity; and efficiently implementing the solution process.

The major objective of this chapter is to identify the issues associated with these topics and put them in the perspective of the whole solution process. Detailed discussion of ordering methods is deferred to Chapters 6, 7, and 8, while detailed discussion of data structures and algorithm implementation is the subject of Chapters 9 and 10.

We approach this overview by looking at features of general-purpose sparse matrix computer programs. This provides an excellent way of considering the key issues without the risk of confusion through too much attention to detail. Furthermore, the use of an existing code is generally the proper first step in solving actual sparse matrix problems. It provides a benchmark for performance even for very specialized problems for which a new code might be written.

For purposes of illustration, we use the MA28 sparse matrix code from Harwell for unsymmetric problems and the SPARSPAK code from the University of Waterloo for symmetric problems. Other codes are available including YSMP (Eisenstat, Gursky, Schultz, and Sherman 1982) from Yale University (primarily for symmetric and near-symmetric problems), Y12M from Denmark (Zlatev, Wasniewski, and Schaumburg 1981, broadly comparable with MA28), and some from Harwell for symmetric and unsymmetric problems. We have selected MA28 (Duff 1977c) and SPARSPAK (Chu, George, Liu, and Ng 1984) because of our familiarity with them, their widespread use, their availability, and the diversity of characteristics which they display.

5.2   Features of a sparse matrix code

We begin by describing some of the features required of a general sparse matrix computer code. The first step is the representation of the sparse matrix problem and here there are a number of choices. The user should not have to write complicated code to force the data into a form that is convenient for the implementation of a sparse solution. An inexpert user may code a faulty or poor sorting algorithm that loses most of the potential gains in efficiency. For this reason a common choice is the requirement to specify the entries as an unordered sequence of triples (i, j, a_{ij}) indicating row index, column index, and numerical value. We call this the coordinate scheme; it was discussed and illustrated in Section 2.6. Sometimes it may be equally convenient to specify the matrix as a collection of column (or row) vectors, as discussed and illustrated in Section 2.7. For example, in a nonlinear problem the Jacobian matrix may be generated column by column using finite differences. Since neither of these schemes is adequate for efficient solution, the original (user's) data structure will have to be transformed to one that is efficient internally.

MA28 (Duff and Reid 1979a) uses the coordinate scheme as its standard interface but also allows for the more sophisticated user to specify the entries column by column. SPARSPAK (George and Liu 1979b) allows these two input forms as options, along with a third option to express the matrix as the sum of dense submatrices using the clique storage scheme discussed in Section 2.15.

Often it is the case that a sequence of systems of equations must be solved where the sparsity pattern is fixed but the numerical values change. One example is the solution of time-dependent nonlinear differential equations. The Jacobian matrix can be accommodated in a single sparsity pattern for every time point, while the numerical values change (though some entries may be zero at the initial point). Another example is the design problem where values of parameters must be chosen to maximize some measure of performance. In these cases it is worthwhile to invest effort in the choice of a good ordering for the particular sparsity pattern and in the development of a good associated data structure, since the costs can be spread over many solutions. Except in the symmetric and positive-definite case, this raises some numerical stability questions when the values of the entries change and these are discussed in Section 5.4.

Finally, we may want to use one matrix factorization to solve a system of equations with many right-hand sides. This is the case when performing iterative refinement (Section 4.11) and may also happen because of different analyses of a mathematical model based on various inputs, or it may come from modifying the method of the last paragraph to use one Jacobian matrix over a number of steps.

Phase		Key features
ANALYSE	1.	Transfer user data to internal data structures.
	2.	Determine a good pivotal sequence.
	3.	Prepare data structures for the efficient execution of the other phases.
	4.	Output of statistics.
FACTORIZE	1.	Transfer new numerical values to internal data structure format.
	2.	Factorize the matrix using the chosen pivotal sequence.
	3.	Estimate condition number of the matrix.
	4.	Monitor stability to ensure reliable factorization.
SOLVE	1.	Perform the appropriate permutation from user order to internal order.
	2.	Perform forward substitution and back-substitution using stored $L\backslash U$ factors.
	3.	Perform the appropriate permutation from internal order to user order and return the solution in user order.

Table 5.2.1. Key features of ANALYSE, FACTORIZE, and SOLVE.

Because of these common uses of a general sparse matrix computer code, it is standard to consider three distinct phases: ANALYSE, FACTORIZE, and SOLVE (although not all codes fit this template exactly). These phases and their features are summarized in Table 5.2.1. For matrices whose pattern is symmetric or nearly so, it is possible to organize the computation so that the first phase works on the sparsity pattern alone and involves no actual computation on real numbers. On the other hand, with unsymmetric problems we usually work on the actual numbers too, so a factorization is a by-product of the analysis in this case. For this reason the first phase for unsymmetric matrices is sometimes termed ANALYSE–FACTORIZE. When we discuss implementation in detail, we separate the two cases into two chapters (Chapters 9 and 10).

5.3 Orderings

Ordering the rows and columns to preserve sparsity in Gaussian elimination was demonstrated to be effective in the example of Section 1.3. Another illustration is shown in Figure 5.3.1 where the reordered matrix preserves all zeros in the factorization but the original order preserves none.

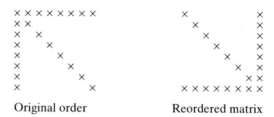

Original order   Reordered matrix

Figure 5.3.1. Reordering can preserve sparsity in factorization.

A very simple but effective strategy for maintaining sparsity is due to Markowitz (1957). At each stage of Gaussian elimination, he selects as pivot the nonzero entry of the remaining reduced submatrix with lowest product of number of other entries in its row and number of other entries in its column. This is the method which was used in Section 1.3, and it is easy to verify that it will accomplish the reordering shown in Figure 5.3.1.

The Markowitz strategy is the ordering method used in MA28 and a symmetric variant is one option in SPARSPAK. This method and some variations are discussed in Chapter 7. Implementation details for the unsymmetric case are considered in Chapter 9 and for the symmetric case in Chapter 10. Surprisingly, the algorithm can be implemented in the symmetric case without explicitly updating the sparsity pattern at each stage, which greatly improves its performance.

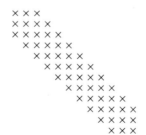

Figure 5.3.2. A band matrix.

A very different approach is to permute the matrix to a form in which the zeros are isolated. One example is the **banded** form illustrated in Figure

5.3.2. If no further interchanges are performed, there is no fill-in outside the band. Some authors have assumed that having a narrow band is the only way that sparsity can be exploited, while here we treat this as one of several alternatives.

Schemes for automatically choosing permutations to band form are discussed in Chapter 8 and data structures for solving band equations are described in Chapter 10. This is one of the options in SPARSPAK.

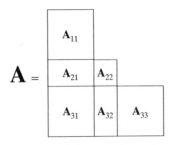

Figure 5.3.3. Block lower triangular matrix.

Another pattern that is worthwhile to seek is the block lower triangular form illustrated in Figure 5.3.3. If we partition **x** and **b** similarly, we may solve the equation **Ax** = **b** by solving the equations

$$\mathbf{A}_{ii}\mathbf{x}_i = \mathbf{b}_i - \sum_{j=1}^{i-1}\mathbf{A}_{ij}\mathbf{x}_j, \quad i=1,2,3, \tag{5.3.1}$$

where the sum is zero for $i=1$. We have to factorize only the diagonal blocks $\mathbf{A}_{ii}$. The off-diagonal blocks $\mathbf{A}_{ij}$, $i>j$, are used only in multiplications $\mathbf{A}_{ij}\mathbf{x}_j$. In particular, all fill-in is confined to the diagonal blocks. Any row and column interchanges needed for the sake of stability and sparsity may be performed within the diagonal blocks and do not affect the block triangular structure. This is an option in MA28 and algorithms for achieving this form are discussed in Chapter 6.

Figure 5.3.4. A dissection of the region and its corresponding matrix.

Another technique for isolating zeros is dissection. If the matrix pattern corresponds to a finite-element or finite-difference approximation over a region, the removal of part of the region (for example region 3 in Figure 5.3.4) may cut the problem into independent pieces. Ordering the variables of the cut (and the corresponding equations) last yields a bordered block diagonal form, as illustrated in Figure 5.3.4. The zero off-diagonal blocks remain zero during factorization if no further interchanges are performed except within blocks.

Methods such as nested dissection and one-way dissection, both options in SPARSPAK, are refinements of this general approach. These and other ordering methods which isolate zeros are the subject of Chapter 8.

The choice of best ordering method is problem dependent. We make some comparisons between methods in both Chapters 7 and 8.

5.4 Numerical stability in sparse computation

The numerical stability of the factorization is not a concern in the symmetric positive-definite case, see Section 4.6. Thus SPARSPAK is not concerned with numerical stability. For the unsymmetric case, however, the stability of the factorization depends on the growth in the size of the matrix entries during Gaussian elimination. The usual strategy for controlling this growth is based on the threshold strategy discussed in Section 4.4. Pivots are selected, on sparsity grounds, from the candidates satisfying the inequality

$$|a_{ij}^{(k)}| \geq u \max_l |a_{lj}^{(k)}| , \tag{5.4.1}$$

where u is a preset parameter in the range $0 < u \leq 1$ (the **threshold parameter**). This limits the growth in a single column during a single step of the reduction according to the inequality

$$\max_i |a_{ij}^{(k+1)}| \leq (1+u^{-1}) \max_i |a_{ij}^{(k)}| \tag{5.4.2}$$

and the growth overall by

$$\max_i |a_{ij}^{(k)}| \leq (1+u^{-1})^{p_j} \max_i |a_{ij}| \tag{5.4.3}$$

where p_j is the number of off-diagonal entries in the j-th column of $U=\{a_{ij}^{(n)}, j \geq i\}$. The inequality (5.4.3) is an immediate consequence of inequality (5.4.2) and the fact that column j changes only p_j times during the reduction. This was discussed by Gear (1975).

The analogous inequality for the dense case has $p_j = j-1$; while this is an attainable bound, it is generally considered over-pessimistic in practice. Inequality (5.4.3) merely indicates that because p_j is usually small in the sparse case, we may take $u < 1$ and avoid too great a restriction on pivot choice. The parameter u is an input parameter for MA28 and $u = 0.1$ is recommended on the basis of extensive testing. The trade-off between sparsity and stability is discussed in Chapter 7, where test results for various values of u are given.

Because of the generally pessimistic nature of the bound (5.4.3), we recommend assessing the stability of the factorization. If a factorization found during ANALYSE is unstable, ANALYSE may be repeated with a larger value of u. If a factorization found during FACTORIZE is unstable, a fresh ANALYSE may be performed with the current set of numerical values.

MA28 can estimate the growth parameter $\rho = \max |a_{ij}^{(k)}|$ using the Erisman and Reid (1974) bound, see Section 4.5, or can compute ρ directly. Duff (1979) has found that in some cases the estimate is too pessimistic, which is why the slightly more expensive direct computation is provided as an alternative.

Another concern in assessing stability for the large, sparse case is the presence of n in the stability bound

$$|h_{ij}| \leqslant 5.01\,\varepsilon\rho n, \tag{5.4.4}$$

(see (4.3.14)). For large problems (say $n = 100\,000$), n could play a substantial role in the bound. Gear (1975) has generalized the bound to the sparse case and obtained

$$|h_{ij}| \leqslant 1.01\,\varepsilon\rho\left(up^3 + (1+u)p^2\right) + O(\varepsilon^2) \tag{5.4.5}$$

where

$$p = \max_j p_j. \tag{5.4.6}$$

While this is now independent of n, p^3 can be large, so the result is not entirely satisfactory.

Another a posteriori way to assess the computational stability is to compute the residual vector, as discussed in Section 4.5. This is an option in MA28.

For the symmetric indefinite case where 2×2 pivots may be used to preserve stability and symmetry, as discussed in Section 4.7, an extension to the sparse case is needed. Duff, Reid, Munksgaard, and Nielsen (1979) recommend

$$u\,|a_{ij}^{(k)}|\,\|\mathbf{E}_k^{-1}\|_1 \leqslant 1, \quad i > k+1, \quad j = k, k+1 \tag{5.4.7}$$

where

$$\mathbf{E}_k = \begin{pmatrix} a_{kk}^{(k)} & a_{k,k+1}^{(k)} \\ a_{k+1,k}^{(k)} & a_{k+1,k+1}^{(k)} \end{pmatrix} \tag{5.4.8}$$

is a 2×2 pivot. This choice maintains the validity of inequality (5.4.3) and again the value $u = 0.1$ will generally be acceptable.

Another issue associated with stability estimation in sparse matrix computation is the proper interpretation of inequality (5.4.4). We have bounded the size of the matrix $\mathbf{H}$ that satisfies the equation

$$\mathbf{A} + \mathbf{H} = \mathbf{LU} \tag{5.4.9}$$

and thus established the successful factorization of a neighbouring matrix. But if **H** has a sparsity pattern different from that of **A**, **A**+**H** is qualitatively different from **A**. This topic is discussed further in Section 12.8.

5.5 Estimating condition numbers in sparse computation

In Chapter 4 we showed that the stability of the factorization is only part of the story in assessing the accuracy of the solution. The other part is the condition of the matrix. Neither MA28 nor SPARSPAK presently provides a direct estimate of the condition of the matrix. Methods (5), (6), or (7) from Section 4.10, however, may be implemented by making use of iterative refinement (MA28) or extra SOLVE entries.

When adapting these methods to the sparse case, one must be cautious. These methods provide an inexpensive assessment of the matrix condition number if SOLVE is inexpensive compared with FACTORIZE. In the dense case SOLVE involves $O(n^2)$ operations compared with $O(n^3)$ for FACTORIZE, but the comparison is often much less striking for the sparse case. Even simple rescaling of the solution can add significantly to SOLVE times. Grimes and Lewis (1981) made a minor modification of the LINPACK condition estimating algorithm (see Appendix B) to reduce the number of times the solution is rescaled (to avoid overflows, the LINPACK code rescales the 'solution' vector whenever the next component to be determined has magnitude greater than one). While this change would be insignificant for the dense case, Table 5.5.1 shows the dramatic impact for the sparse case. The timings were done on a CYBER 175 computer using code which interfaced with SPARSPAK.

Factorization time	Standard condition estimation	Modified condition estimation
0.075 sec.	2.7 sec.	0.19 sec.

Table 5.5.1. The impact of scaling in sparse condition estimation for a test problem of order 2000.

Having looked at ordering and numerical computation adapted to the sparse case, we now look at the three phases of the computer code, again from a user viewpoint. The purpose is not to discuss implementation details, the subject of Chapters 9 and 10, but to provide insight into critical operations for the user of a sparse matrix code.

5.6 The ANALYSE phase

It is in the ANALYSE phase that the fundamental differences between MA28 and SPARSPAK are most evident. MA28 must deal with both numerical and structural data in this phase, and it actually produces a factorization as a by-product. In the course of selecting the pivot order, the updated elimination matrix with the numerical and structural information, including fill-in, must be held. This means that an unknown amount of additional storage is required during the ANALYSE step. The user is required to provide work space with three possible outcomes: the solution is achieved efficiently, the solution is achieved less efficiently because the small amount of storage requires a great deal of internal data movement, or no solution is found but the user is given an estimate of how much storage is required.

SPARSPAK assumes that pivots may be selected from the diagonal without risk of instability. The pivot selection is accomplished without numerical values and fill-in is tracked implicitly. Thus no additional storage is needed during ANALYSE, and the precise requirements for FACTORIZE can be reported to the user. Similar features are provided by MA27/MA37 from Harwell and YSMP from Yale.

Order of matrix	199	822	900
Number of nonzeros	701	4841	4380
ANALYSE	0.24	1.76	11.40
FACTORIZE	0.05	0.26	1.52
SOLVE	0.01	0.04	0.09

Table 5.6.1. Times for the three phases of MA28 (seconds on an IBM 370/168).

In the case of MA28, the FACTORIZE step is much less costly than the ANALYSE step, as is illustrated by the times in Table 5.6.1. In SPARSPAK, since no numerical data is present in ANALYSE, even the first numerical factorization is carried out in FACTORIZE. Because of some critical implementation details in ANALYSE, discussed in Chapter 10, ANALYSE can be less costly than FACTORIZE, as we can see from the results in Table 5.6.2. Because the problems and the computers of Tables 5.6.1 and 5.6.2 are different, no comparison between the codes is implied.

All the competitive codes depend for their efficiency on careful handling of implementation details. Many of these are discussed in Chapters 9 and 10.

Order of matrix	147	1176	778	503	1005	1723	5300
Number of nonzeros	2449	18552	5272	6027	8621	6511	21842
ANALYSE	0.05	0.37	0.21	0.21	0.51	0.48	2.00
FACTORIZE	0.05	0.29	0.30	0.28	0.69	0.08	0.48
SOLVE	0.01	0.03	0.04	0.03	0.06	0.03	0.13

Table 5.6.2. Times for different phases of SPARSPAK (seconds on an IBM 3033).

5.7 The FACTORIZE and SOLVE phases

For the FACTORIZE and SOLVE phases, the different types of computer codes represented by MA28 and SPARSPAK are much more alike. In FACTORIZE, the numerical values for **A** are usually supplied in exactly the same order as they were supplied to ANALYSE. It is assumed that static data structures can now be used because the pivotal sequence, fill-in pattern, and the details of the transformation of the user's data to internal format have been preserved.

The distinguishing feature of MA28 for this stage is the question about the stability of the factorization for the new set of numerical values. If instability is indicated, the ANALYSE phase may have to be repeated. It is our experience, however, that problems in a sequence with the same sparsity pattern are often related closely enough for the subsequent factorizations to remain stable.

```
DO 10 K = K1,K2
   J = JCN(K)
   W(J) = W(J) - ALPHA*VAL(K)
10 CONTINUE
```

Figure 5.7.1. The innermost loop of FACTORIZE.

The key operation in the FACTORIZE step of both codes involves a loop like that illustrated in Figure 5.7.1. A detailed discussion of how this arises is left to Chapters 9 and 10, but here we make two important observations which may be helpful to the user:

(i) we have been able to avoid searching and testing in the innermost loop, and

(ii) W is addressed indirectly through JCN, which means that the store is accessed more often and in an irregular manner.

The absence of searching is critical because it reduces the overheads to the difference between the use of direct and indirect addressing. Dembart and

Neves (1977) did a detailed study of the implementation of this loop on vector architectures and showed regions of best performance for various loop lengths. More recently (1985), the CRAY X-MP and FACOM VP computers have added hardware instructions which largely overcome the degradation associated with indirect addressing. This represents an excellent, but all too infrequent, example of hardware responding to software needs.

In Chapter 3 we described forward substitution and back-substitution in terms of accessing rows of **L** and rows of **U**, but the adaptation to columns of one or both is quite easy (see Exercise 5.1). The variations are insignificant unless advantage is to be taken of sparsity in **b** or a requirement for only a partial solution. Again, details of implementation for SOLVE are deferred to Chapters 9 and 10.

Both MA28 and SPARSPAK provide the user with the option of entering new numerical values for **b** and getting a new solution using the old factors. For a fresh factorization, SPARSPAK requires that **A** be provided in the same order as the original non-numerical data. MA28, however, permits the order to be changed and indeed a different pattern can be input as long as it lies within the pattern of the **L\U** factors. The user should note that this flexibility is bought at the cost of a more expensive sort. For both codes, SOLVE does not require the user to perform any prior permutations of the right-hand side vector and returns the solution in the user's order. The distinguishing feature for SOLVE in MA28 is the facility to solve either $\mathbf{Ax} = \mathbf{b}$ or $\mathbf{A}^T\mathbf{x} = \mathbf{b}$ according to user choice. Since SPARSPAK primarily solves symmetric problems, this feature is not needed there.

Another comment about SOLVE is important to the user. While in the dense case SOLVE is usually trivial compared with FACTORIZE (involving $O(n^2)$ compared with $O(n^3)$ operations), this may not be the case for sparse problems. The last two columns of Table 5.6.2 illustrate that SOLVE time can be more than a quarter of the corresponding FACTORIZE time. These very sparse problems from power systems load flow calculations were discussed by Lewis and Poole (1980). Another example where SOLVE is significant was given in Section 5.5, where multiple SOLVE entries were used to estimate condition numbers.

5.8 Writing compared with using sparse matrix software

If a large sparse matrix problem had to be solved in 1970, the software needed to be written for the problem. Today the situation is very different. Very useful sparse matrix software is generally available and actively being improved, see Duff (1984b), and for almost any problem it should provide a starting point. While it may be true that a particular problem has many special characteristics which can be exploited, use of general-purpose software can provide a benchmark against which improvements can be measured.

Often the work involved in performing operations on a sparse matrix of order n with $\tau = cn$ nonzeros is only $O(c^2 n)$ (see Exercise 5.5). One of the subtleties in writing sparse matrix software lies in avoiding $O(n^2)$ or more operations. This can happen in very innocent ways and then dominate the computation. This was illustrated in Table 5.5.1 in estimating the condition number of a matrix. Another example lies in the implementation of the pivot selection in ANALYSE. A naïve strategy for the algorithm of Markowitz would require checking all remaining entries at every step, which would require at least $O(n^2)$ comparisons in all and would be likely to dominate the computation. A number of other $O(n^2)$ traps were identified by Gustavson (1978).

If only a few problems of moderate order are to be solved, general-purpose codes are very adequate. If problems of interest are infeasible using a standard code, then special characteristics should be sought which will allow their solution. Often, however, a general-purpose sparse matrix code may serve as a good starting point for developing a more specialized one.

Exercises

5.1 Rewrite the algebraic description of back-substitution and forward substitution, equations (3.2.5) and (3.2.6), to access both **L** and **U** by columns.

5.2 Write the pattern for the matrix corresponding to a 9-point operator (where each equation involves a node and its neighbours in the directions N, S, E, W, NW, NE, SE, SW) on a 4×4 square grid. Try two different initial node orderings to see their effect.

5.3 With columns of **L** stored in packed form, write a Fortran program to perform forward substitution.

5.4 Using a standard sparse matrix code, such as one of those mentioned in this chapter, gain some experience as a user by trying several different problems (perhaps from the test collection). Experiment with different features and try to assess the validity of your results.

5.5 Give a plausible argument to illustrate that a sparse **LU** factorization may be performed in $O(c^2 n)$ operations, if n is the order and cn is the number of nonzeros.

6 Reduction to block triangular form

We consider algorithms for reducing a general sparse matrix to
block triangular form. This form allows the corresponding set of
linear equations to be solved as a sequence of subproblems. We
discuss the assignment problem of placing entries on the
diagonal as a part of the process of finding the block triangular
form, though this problem is of interest in its own right.

6.1 Introduction

If we are solving a set of linear equations

$$\mathbf{A x} = \mathbf{b} \qquad (6.1.1)$$

whose matrix has a block triangular form, savings in both computational
work and storage may be made by exploiting this form. Our purpose in this
chapter is to explain how a given matrix may be permuted to this form, and
to demonstrate that the form is (essentially) unique. Remarkably
economical algorithms exist for this task, typically requiring $O(n) + O(\tau)$
operations for a matrix of order n with τ entries.

We find it convenient to consider permutations to the block lower
triangular form

$$\mathbf{P A Q} = \begin{bmatrix} \mathbf{B}_{11} & & & & & \\ \mathbf{B}_{21} & \mathbf{B}_{22} & & & & \\ \mathbf{B}_{31} & \mathbf{B}_{32} & \mathbf{B}_{33} & & & \\ \cdot & \cdot & \cdot & \cdot & & \\ \cdot & \cdot & \cdot & & \cdot & \\ \cdot & \cdot & \cdot & & & \cdot \\ \mathbf{B}_{N1} & \mathbf{B}_{N2} & \mathbf{B}_{N3} & \cdot & \cdot & \cdot & \mathbf{B}_{NN} \end{bmatrix}, \qquad (6.1.2)$$

though with only minor modification the block upper triangular form could
be obtained (see Exercise 6.1).

A matrix which can be permuted to the form (6.1.2), with $N > 1$, is said
to be **reducible**. If no block triangular form other than the trivial one
($N = 1$) can be found, the matrix is called **irreducible**. We expect each $\mathbf{B}_{ii}$ to
be irreducible, for otherwise a finer decomposition is possible. Some
authors reserve the terms reducible and irreducible for the case $\mathbf{Q} = \mathbf{P}^T$.
For $\mathbf{Q} \neq \mathbf{P}^T$, Harary (1971) uses the terms 'bireducible' and 'bi-irreducible'
and Schneider (1977) uses the term 'fully indecomposable'.

The advantage of using block triangular forms such as (6.1.2) is that the
set of equations (6.1.1) may then be solved by the simple forward
substitution process

$$\mathbf{B}_{ii}\mathbf{y}_i = (\mathbf{P}\,\mathbf{b})_i - \sum_{j=1}^{i-1} \mathbf{B}_{ij}\mathbf{y}_j, \ i = 1, 2, \dots, N \qquad (6.1.3)$$

(where the sum is zero for $i=1$) and the permutation

$$\mathbf{x}=\mathbf{Q}\,\mathbf{y}. \qquad (6.1.4)$$

We have to factorize only the diagonal blocks $\mathbf{B}_{ii}$, $i = 1, 2, \dots, N$. The off-diagonal blocks $\mathbf{B}_{ij}$, $i > j$, are used only in the multiplications $\mathbf{B}_{ij}\mathbf{y}_j$. Notice in particular that all fill-in is confined to the diagonal blocks. If row and column interchanges are performed within each diagonal block for the sake of stability and sparsity, this will not affect the block triangular structure.

There are classes of problems for which the algorithms of this chapter may not be useful. For some applications it may be known a priori that the matrix is irreducible. Matrices arising from the discretization of partial differential equations are often of this nature. If $\mathbf{A}$ is symmetric with nonzero diagonal entries, reducibility means that $\mathbf{A}$ may be permuted to block diagonal form. Because of the underlying physical structure, this decomposition may be known in advance and again the automatic methods are not useful.

On the other hand, many areas produce reducible systems, including chemical process design (Westerberg and Berna 1979), linear programming (Hellerman and Rarick 1971), and economic modelling (Szyld 1981). Because of the speed of the block triangularization algorithms and the savings that result from their use in solving reducible systems of equations, they can be very beneficial.

Techniques exist for permuting directly to the block triangular form (6.1.2) but we do not know of any advantage over the usual two-stage approach:

(1) permute entries onto the diagonal (usually called finding a **transversal**), and

(2) use symmetric permutations to find the block form itself.

In this chapter we discuss the two stages separately (Sections 6.2 to 6.5 and 6.6 to 6.9). The essential uniqueness of the form is discussed in Section 6.10, the benefits of using a block triangular form are illustrated in Section 6.11, and the use of the transversal algorithms is considered in a wider context in Section 6.12.

6.2 Background for finding a transversal

Our purpose in this section is to motivate an algorithm for finding a transversal based on the work of Hall (1956). The algorithm which appeared in Hall's paper was based on a breadth-first search, while the algorithm we discuss here follows the lines of the work of Kuhn (1955) and uses a depth-first search. As we shall see in the following text, however, it is the implementation of the search which is of crucial importance to the efficiency of the algorithm. The algorithm we describe follows closely the work of Duff (1972, 1981c) and Gustavson (1976), who have discussed this algorithm extensively, including algorithmic details, heuristics, implementation, and testing. Duff (1981d) provides a Fortran implementation.

The algorithm can be described in either a row or a column orientation; we follow the variant that looks at the columns one by one and performs row permutations. Suppose that permutations have been found that place entries in the first $k-1$ diagonal positions. We then examine column k, and seek a row permutation which will

(1) preserve the presence of entries in the first $k-1$ diagonal positions, and

(2) result in column k having an entry in row k.

The algorithm continues in this fashion extending the transversal by one at each stage. Success in this extension is called an **assignment**.

Sometimes this transversal extension step is trivial. If column k has an entry in row k or beyond, say in row i, a simple interchange between rows i and k will place an entry in the (k,k) position: a **cheap assignment**. This is illustrated in Figure 6.2.1, where the assignment in column six is made by the interchange of rows six and seven.

$$\begin{bmatrix} \times & & \times & & & & \\ & \times & & & \times & & \\ & & \times & & & \times & \\ \times & & \times & & & & \\ & \times & & \times & & & \\ & & & & & \times & \\ & & & \times & & & \end{bmatrix}$$

Figure 6.2.1. A cheap assignment for column 6.

Figure 6.2.2 shows a simple case where a cheap assignment is not available, but the interchange of rows one and seven is adequate to preserve the (1,1) entry and make a cheap assignment possible in column six.

Sometimes the transversal cannot be extended because the matrix is singular for all numerical values of the entries. Such a matrix is called **symbolically singular** or **structurally singular**. This is illustrated in Figure

$$\begin{bmatrix} \times & & & & \times & & \\ & \times & & & & & \times \\ \times & & \times & & & & \\ & & & & \times & & \\ & & & & & \times & \\ & & & \times & & & \\ \times & & & & & & \times \end{bmatrix}$$

Figure 6.2.2. A single preliminary row interchange is needed.

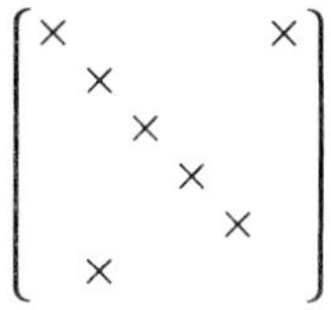

$$\begin{bmatrix} \times & & & & \times & \\ & \times & & & & \\ & & \times & & & \\ & & & \times & & \\ & & & & \times & \\ & \times & & & & \end{bmatrix}$$

Figure 6.2.3. A symbolically singular case.

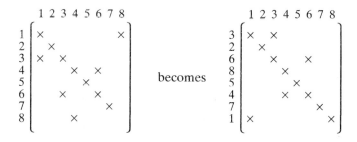

Figure 6.2.4. A more complicated transversal extension.

6.2.3. Observe that columns one and six (or rows two and six) must be multiples of each other irrespective of the numerical values of the entries.

Finally the transversal may be extendible, but the permutation that accomplishes this may be rather more complicated. Such a matrix pattern is shown in Figure 6.2.4. It may be verified that the sequence of row interchanges (1,3), (3,6), (6,4), (4,8) succeeds in placing an entry in the last diagonal position while keeping entries in the other diagonal positions.

The goal of the algorithm described in the next section is to accomplish the transversal extension step quickly and reliably when it can be done, and to identify symbolically singular cases. We show in Section 6.4 that it is sufficient to consider the columns one at a time.

6.3 A depth-first search algorithm for transversal extension

Our algorithm provides a systematic means of permuting the rows so that the presence of entries in the first $k-1$ diagonal positions is preserved while an entry is moved to position (k,k). If column k has an entry in row k or beyond, say in row i, a simple interchange of rows i and k suffices. Otherwise we proceed as follows.

Suppose the first entry in column k is in row j. If column j has an entry in row k or beyond, say in row p, this means that an interchange between rows p and j will preserve the (j,j) entry and move the entry in column k to row $p \geqslant k$. This is illustrated in Figure 6.2.2, where k=6, j=1, p=7. If column j does not have an entry in row k or beyond, any off-diagonal entry in column j, say (p,j), will allow the row interchange (p,j) to preserve the (j,j) entry while providing an opportunity to look in column p for an entry in row k or beyond. This sequence is illustrated in Figure 6.3.1 where k=6, j=2, p=4, and the pair of interchanges (2,4) and (4,6) suffices.

$$
\begin{array}{c}
\\ 1\\2\\3\\4\\5\\6
\end{array}
\begin{array}{c}
\begin{array}{cccccc}1&2&3&4&5&6\end{array}\\
\left[\begin{array}{cccccc}
\times&&&&&\\
&\times&&&&\times\\
&&\times&&&\\
&\times&&\times&&\\
&&&&\times&\\
&&&\times&&
\end{array}\right]
\end{array}
\quad \text{becomes} \quad
\begin{array}{c}
\\ 1\\4\\3\\6\\5\\2
\end{array}
\begin{array}{c}
\begin{array}{cccccc}1&2&3&4&5&6\end{array}\\
\left[\begin{array}{cccccc}
\times&&&&&\\
&\times&&\times&&\\
&&\times&&&\\
&&&\times&&\\
&&&&\times&\\
&\times&&&&\times
\end{array}\right]
\end{array}
$$

Figure 6.3.1. Two row interchanges needed in the transversal algorithm.

In general we seek a sequence of columns $c_1, c_2, \ldots, c_j$ with $c_1 = k$ having entries in rows $r_1, r_2, \ldots, r_j$, with $r_i = c_{i+1}$, $i=1,2,\ldots,j-1$, and $r_j \geqslant k$. Then the sequence of row interchanges $(r_1, r_2), \ldots, (r_{j-1}, r_j)$ achieves what we need. To find this we make use of a ***depth-first search*** with a look-ahead feature. Starting with the first entry in column k we take its row number to indicate the next column and continue letting the first off-diagonal entry in each column indicate the subsequent column. In each column we look for an entry in row k or beyond. Always taking the next column rather than trying other rows in the present column is the depth-first part. Looking to see if there is an entry in row k or beyond is the look-ahead feature. This sequence has one of the following possible outcomes:–

(1) we find a column with an entry in row k or beyond (some authors call this a cheap assignment but we prefer to reserve the term for immediate assignments in column k),

(2) we reach a row already considered (which is not worth taking because it leads to a column already considered), or

(3) we come to a dead end (that is a column with no off-diagonal entries

or one whose off-diagonal entries have all already been considered).

In case (1) we have the sequence of columns that we need. In case (2) we take the next entry in the current column. In case (3) we return (backtrack) to the previous column and start again with the next entry there. An example where a dead end is reached is shown in Figure 6.3.2 and an example where we reach a row already considered is shown in Figure 6.3.3. A more complicated example, containing both features, is shown in Figure 6.3.4.

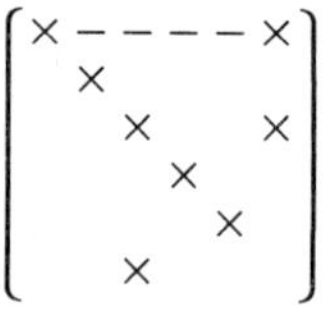

$$\begin{bmatrix} \times - - - - \times \\ \quad \times \\ \qquad \times \quad \times \\ \qquad\quad \times \\ \qquad\qquad \times \\ \qquad \times \end{bmatrix}$$

Figure 6.3.2. Encountering a dead end.

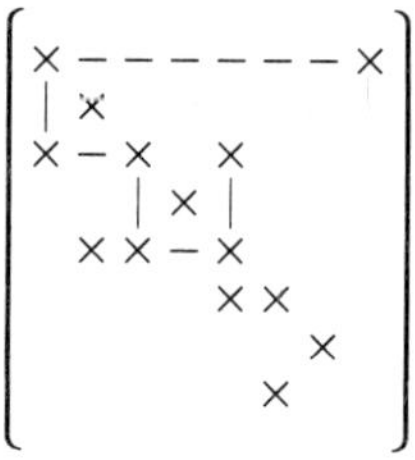

$$\begin{bmatrix} \times - - - - - \times \\ | \times \\ \times - \times \quad \times \\ \quad | \times | \\ \times \times - \times \\ \qquad \times \times \\ \qquad\quad \times \\ \qquad\quad \times \end{bmatrix}$$

Figure 6.3.3. Reaching a row already considered.

$$\begin{bmatrix} \times - - - - - \times \\ | \times \\ \times - \times \quad \times \\ \quad | \times \\ \times \times - \times \\ \quad \times \quad \times \\ \qquad\quad \times \\ \qquad\quad \times \end{bmatrix}$$

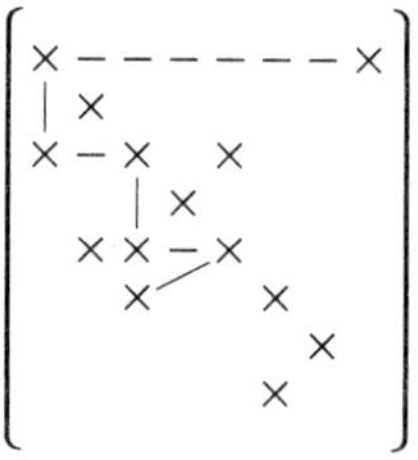

Figure 6.3.4. Returning to a previous column.

Note that in the case of backtracking, the paths involved in the backtrack do not contribute to the final column sequence. The row interchanges for Figure 6.3.3 are (1,3), (3,5), (5,6), (6,8). The interchanges for Figure 6.3.4 are (1,3), (3,6), (6,8).

The combined effect of the row interchanges at stage k is that the new rows 1 to k consist of the old rows 1 to $k-1$ and one other. Therefore any entry in rows 1 to $k-1$ before the interchanges will be in rows 1 to k afterwards. It follows that for the look-ahead feature each entry need be tested only once during the whole algorithm to see if it lies in or beyond the current row k. We therefore keep pointers for each column to record how far the look-ahead searches have progressed. In particular note that the columns in the interior of the depth-first sequence have all been fully examined in look-ahead searches and never need to be so examined again. This simple observation has a significant effect on the efficiency of the implementation.

6.4 Analysis of the depth-first search transversal algorithm

In this section, we examine the behaviour of the transversal algorithm when the matrix is symbolically singular, look at its computational complexity, and comment on another transversal finding algorithm.

Suppose the algorithm has failed to find a path from column k to row k or a row beyond it. If the total number of rows considered is p, we will have looked at p of the first $k-1$ columns and also at column k and found that none of these $p+1$ columns contains an entry outside the p rows. Thus the $p+1$ columns have entries in only p rows and the matrix is symbolically singular. A simple example of this is in Figure 6.2.3 and another is in Exercise 6.2. We may continue the algorithm from column $k+1$ to get other entries onto the diagonal, but we must leave a zero in the (k,k) position.

From the description of the algorithm, we observe that there are n major steps, if n is the matrix order. At each step we consider each entry at most once in the depth-first search and at most once in the look-ahead part, so if the matrix has τ entries the number of elementary operations is at most proportional to $n\tau$. Duff (1981c) provides a detailed discussion of this and includes an example, see Figure 6.4.1 and Exercise 6.3, which shows that this bound may be attained. In general, however, we would expect far fewer steps and Duff's test results confirm this.

Duff (1981c) also examines other algorithms for finding a transversal, including the algorithm of Hopcroft and Karp (1973) which requires at most $O(n^{\frac{1}{2}}\tau)$ operations. Based on timing comparisons of his implementations of the two algorithms, Duff reports that the depth-first search algorithm often performs faster in spite of the poorer complexity bound. Efficient implementations of the algorithm of Hopcroft and Karp are much more complicated, result in much longer code, and need more storage.

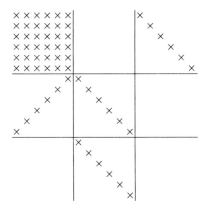

Figure 6.4.1. An example requiring $O(n\tau)$ accesses.

6.5 Implementation of the transversal algorithm

As described in its column orientation, the transversal algorithm may be implemented with the positions of the entries stored as a collection of sparse column vectors (Section 2.7). An array is needed to keep track of the row permutations, since all interchanges should be done implicitly. An array is needed to record the column sequence $c_1, c_2, \ldots$ and we need arrays to record how far we have progressed in each column in the look-ahead search and in the local depth-first search. Finally we require an array that records which rows have already been traversed at this stage. It need not be reinitialized if the value k is used as a flag at stage k.

Other implementation heuristics have been considered to increase the speed of the algorithm. Perhaps the most obvious is to try to maximize the number of cheap assignments. Duff (1972) and Gustavson (1976) suggest a number of heuristics, including ordering the columns initially by increasing numbers of entries. Figure 6.5.1 illustrates why this appears to be a good idea. As shown, only $\frac{1}{2}n$ cheap assignments could be made followed by $\frac{1}{2}n$ row interchanges. If the columns are reordered by increasing numbers of entries, the entire transversal selection can be made with cheap assignments. The asymptotic bound of the algorithm does, however, remain the same (see Exercise 6.4). Note that each column is searched only once for a cheap assignment, so reordering the rows provides little scope for improving the efficiency of the searches for cheap assignments.

Another possibility is to interchange a later column with column k whenever a cheap assignment is not available. In this way we may assign cheaply at least half the transversal of a nonsingular matrix (see Exercise 6.5). Note that our algorithm does not always get as many as $\frac{1}{2}n$ cheap assignments (see Exercise 6.6).

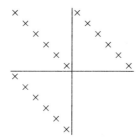

Figure 6.5.1. Illustration of benefit of preordering columns.

In spite of the apparent benefits, Duff (1981c) found the implementation of even simple heuristics caused overheads which usually could not be overcome by being able to do a faster assignment on the resulting matrix. For this reason, he discourages the use of such heuristics. A possible explanation for this is that it is common for the given matrix to have entries on all or most of its diagonal. The initial permutation will, in general, destroy this property, although the main algorithm will not.

6.6 Symmetric permutations to block triangular form

We assume that a row permutation $\mathbf{P}_1$ has been computed so that $\mathbf{P}_1\mathbf{A}$ has entries on the whole of its diagonal (unless $\mathbf{A}$ is symbolically singular), and we now wish to find a symmetric permutation which will put the matrix into block lower triangular form. That is, we wish to find a permutation matrix $\mathbf{Q}$ such that $\mathbf{Q}^T(\mathbf{P}_1\mathbf{A})\mathbf{Q}$ has the form

$$\mathbf{Q}^T(\mathbf{P}_1\mathbf{A})\mathbf{Q} = \begin{bmatrix} \mathbf{B}_{11} & & & & & \\ \mathbf{B}_{21} & \mathbf{B}_{22} & & & & \\ \mathbf{B}_{31} & \mathbf{B}_{32} & \mathbf{B}_{33} & & & \\ \cdot & \cdot & \cdot & \cdot & & \\ \cdot & \cdot & \cdot & & \cdot & \\ \cdot & \cdot & \cdot & & & \cdot \\ \mathbf{B}_{N1} & \mathbf{B}_{N2} & \mathbf{B}_{N3} & \cdot & \cdot & \cdot & \mathbf{B}_{NN} \end{bmatrix}, \qquad (6.6.1)$$

where each $\mathbf{B}_{ii}$ cannot itself be symmetrically permuted to block triangular form. Then with $\mathbf{P} = \mathbf{Q}^T\mathbf{P}_1$, we will have achieved the desired permutation (6.1.2).

It is convenient to describe algorithms for this process with the help of the digraphs (directed graphs) associated with the matrices (see Section 1.2). With only symmetric permutations in use, the diagonal entries play no role so we have no need for self-loops associated with diagonal entries. Applying a symmetric permutation to the matrix causes no change in the associated digraph except for the relabelling of its nodes. Thus we need only consider relabelling the nodes of the digraph, which we find easier to explain than permuting the rows and columns of the matrix.

If we cannot find a closed path through all the nodes of the digraph, then we must be able to divide the digraph into two parts which are such that there is no path from the first part to the second. Renumbering the first group of nodes $1, 2, ..., k$ and the second group $k+1, ..., n$ will produce a corresponding (permuted) matrix in block lower triangular form. An example is shown in Figure 6.6.1, where there is no connection from nodes (1,2) to nodes (3,4,5). The same process may now be applied to each resulting block until no further subdivision is possible. The resulting sets of nodes corresponding to diagonal blocks are called **strong components**. For each there is a closed path passing through all its nodes but no closed path includes these and other nodes. The digraph of Figure 6.6.1 contains just two strong components, which correspond to the two irreducible diagonal blocks.

Figure 6.6.1. A 5×5 matrix and its digraph.

A triangular matrix may be regarded as the limiting case of the block triangular form in the case when every diagonal block has size 1×1. Conversely, the block triangular form may be regarded as a generalization of the triangular form with strong components in the digraph corresponding to generalized nodes. This observation forms the basis for the algorithms in the next three sections.

6.7 The algorithm of Sargent and Westerberg

Algorithms for finding the triangular form may be built upon the observation that if A is a symmetric permutation of a triangular matrix, there must be a node in its digraph from which no path leaves. This node should be ordered first in the relabelled digraph (and the corresponding row and column of the matrix permuted to the first position). Eliminating this node and all edges pointing into it (corresponding to removing the first row and column of the permuted matrix) leaves a remaining subgraph which again has a node from which no paths leave. Continuing in this way, we eventually permute the matrix to lower triangular form.

To implement this strategy for a matrix that may be permuted to triangular form, we may start anywhere in the digraph and trace a path until we encounter a node from which no paths leave. This is easy since the digraph contains no closed paths (such a digraph is called acyclic); any available choice may be made at each node and the path can have length at most $n-1$, where n is the matrix order. We number the node at the end of the path first and eliminate it and all edges pointing to it from the digraph, then continue from the previous node on the path (or choose any remaining node if the path is now empty) until once again we reach a node with no path leaving it. In this way the triangular form is identified and no edge is inspected more than once, so the algorithm is economical. We illustrate this with the digraph of Figure 6.7.1. The sequence of paths is illustrated in Figure 6.7.2.

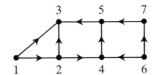

Figure 6.7.1. A digraph corresponding to a triangular matrix.

$$
\text{Path} \left\{
\begin{array}{ccccccccccc}
 & & & \mathbf{5} & & & & & & & \\
 & & \mathbf{3} & 4 & 4 & \mathbf{4} & & & & & \\
 & 2 & 2 & 2 & 2 & 2 & 2 & & & \mathbf{7} & \\
\mathbf{1} & 1 & 1 & 1 & 1 & 1 & 1 & \mathbf{1} & 6 & 6 & \mathbf{6}
\end{array}
\right.
$$

Step   1   2   3   4   5   6   7   8   9   10   11

Figure 6.7.2. The sequence of paths used for the Figure 6.7.1 case, where nodes selected for ordering are shown in bold.

It may be verified (see Figure 6.7.3) that the relabelled digraph $3\rightarrow 1$, $5\rightarrow 2$, $4\rightarrow 3$, $2\rightarrow 4$, $1\rightarrow 5$, $7\rightarrow 6$, $6\rightarrow 7$ corresponds to a triangular matrix. Observe that at step 5 there was no path from node 5 because node 3 had already been eliminated. Note also that at step 9 the path tracing was restarted because the path became empty.

$$
\begin{bmatrix}
\times & \times & \times & & & & \\
 & \times & \times & \times & & & \\
 & & \times & & & & \\
 & & & \times & \times & & \\
 & & \times & & \times & & \\
 & & & \times & & \times & \times \\
 & & & & \times & & \times
\end{bmatrix}
\quad \text{becomes} \quad
\begin{bmatrix}
\times & & & & & & \\
\times & \times & & & & & \\
 & \times & \times & & & & \\
\times & & \times & \times & & & \\
\times & & & \times & \times & & \\
 & & & \times & & \times & \\
 & & & & \times & & \times & \times
\end{bmatrix}
$$

Figure 6.7.3. The matrices before and after renumbering.

Sargent and Westerberg (1964) generalized this idea to the block case. They define as a **composite node** any group of nodes through which a closed path (cycle) has been found. Starting from any node, a path is followed through the digraph until

(1) a closed path is found (identified by encountering the same node or composite node twice), or

(2) a node or composite node is encountered with no edges leaving it.

In case (1), all the nodes on the closed path must belong to the same strong component and the digraph is modified by collapsing all nodes on the closed path into a single composite node. Edges within a composite node are ignored and edges entering or leaving any node of the composite node are regarded as entering or leaving the composite node. The path is now continued from the composite node.

In case (2), as for ordinary nodes in the triangular case, the composite node is numbered next in the relabelling. It and all edges connected to it are eliminated and the path now ends at the previous node or composite node, or starts from any remaining node if it would otherwise be empty.

Thus the blocks of the required form are obtained successively. This generalization of the triangularization algorithm shares the property that each edge of the original digraph is inspected at most once.

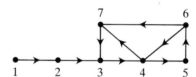

Figure 6.7.4. A digraph illustrating the algorithm of Sargent and Westerberg.

We illustrate with the example shown in Figure 6.7.4. Starting the path at node 1, it continues $1 \rightarrow 2 \rightarrow 3 \rightarrow 4 \rightarrow 5 \rightarrow 6 \rightarrow 4$, then $(4,5,6)$ is recognized as a closed path and nodes 4, 5, and 6 are relabelled as composite node $4'$. The path is continued from this composite node to become $1 \rightarrow 2 \rightarrow 3 \rightarrow 4' \rightarrow 7 \rightarrow 3$. Again a closed path has been found and $(3,4',7)$ is labelled as $3'$ and the path becomes $1 \rightarrow 2 \rightarrow 3'$. Since there are no edges leaving $3'$, it is numbered first and removed. The path is now $1 \rightarrow 2$ and no edges leave node 2, so this is numbered second. Finally node 1 is numbered as the last block. The corresponding original and reordered matrices are shown in Figure 6.7.5.

The difficulty with this approach is that there may be large overheads associated with the relabelling in the node collapsing step. A simple scheme such as labelling each composite node with the lowest label of its constituent nodes can result in $O(n^2)$ relabellings. For instance, in Figure

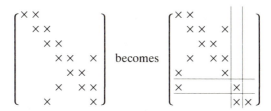

Figure 6.7.5. The matrices before and after renumbering.

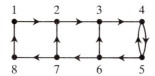

Figure 6.7.6. A case causing many relabellings.

6.7.6 successive composite nodes are (4,5), (3,4,5,6), (2,3,4,5,6,7), (1,2,3,4,5,6,7,8); in general, such a digraph with n nodes will involve $2+4+6+\ldots+n = n^2/4+n/2$ relabellings. Various authors (for example Munro 1971a, 1971b and Tarjan 1975) have proposed schemes for simplifying this relabelling, but the alternative approach of Tarjan (1972) eliminates the difficulty. This is described in the next section.

6.8 Tarjan's algorithm

The algorithm of Tarjan (1972) follows the same basic idea as that of Sargent and Westerberg, tracing paths and identifying strong components. It eliminates the relabelling through the clever use of a stack very like that in Figure 6.7.2, which was used to find the triangular form. The stack is built using a depth-first search and records both the current path and all the closed paths so far identified. Each strong component eventually appears as a group of nodes at the top of the stack and is then removed from it. We first illustrate this algorithm with two examples and then explain it in general.

Figure 6.8.1 shows the stack at all steps of the algorithm for the digraph of Figure 6.7.4 starting from node 1. In the first six steps, the stack simply records the growing path $1 \to 2 \to 3 \to 4 \to 5 \to 6$. At step 6 we find an edge connecting the node at the top of the stack (node 6) to one lower down (node 4). This is recorded by adding a link, shown as a subscript. Since we know that there is a path up the stack, $4 \to 5 \to 6$, this tells us that (4,5,6) lie on a closed path, but we do not relabel the nodes. Similarly at step 7, we record the link $7 \to 3$ and this indicates that (3,4,5,6,7) lie on a closed path. There are no more edges from node 7 so it is removed from

Stack	Step 1	2	3	4	5	6	7	8	9	10	11	12	13
									7_3	**7**	**7**	**7**	**7**
						6_4	6_4	6_3	**6**	**6**	**6**		
					5	5	5	5	5_3	**5**	**5**		
				4	4	4	4	4	4	4_3	**4**		
			3	3	3	3	3	3	3	3	3		
		2	2	2	2	2	2	2	2	2	2	2	
	1	1	1	1	1	1	1	1	1	1	1	1	1

Figure 6.8.1. The stack corresponding to Figure 6.7.4.

the path. However it is not removed from the stack because it is part of the closed path 3–7. We indicate this in Figure 6.8.1 by showing 7 in bold. Now node 6, which is immediately below node 7 on the stack, is at the path end so we return to it, but the only relabelling we do is to make its link point to 3 and we discard the link from 7, thereby recording the closed path 3–7.

We now look for unsearched edges at node 6 and find that there are none so we label 6 in bold, set the link from 5 to 3, and discard the link from 6 (see column 9 of Figure 6.8.1). At the next step we treat node 5 similarly and in the following loop we label 4 in bold and discard its link, but no change is made to the link at 3 which may be regarded as pointing to itself. Column 11 of Figure 6.8.1 still records that nodes 3, 4, 5, 6, and 7 lie on a closed path. Next we find that 3 has no unsearched edges. Since it does not have a link to a node below it, there cannot be any path from it or any of the nodes above it to a node below it. We have a 'dead-end', as in the triangular case. The nodes 3–7 constitute a strong component and may be removed from the stack. The trivial strong components (2) and (1) follow. The algorithm also works starting from any other node. Notice that the strong component was built up gradually by relabelling stack nodes in bold once their edges have been searched and each addition demands no relabelling of nodes already in the composite node.

This example was carefully chosen to be simple in the sense that the path always consists of adjacent nodes in the stack. This is not always so, and we illustrate the more general situation in the next example.

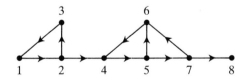

Figure 6.8.2. An example with two nontrivial strong components.

```
                                          8
                                7   7   7₆  7   7
                          6₄  6   6   6   6   6   6
                       5   5   5₄  5₄  5₄  5₄  5₄  5
Stack                4   4   4   4   4   4   4   4   4
                3₁  3   3   3   3   3   3   3   3   3   3   3   3
            2   2   2₁  2₁  2₁  2₁  2₁  2₁  2₁  2₁  2₁  2₁  2₁  2
        1   1   1   1   1   1   1   1   1   1   1   1   1   1   1

Step    1   2   3   4   5   6   7   8   9   10  11  12  13  14  15
```

Figure 6.8.3. The stack corresponding to Figure 6.8.2.

Consider the digraph shown in Figure 6.8.2. The stack, starting from node 1, is shown in Figure 6.8.3. At step 3 node 3 is removed from the path because it has no unsearched edges, but is not removed from the stack because of its link to node 1. Node 4 is added at step 5 because of the edge (2,4) and the path is $1 \rightarrow 2 \rightarrow 4$. Similarly node 7 is added because an edge connects to it from node 5. At step 10 node 8 is identified as a strong component because it has no edges leading from it and it does not have a link pointing below it in the stack. At step 11 node 7 has no more edges and is recognized as a member of a strong component because of its link to node 6 and is therefore labelled in bold. The strong components (4,5,6,7) and (1,2,3) are found in steps 13 and 15. A useful exercise is to carry through this computation from another starting node.

As with the Sargent and Westerberg algorithm, we start with any node and if ever the stack becomes empty we restart with any node not so far used. It is convenient to give such starting nodes links to themselves. At a typical step we look at the next unsearched edge of the node at the end of the path (we will call this node the current node) and distinguish these cases:

(1) The edge points to a node not on the stack, in which case this new node is added to the top of the stack and given a link that points to itself.

(2) The edge points to a node lower on the stack than the node linked from the current node, in which case the link is reset to point to this lower node.

(3) The edge points to a node higher on the stack than the node linked from the current node, in which case no action is needed.

(4) There are no unsearched edges from the current node and the link from the current node points below it. In this case the node is left on the stack but removed from the path. The link for the node before it

on the path is reset to the lesser of its old value and that of the current node.

(5) There are no unsearched edges from the current node and the link from the current node does not point below it. In this case the current node and all those above it on the stack constitute a strong component and so are removed from the stack.

It is interesting (see Exercise 6.9) to see that the excessive relabelling associated with the digraph of Figure 6.7.6 is avoided.

A formal proof of the validity of the Tarjan algorithm can readily be developed from the following observations:−

(i) at every step of the algorithm, there is a path from any node on the stack to any node above it on the stack;

(ii) for any contiguous group of nodes on the stack but not on the path (nodes marked in bold) there is a closed path between these nodes and the next node on the stack below this group;

(iii) if nodes α and β are a part of the same strong component, α cannot be removed from the stack by the algorithm unless β is removed at the same time.

Together, (i) and (ii) say that nodes removed from the stack at one step must be a part of the same strong component while (iii) says that all of the strong component must be removed at the same step.

6.9   Implementation of Tarjan's algorithm

To implement Tarjan's algorithm, it is convenient to store the matrix as a collection of sparse row vectors, since following paths in the digraph corresponds to accessing entries in the rows of the matrix.

It was convenient for illustration in the last section to label bold those nodes on the stack whose edges have all been searched, but for efficiency we need a rapid means of backtracking from the current node to the previous node on the path. This is better done by storing a pointer with each path node. Hence the following arrays of length n, the matrix order, are needed:

(1) one holding the nodes of the stack,

(2) one holding, for each path node, a link to the lowest stack node to which a path has so far been found,

(3) one holding, for each path node, a pointer to the previous path node,

(4) one holding for each node, its position on the stack if it is there, its position in the final order if it has been ordered, or zero if neither,

(5) one holding, for each path node, a record of how far the search of its edges has progressed.

In addition, the starts of the blocks must be recorded.

Note that each edge is referenced only once and that each of the steps (1) to (5) of Section 6.8 that is associated with an edge involves an amount of work that is bounded by a fixed number of operations. There will also be some $O(n)$ costs associated with initialization and accessing the rows, so the overall cost is $O(\tau) + O(n)$ for a matrix of order n with τ entries.

Further details of the algorithm and its implementation are provided by Tarjan (1972) and Duff and Reid (1978a). Duff and Reid (1978b) also provide an implementation in Fortran; it is interesting to note that their implementation involves less than 75 executable statements.

6.10 Essential uniqueness of the block triangular form

In the last three sections we have described two alternative algorithms for finding a block triangular form given a matrix with entries on its diagonal. There remains, however, some uncertainty as to whether the final result may depend significantly on the algorithm or the set of entries placed on the diagonal in the first stage. We show in this section that for nonsingular matrices, the result is essentially unique in the sense that any one block form can be obtained from any other by applying row permutations that involve the rows of a single block row, column permutations that involve the columns of a single block column, and symmetric permutations that reorder the blocks (see also Duff 1977b).

We begin by considering the variety of forms possible from symmetric permutations of a matrix. These correspond to no more than different relabellings of the same digraph. Since the strongly connected components of the digraph are uniquely defined, it follows that the diagonal blocks of the block triangular form are uniquely defined. The order in which these blocks appear on the block diagonal must be such that block α must precede block β for every pair of blocks α, β which are such that the component corresponding to β is connected to the component corresponding to α. It follows that often there is scope for permuting the blocks as wholes. Also, of course, we are free to use any permutation within each block. This describes the full freedom available and what is meant by 'essential uniqueness'.

It should be noted that this result is true for symmetric permutations whether or not the diagonal consists entirely of entries. However, it can only be extended to unsymmetric permutations if the diagonal does consist entirely of entries. Otherwise we may be able to find a block form with smaller diagonal blocks by first permuting entries onto the diagonal (see Exercise 6.10).

We now suppose that $\mathbf{B} = \mathbf{P}\mathbf{A}\mathbf{Q}$ is in the block triangular form (6.1.2) and show that any set of entries that may be permuted onto the diagonal must consist of entries from the diagonal blocks $\mathbf{B}_{ii}$, $i = 1, 2, ..., N$. Such a set of entries must include exactly one from each row of the matrix and exactly one from each column. Those from the n_1 rows corresponding to $\mathbf{B}_{11}$ must come from the first n_1 columns, that is from $\mathbf{B}_{11}$ itself. The remainder must all come from the last $n-n_1$ rows and columns. The submatrix of the last $n-n_1$ rows and columns has the same triangular form, so we can use the same argument to show that n_2 of the entries must come from $\mathbf{B}_{22}$. Continuing, we find that the entries all come from diagonal blocks.

It follows that column permutations within diagonal blocks can be used to place the new set of entries on the diagonal. Such permutations make no change to the block structure.

We may deduce that, if $\mathbf{A}$ has entries on its diagonal and $\mathbf{P}\mathbf{A}\mathbf{Q} = \mathbf{B}$ has the block form (6.1.2), then there is a symmetric permutation $\mathbf{P}_1\mathbf{A}\mathbf{P}_1^T$ having the same form, because we can find a column permutation $\mathbf{B}\mathbf{Q}_1$ of $\mathbf{B}$ that preserves its form while placing the diagonal entries of $\mathbf{A}$ on the diagonal of $\mathbf{B}\mathbf{Q}_1$. Therefore $\mathbf{B}\mathbf{Q}_1$ is a symmetric permutation of $\mathbf{A}$, say $\mathbf{B}\mathbf{Q}_1 = \mathbf{P}_1\mathbf{A}\mathbf{P}_1^T$. We have therefore established that once entries have been placed on the diagonal there is no loss of generality in considering symmetric permutations only. This establishes the essential uniqueness, as defined at the end of the first paragraph of this section.

6.11 Experience with block triangular forms

Having described an efficient algorithm for finding the reducible form of a matrix, we now discuss experience with it on practical problems. Since this capability has been incorporated as a user option in the Harwell general unsymmetric code MA28, we make reference to results using this code in describing timing and performance.

As we have already remarked, we would not usually expect gains for symmetrically structured matrices, since reducibility then corresponds to independent diagonal blocks. This leads to the conjecture that 'almost symmetric' matrix structures show less benefit from block triangularization attempts than 'very unsymmetric' ones.

Erisman, Grimes, Lewis, Poole, and Simon (1987) defined a measure of structural symmetry as the proportion of off-diagonal entries for which there is a corresponding entry in the transpose. They collected unsymmetric matrices from chemical engineering, linear programming, simulation, partial differential equation grids, and other applications to evaluate the effectiveness of block triangularization on real problems. Their results are summarized in Table 6.11.1. They also quantified the

Matrix origin	Number of matrices	Average symmetry measure	Reducibility		
			None	Some	Much
Chemical engineering	16	0.05	0	2	14
Linear programming	16	0.01	0	0	16
Simulations	11	0.46	7	4	0
PDE grids	7	0.99	5	0	2
Miscellaneous	6	0.54	3	2	1

Table 6.11.1. Reducibility of test matrices compared with the measure of symmetry of their structures.

Matrix origin	Number of matrices	Average order	Average number of diagonal blocks	Average size of nontrivial diagonal blocks
Chemical engineering	16	506.4	6.9	122.5
Linear programming	16	677.4	21.8	19.0
Simulations	11	441.6	1.4	441.3
PDE grids	7	959.0	1.7	694.8
Miscellaneous	6	97.5	2.2	69.4

Table 6.11.2. Benefits from block triangularization.

amount of reducibility and average figures for these results are summarized in Table 6.11.2. These results tend to support the conjecture that very unsymmetric matrices are the ones which are likely to benefit most.

Duff and Reid (1979a) looked at the timing impact of block triangularization. Their results are summarized for four test problems in Table 6.11.3. Cases one and three were irreducible, so this extra step represented simply an added cost. The structural gain on the second problem was slight, and approximately balanced the cost of finding the block form. The FACTORIZE time was essentially unchanged. The fourth case, however, showed significant benefit from the block triangularization; the ANALYSE time was approximately halved and the FACTORIZE time was reduced by about 20 per cent. Here there were 450 blocks and the largest was of order 217.

We conclude that, especially for very unsymmetric problems, the likely

Matrix order	No. of nonzeros	Block triangular- ization	ANALYSE without block option	ANALYSE with block option
147	2449	0.07	1.54	1.62
199	701	0.03	0.24	0.25
292	2208	0.10	0.83	0.92
822	4841	0.25	1.76	0.94

Table 6.11.3 Timings on an IBM 370/168 (seconds) of block triangularization and of ANALYSE, with and without block triangularization.

benefits from block triangularization can be significant and the modest penalty for this step is worthwhile. Although the overheads are slight, several applications yield matrices (usually near-symmetric ones) which are not amenable to block triangularization. In such cases, we would advise that the option is not invoked.

6.12 Maximum transversals

Throughout this chapter we have been concerned with developing efficient algorithms for block triangularization, so that subsequent Gaussian elimination operations can be confined to the blocks on the diagonal and the original problem can be solved as a sequence of subproblems. The algorithms of this chapter are, however, of interest in other contexts and, in particular, the transversal selection problem has a long and varied history.

Transversal selection corresponds to the assignment or matching problem, which has been of interest to management scientists for many years. They are concerned with allocating people to tasks, activities to resources, etc. This area has given rise to many algorithms, most being based on variants of breadth- or depth-first search. A selection of these algorithms is discussed by Duff (1981c). Other disciplines for which transversal selection is important include graph theory, games theory, and nonlinear systems. A different language is used in each area and a brief summary of the terms involved is given by Duff (1981c).

Often it is important to obtain a maximum transversal (or maximum assignment) which, in matrix terms, is equivalent to determining a permutation that places the maximum number of entries on the diagonal. We now show that the algorithm that we have described achieves this if we continue searching the columns after encountering one from which no assignment is possible. This leads to the situation shown in Figure 6.12.1. The bottom $(n-k) \times (n-k)$ block is zero and for $i = 1, 2, \ldots, k$, if there is an

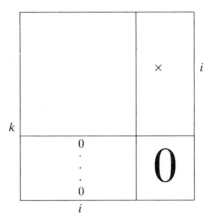

Figure 6.12.1. The situation at the end of the transversal selection algorithm.

entry in positions $k+1$ to n of row i then there can be no entries in positions $k+1$ to n of column i since otherwise our algorithm would have found an augmenting path. Therefore, if Gaussian elimination is applied to this matrix, no fill-in ever takes place in the bottom $(n-k) \times (n-k)$ zero submatrix. The resulting upper triangular matrix therefore cannot have rank greater than k and, since the elementary operations of Gaussian elimination do not change the rank, it follows that the rank of $\mathbf{A}$ is at most k. Therefore a transversal with more than k entries can never be found.

Because the rank of $\mathbf{A}$ will be k if the transversal entries have the value one and all other entries have the value zero, we have thus shown that k is equal to the maximum rank which the matrix can have over all choices of values for the entries (the symbolic rank).

Exercises

6.1 Given a block lower triangular matrix $\mathbf{A}$, find permutations $\mathbf{P}$, $\mathbf{Q}$ such that $\mathbf{P A Q}$ is block upper triangular.

6.2 Show that the matrix

$$\begin{bmatrix} \times & & & & \times & \times \\ & \times & \times & & & \\ & \times & \times & & & \\ \times & & & \times & & \times \\ & & & & \times & \\ & & & \times & & \times & \times \\ & & \times & & \times & \end{bmatrix}$$

is symbolically singular by attempting to extend the transversal and identifying columns which must be linearly dependent.

6.3 Show that the transversal algorithm, as described in Section 6.3, requires $O(n\tau)$ = $O(n^3)$ operations when applied without prior column ordering to the matrix

$$\begin{pmatrix} F & 0 & I \\ I & I & I \\ 0 & I & 0 \end{pmatrix},$$

where each block is of size $\frac{1}{3}n \times \frac{1}{3}n$ and F is full. What is the effect of prior column ordering by numbers of entries?

6.4 Construct a matrix whose columns are in order of increasing number of entries and for which the transversal algorithm needs $O(n^3)$ operations.

6.5 Consider the transversal extension algorithm that uses only cheap assignments and switches a later column with column k whenever a cheap assignment is not possible in column k. Show that this obtains at least $\frac{1}{2}n$ cheap assignments for a nonsingular matrix of order n.

6.6 By considering the example

$$\begin{pmatrix} \times & \times & \times \\ \times & & \times \\ \times & & \end{pmatrix},$$

show that when the transversal algorithm of Sections 6.2 to 6.5 is applied to a nonsingular matrix of order n, it may obtain less than $\frac{1}{2}n$ assignments cheaply.

6.7 Apply the Tarjan algorithm to the digraph in Figure 6.7.4 starting from node 7.

6.8 Apply the Tarjan algorithm to the digraph in Figure 6.8.2 starting from node 7.

6.9 Apply the Tarjan algorithm to the digraph of Figure 6.7.6 starting at node 1 and comment on the work.

6.10 Illustrate that, if the diagonal of A does not consist entirely of entries, a finer block triangular structure may be found by using unsymmetric permutations than is possible by using symmetric permutations.

7 Local pivotal strategies for sparse matrices

A look at research results and open questions on various methods of reordering equations from the viewpoint of 'locally' optimizing certain sparsity objectives, including dealing with numerical concerns. Discussion includes obvious algorithm extensions which do not work.

7.1 Introduction

In this and in following chapters we address the problem of finding the best (or more realistically a good) ordering for sparse equations. It is difficult to define what we mean by best and, given a definition, it is difficult to find such an ordering.

To put this in perspective, suppose we use the objective of finding the ordering which introduces the least new entries in the LU factorization. This will be referred to as the minimum fill-in objective. Even if we start with entries on the diagonal and restrict the pivot choice to the diagonal, this problem has been shown by Rose and Tarjan (1978) in the unsymmetric case and by Yannakakis (1981) in the symmetric case to be 'NP complete'. This means (see Karp 1986) that the problem is very hard (like the travelling salesman problem), and only heuristic algorithms are computationally feasible for large problems.

Going back to defining a 'best' ordering, we see that a solution to the minimum fill-in problem may not be best for a number of reasons. It may be very costly to compute, so that the total cost of allowing more fill-in but computing the ordering more economically may be less. It may be numerically unstable. The ordering may require a very sophisticated data structure and solution program, the cost of which could override any computational savings. It may not permit as much exploitation of vector and parallel architectures (see Section 1.4) as some ordering allowing more fill-in. The user may need to solve just one sparse system of equations or, as is often the case, may need to solve many systems with the same sparsity structure. These are just a few of the factors which demonstrate that 'best' cannot be defined absolutely.

For these reasons we will not seek a single best ordering. We will discuss a number of proposed orderings, using both analysis and computational experiments to retain some and reject others. For those which we recommend, we identify the demands made on the data and the conditions under which they may be useful. Implementation details are deferred to Chapters 9 and 10 where alternative data structures and solution algorithms are discussed.

All the ordering strategies have as at least a partial objective that of controlling fill-in. The strategies fall into two basic categories: those which at each pivotal step minimize some objective for that step without reference to effects on later steps (**local** strategies) and those that confine the fill-in within some **desirable form** (for example within a band or within some small number of columns). We examine the local strategies in this chapter and other strategies in the next chapter. We make comparative statements between these classes when we can.

7.2 The Markowitz criterion

The ordering strategy introduced at the beginning of Section 5.3 is due to Markowitz (1957) and has proved extremely successful for general-purpose use. We consider it in more detail here so that it may be compared with other local strategies.

Suppose Gaussian elimination has proceeded through the first k stages. For each row i in the active $(n-k)\times(n-k)$ submatrix let $r_i^{(k)}$ denote the number of entries. Similarly let $c_j^{(k)}$ be the number of entries in column j. Then the Markowitz criterion is to select the entry $a_{ij}^{(k)}$ from the $(n-k)\times(n-k)$ submatrix that is not too small numerically (see Sections 5.4 and 7.8) and that minimizes the expression

$$(r_i^{(k)} - 1)(c_j^{(k)} - 1). \tag{7.2.1}$$

Observe that it is better to use $(r_i^{(k)}-1)(c_j^{(k)}-1)$ than $r_i^{(k)}c_j^{(k)}$ because this forces the algorithm to select a row singleton $(r_i^{(k)}=1)$ or a column singleton $(c_j^{(k)}=1)$ if either is present. Such a choice produces no fill-in at all.

Markowitz interpreted this strategy as finding the pivot which, given that the first k pivots have been chosen, modifies the least coefficients in the remaining submatrix. It may also be regarded as an approximation to the local minimum multiplication count, since using $a_{ij}^{(k)}$ as pivot requires $r_i^{(k)}(c_j^{(k)}-1)$ multiplications. Finally, we may think of this as an approximation to the choice of pivot which introduces the least fill-in at this stage, since that would be the case if all $(r_i^{(k)}-1)(c_j^{(k)}-1)$ modifications yielded fill-in.

To implement the Markowitz strategy generally requires knowledge of the updated sparsity pattern of the reduced $(n-k)\times(n-k)$ submatrix at each stage of elimination. It requires access to the rows **and** columns, since the positions of nonzero entries in the pivotal column are needed to determine which rows change. It requires the numerical values to judge the acceptability of the pivot size. If the search is not cleverly limited (see Section 9.2, for example), it may require examining all the entries in the active submatrix at every stage. It should be observed that the Markowitz count (7.2.1) may be greater than

$$\min_t(r_t^{(k)}-1)\ \min_t(c_t^{(k)}-1),\tag{7.2.2}$$

for there may be no entries, or no acceptable entries, in the intersection between rows with minimum $r_i^{(k)}$ and columns with minimum $c_j^{(k)}$ (see Exercise 7.1).

7.3 Minimum degree (Tinney scheme 2)

Before examining some related local algorithms and comparing them with the Markowitz strategy, we look at the case where the pattern of **A** is symmetric and we can be sure that diagonal pivots produce a stable factorization (the most important example is when **A** is symmetric and positive definite). Two things simplify. We do not have to carry numerical values to check for stability, and the search for the pivot is simplified to finding i such that

$$r_i^{(k)} = \min_t r_t^{(k)}\tag{7.3.1}$$

and using $a_{ii}^{(k)}$ as pivot. This special case was not considered by Markowitz (1957). It was introduced by Tinney and Walker (1967) as their **scheme 2**. It is called the **minimum degree** algorithm because of its graph theoretic interpretation: in the graph associated with a symmetric sparse matrix, this strategy corresponds to choosing that node for the next elimination which has the least edges connected to it.

Notice that the diagonal entry of minimum degree will always have Markowitz count equal to the minimum for any diagonal or off-diagonal entry. This property is maintained by choosing pivots from the diagonal, since symmetry is preserved.

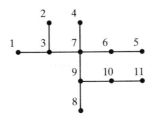

Figure 7.3.1. A tree graph.

The minimum degree strategy has been very successful in practice. It is easy to verify that for a tree graph (a graph with no loops, see the Figure 7.3.1 example), it introduces no fill-in (Exercise 7.2). In this case, it is certainly best in the sense of least overall fill-in, but it does not always produce an ordering which minimizes the amount of fill-in introduced. We demonstrate this by the example shown in Figure 7.3.2. The minimum

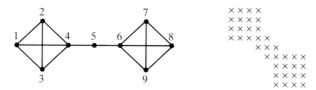

Figure 7.3.2. Minimum degree not optimal.

degree algorithm selects node 5 first, introducing fill-in between nodes 4 and 6. The given order produces no fill-in. Since minimum degree is a Markowitz ordering applied to a symmetric problem, we conclude that the Markowitz ordering need not be optimal in the sense of least fill-in.

A rather surprising result about this strategy is that it may be implemented without the explicit updating of the sparsity pattern at each stage. Because we are not concerned about the numerical sizes of the pivots, we can safely do this. The result is that the symmetric version of the Markowitz strategy runs significantly faster than the unsymmetric version. For this reason the minimum degree algorithm is often applied to unsymmetric systems whose pattern is symmetric or nearly so (see Section 10.8).

7.4 Simpler strategies

We have now identified two basic difficulties with the Markowitz ordering: it does not always produce the best ordering and may be costly to implement. We will show in Chapters 9 and 10 that both the pivot search and the updating may be performed efficiently by a careful implementation. Nevertheless, we are led to examine two alternatives. In this section, we consider simpler orderings in the hope of reducing the work of finding the ordering without greatly increasing the fill-in. In Section 7.5, we try to improve the performance by using more sophisticated strategies.

To eliminate the costly update of the sparsity pattern (or its implicit equivalent) at each stage, we could estimate the Markowitz count (7.2.1) by ignoring changes in the sparsity pattern. One such method would be to order the columns of the original matrix by increasing numbers of entries and keep to this order during the elimination. The k-th pivot itself could then be chosen from the reduced part of the k-th column to minimize its original row count. This is not really a local strategy at all, since the column order is chosen a priori. We include it in this discussion because of its relationship to other local strategies. The symmetric version of this algorithm is scheme 1 of Sato and Tinney (1963): order all of the nodes in the graph in increasing degree order a priori.

While these schemes eliminate the need to update the sparsity counts and search for a pivot at each stage, they often significantly increase the

Order	100	100	54	57	199
Number of nonzeros	394	297	291	281	701
Fill-in					
Markowitz	584	199	90	34	686
A priori pivot choice	1035	363	184	127	2066
Minimum row count within a priori column choice	886	332	154	120	1337
Multiplications and divisions					
Markowitz	3526	830	737	505	3189
A priori pivot choice	8105	1532	1243	871	16069
Minimum row count within a priori column choice	5869	1390	1044	821	7640

Table 7.4.1. Experimental comparison of some orderings on unsymmetric sparse matrices.

cost of solving the problem due to increased fill-in. This is shown in Table 7.4.1, which contains a subset of the results of experiments performed by Duff and Reid (1974). We label the algorithm of the last paragraph as 'a priori pivot choice', and it can be seen that both the fill-in and the arithmetic are markedly greater than with the Markowitz ordering. Since the overall cost involves both the cost of ordering and the cost of factorizing the ordered matrix, in general we conclude that this is too high a price to pay for a faster ordering. Curtis and Reid (1971a) discuss an application where most of the matrix must be in secondary storage, making this ordering more attractive. Sherman (1978) has developed a code, NSPIV, which uses an a priori row ordering and then chooses the pivot from each row in turn on numerical stability grounds. In a more recent code, NSPFAC, he permits threshold pivoting (see equation 4.4.7) but adheres to the a priori column ordering if the pivot candidate satisfies the threshold test. This code has been incorporated in the PORT library by Kaufman (1982). To overcome the difficulties we have observed in choosing an a priori ordering, Sherman has suggested using a Markowitz approach to obtain it. The benefits of his codes are that they are very short and, if a good ordering is supplied, can be efficient.

Figure 7.4.1. Difficulty with a priori ordering.

One difficulty with a priori orderings is illustrated by the simple graph in Figure 7.4.1. Without some special tie-breaking strategy we have no reason to renumber the nodes from the order shown. This leads to a fill-in between nodes 4 and 5. In spite of this, Sato and Tinney (1963) cite 4:1

savings in using this a priori ordering compared with not reordering for sparsity at all.

If we can update the row counts as the computation progresses, then we could perhaps improve the ordering by choosing the minimum row count at each stage while keeping the a priori column ordering. This does indeed provide an improved ordering as shown in Table 7.4.1. We now have a reduced pivot search, but must update row counts. There is still a significant penalty in fill-in compared with the Markowitz ordering.

Our final alternative involves keeping an updated sparsity pattern but limiting the pivot search: select the column with *updated* minimum column count and choose the entry in that column in the row with minimum updated row count among entries in that column. This is referred to as the **min. row in min. col.** strategy (and there is an analogous **min. col. in min. row** strategy). For the symmetric case when only diagonal pivoting is permitted, it is equivalent to the minimum degree algorithm. In practice, it performs better than the a priori or partial a priori orderings discussed previously. But now the savings compared with the Markowitz ordering are only in the pivot search. Tosovic (1973) and Duff (1979) give examples which show the performance to be somewhat poorer (typically the number of entries in the factorized form is greater in the range 0 to 20 per cent than with the strategy of Markowitz). The danger is that, although the pivot is in a column with few entries, it may be in a row with many entries. An example of this behaviour is shown in Table 7.4.2; although the Markowitz and min. row in min. col. algorithms perform similarly on the matrix and Markowitz performs similarly on its transpose, min. row in min. col. performs significantly worse on the transpose.

Matrix	$\mathbf{A}$	$\mathbf{A}^T$
Fill-in		
Markowitz	729	747
Min. row in min. col.	809	4889
Operations in factorization		
Markowitz	6440	7865
Min. row in min. col.	7811	51375

Table 7.4.2. Markowitz and min. row in min. col. ordering on a matrix of order 363 with 3157 nonzeros and on its transpose.

A further suggestion for simplifying the Markowitz search was made by Zlatev (1980) and is now incorporated in the MA28 code. This strategy restricts the search to a predetermined number of rows of lowest row count, choosing entries with best Markowitz count and breaking ties on numerical grounds. This can suffer from some of the difficulty of the min. col. in min. row strategy, but has proved very useful on regular matrices from grid-based problems. The performance is quite dependent on the number of rows to which the search is restricted and that in turn depends on the problem. Zlatev (1980) recommends searching two or three rows. We will see in Section 9.2 that an unrestricted Markowitz strategy often does not need any more searching.

The conclusions of this section are that attempts to simplify the Markowitz ordering strategy generally do not succeed. The observation of Sato and Tinney (1963) that the a priori ordering produced a significant improvement over not ordering at all should be taken as an argument in favour of ordering for sparsity, not in favour of a priori orderings. In the case where the matrix is mostly out-of-core and only one solution for that sparsity structure is wanted, then the simple strategies may be considered, but an increased number of arithmetic operations while factorizing the matrix must be expected. There are, however, some occasions when the gains from working with a simpler data structure can outweigh the extra arithmetic. An example is given by the variable-band techniques in Section 8.3. The balance between increased arithmetic and decreased data handling will be influenced by the machine. For example, IBM mainframes are usually more efficient in data handling (relative to arithmetic) than vector machines like the CRAY-1.

7.5 Local minimum fill-in

Many efforts have been made to improve the Markowitz ordering by trying to reduce the overall fill-in further. One such strategy is to replace the Markowitz criterion with a local minimum fill-in criterion. That is, at the k-th stage of Gaussian elimination select as pivot the nonzero entry (which is not too small numerically) that introduces the least amount of fill-in at this stage. Markowitz (1957) actually suggested this algorithm but rejected using it on the grounds of cost. For the symmetric case, this is scheme 3 of Tinney and Walker (1967). It is also sometimes called the minimum deficiency algorithm.

We readily observe that this minimum fill-in criterion is considerably more expensive than that of Markowitz. We not only require the updated sparsity pattern but must identify the pivot candidate which introduces the least fill-in. It is not necessary, in a careful implementation of the minimum fill-in algorithm, to recompute the fill-in produced by all pivot candidates at each stage since only a small part of a sparse matrix changes with each elimination step. Even with careful implementation, however, we need a significant improvement in performance to make this extra effort worthwhile.

We have already remarked that no fill-in results from using the minimum

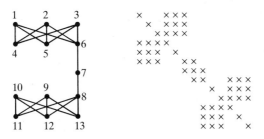

Figure 7.5.1. Minimum fill-in not optimal.

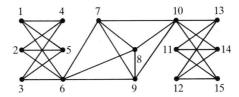

Figure 7.5.2. Markowitz optimal, minimum fill-in not optimal.

fill-in criterion on the case shown in Figure 7.3.2, whereas one fill-in was generated with a Markowitz ordering. However, the graph in Figure 7.5.1 (where the given ordering is globally optimal) illustrates that local minimum fill-in does not always produce a globally optimal result. Selecting node 7 first produces one immediate fill-in while any other selection leads to at least two. However, this fill-in is in addition to those already resulting from the given order. Compounding the problem further is the example in Figure 7.5.2, where the Markowitz algorithm leads to an optimum order (the order given), while the minimum fill-in algorithm introduces extra fill-in. These examples, and others, are discussed by Ogbuobiri, Tinney, and Walker (1970), Duff (1972), Rose (1972), Dembart and Erisman (1973), and Duff and Reid (1974). They serve to demonstrate that a locally best decision is not necessarily best globally and to illustrate the general difficulty with local algorithms.

A more extensive comparison between the Markowitz ordering and the minimum fill-in ordering is shown in Table 7.5.1 (Duff and Reid 1974). Notice that minimum fill-in is often marginally better than the Markowitz ordering, but is sometimes worse. Also observe, in the second case, that we have less fill-in (overall) but more arithmetic (overall). In general, we would expect only small differences of this kind. This lack of significant improvement has been supported by further testing by the authors and others.

Order	100	100	54	57	199	64
Number of nonzeros	394	297	291	281	701	352
Fill-in						
Markowitz	584	199	90	34	686	652
Minimum local fill-in	551	196	70	10	671	692
Multiplications and divisions in factorization						
Markowitz	3526	830	737	505	3189	5117
Minimum local fill-in	3198	846	673	439	3078	5589

Table 7.5.1. Comparison between Markowitz and minimum fill-in criteria (unsymmetric matrices).

7.6 Other local strategies

Numerous other extensions and modifications of the Markowitz and minimum fill-in strategies have been proposed in the literature and are surveyed by Duff (1977a). They include tie-breaking strategies for both algorithms. While many of them can work well on a particular problem, none offers an overall worthwhile improvement over the Markowitz ordering. Hsieh and Ghausi (1972) propose a strategy for choosing the pivot based on a probabilistic prediction of fill-in. Although Duff (1974) has shown that their calculation of probability is flawed, the idea is rather intriguing and the pivot selection process is almost as simple as the Markowitz strategy. Dembart and Erisman (1973) showed that in practice it was usually simply a tie-breaker for Markowitz and gave little practical improvement because it is still a local algorithm.

As a final example of the futility of refining these local strategies, Dembart and Erisman cited an example of order 50 with 375 nonzeros. The Markowitz strategy introduced 460 fill-ins, the probabilistic method introduced 469 fill-ins. Choosing at each stage the probabilistic or Markowitz candidate, depending on which introduced the least fill-ins at that stage, led to 459 fill-ins. But choosing the probabilistic or Markowitz candidate, depending on which introduced the most fill-ins at that stage, led to 455 fill-ins.

Our conclusion is that the strategies of Markowitz and minimum degree are the best local algorithms. This conclusion is based on experimental results. Attempts to simplify the heuristics produce significantly worse results, while more complicated heuristics greatly add to the algorithm time and have little impact on performance. Unfortunately, attempts to establish theoretically the effectiveness of these algorithms have not been successful to date, and one reason for this appears to be the effect of tie-breaking. This is discussed in the next section.

7.7 The effect of tie-breaking on the minimum degree algorithm

Because of the excellent results obtained with the minimum degree algorithm in practice, many people have attempted to analyse its performance formally, sometimes on limited problem classes. For example, it is optimal in the minimum fill-in sense for tree graphs (see Section 7.3), but this is a very limited problem set.

Matrices from 5-point and 9-point operators on a square grid (see Section 1.6) provide a natural class to analyse. Their regular pattern and frequent use in applications motivate such an investigation. In Section 8.7 we introduce the nested dissection algorithm which has been shown to be optimal (in the sense that it involves $O(n^{3/2})$ operations and no ordering involves $O(n^{\alpha})$ operations, with $\alpha < \frac{3}{2}$) on these matrices, and we compare the minimum degree and nested dissection algorithms there.

The difficulty in analysing the minimum degree algorithm for these matrices is the lack of a well-defined tie-breaking strategy. We provide some experimental data in this section to demonstrate the effect of tie-breaking. This raises questions as to the effect of tie-breaking in general.

To see why tie-breaking is so important for these matrices, observe that for a 9-point operator on a two-dimensional grid with N subdivisions in each direction (the matrix order is $(N+1)^2$), four nodes have degree 3, $4(N-1)$ nodes have degree 5, and all the rest have degree 8. Similarly for a 5-point operator on the same grid, the matrix order is $(N+1)^2$, four nodes have degree 2, $4(N-1)$ nodes have degree 3, and the rest have degree 4. We would expect that after a few steps of the algorithm, the arbitrary choices for tie-breaking could produce a wide range of results.

This range is evident in Figures D.13 to D.15 of Appendix D, which show the filled-in matrix patterns from three different efficient implementations of the minimum degree algorithm applied to the matrix corresponding to a 5-point operator on a square region with $N = 19$ and $n = 400$. There is no substantial difference in the number of fill-ins between the different orderings, but the resulting patterns indicate that the tie-breaking strategies differ.

Duff, Erisman, and Reid (1976) did some other experiments to illustrate the effect of tie-breaking in the minimum degree algorithm. They used a variety of initial orders on an implementation of the algorithm which broke ties by selecting the candidate that was first in the initial order. Two initial orders which displayed a large difference in results are illustrated in Figure 7.7.4. The spiral ordering was found by experimentation while the pagewise ordering is a natural one. The results using these two initial orderings are displayed in Table 7.7.1. The normalization factor is proportional to the asymptotic behaviour of nested dissection and is included to aid the comparison for varying sizes of grid.

```
1 16 15 14 13          1  2  3  4  5
2 17 24 23 12          6  7  8  9 11
3 18 25 22 11         11 12 13 14 15
4 19 20 21 10         16 17 18 19 20
5  6  7  8  9         21 22 23 24 25
```

Figure 7.7.4. Spiral and pagewise ordering on a 5×5 grid.

N	η, number of off-diagonal entries in $\mathbf{L}$		$\dfrac{\eta}{N^2 \log_2 N}$	
	Pagewise	Spiral	Pagewise	Spiral
4	111	111	3.5	3.5
6	321	359	3.4	3.9
8	621	662	3.2	3.4
11	1 421	1 523	3.4	3.6
16	3 954	4 334	3.9	4.2
21	7 381	8 613	3.8	4.4
26	12 794	16 190	4.0	5.1
32	23 142	27 903	4.5	5.4
64	129 046	184 565	5.3	7.5

Table 7.7.1. Comparison between different tie-breaking strategies for the minimum degree algorithm.

Eisenstat (1982, private communication) did some experiments which show the critical nature of tie-breaking from a different angle. For 9-point operators on grids ranging from $N = 20$ to $N = 199$ (matrix order 441 to 40 000) using the YSMP program produces the normalized scatter plots shown in Figures 7.7.5 and 7.7.6. The expected monotonic behaviour (the larger the grid the greater the computational and storage requirements) is not present because of the effect of tie-breaking.

The points at (32, 5.4) and (64, 7.5) in Figure 7.7.5 illustrate that the scatter is not as great as it might be. These points correspond to the extremes of Table 7.7.1.

We tentatively hypothesize that analysis of the minimum degree algorithm on any broad class of problems will depend on defining a tie-breaking strategy. The results in Table 7.7.1 show that local tie-breaking strategies are ineffective. Thus it would appear that general analysis of the minimum degree algorithm, except in conjunction with one of the methods of Chapter 8, will at best provide a wide performance band.

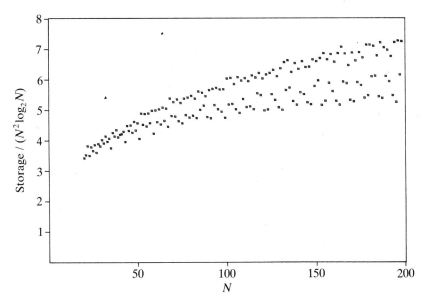

Figure 7.7.5. Normalized storage requirements for 9-point grids.

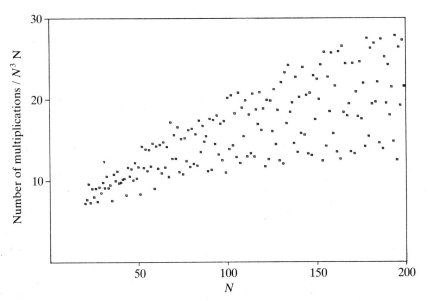

Figure 7.7.6. Normalized computational requirements for 9-point grids.

7.8 Numerical pivoting

With the recommendation for the Markowitz ordering strategy comes the need in the unsymmetric case to establish a suitable threshold parameter u for numerical stability. In particular, we restrict the Markowitz selection to those pivot candidates satisfying the inequality

$$|a_{kk}^{(k)}| \geq u |a_{ik}^{(k)}|, \quad i > k, \tag{7.8.1}$$

where u is a preset parameter in the range $0 < u \leq 1$, as discussed in Section 5.4.

Recommendations for suitable values for u have varied quite widely. Tomlin (1972) recommended the rather low figure of 0.01 for linear programming cases, where the number of entries in column j of U is usually very small (for example 3, 4, or 5). Curtis and Reid (1971a) recommended the value $\frac{1}{4}$ on the basis of experiments that now appear to be of very modest order.

Intuitively it might seem that the smaller the value of u the greater freedom there would be to choose pivots that are satisfactory from a sparsity point of view, so that the limit u is needed only for stability reasons. Duff (1979) found experimentally that this is not the case. We reproduce some of his experimental results in Table 7.8.1.

Order	147		199		822		541	
Nonzeros	2441		701		4790		4285	
u	m	Error	m	Error	m	Error	m	Error
1×10^{-10}	4881	1×10^2	1350	4×10^{-9}	6474	1×10^{-8}	16553	5×10^{15}
1×10^{-4}	5028	3×10^{-9}	1382	4×10^{-9}	6474	1×10^{-8}	16198	5×10^{-2}
1×10^{-2}	5867	4×10^{-10}	1429	2×10^{-11}	6495	2×10^{-10}	15045	4×10^{-6}
1×10^{-1}	5095	2×10^{-12}	1478	1×10^{-12}	6653	1×10^{-11}	13660	3×10^{-9}
0.25	6449	3×10^{-12}	1598	8×10^{-13}	6910	4×10^{-12}	14249	8×10^{-11}
0.5	6381	2×10^{-12}	1728	5×10^{-13}	7231	1×10^{-12}	14109	8×10^{-11}
1.0	6772	2×10^{-12}	1915	3×10^{-13}	8716	6×10^{-12}	16767	2×10^{-10}

Table 7.8.1. Varying the threshold parameter u. m is the number of entries in the factors. Error is the l_2 error norm, when running in double precision on an IBM 370/168.

The case for $n = 541$ is particularly interesting, since fill-in increases significantly for values of u smaller than 0.1. The explanation is that with smaller values of u the spread of numerical values within the reduced submatrix becomes very wide. This wide spread causes inequality (7.8.1) to be very restrictive at later stages.

The experimental recommendation of $u = 0.1$ comes from the balance that this choice seems to allow between sparsity and numerical error. Clearly the best choice is problem dependent, and other experiments may determine a better choice for a particular problem class.

7.9 Sparsity in the right-hand side and partial solution

Often in solving $Ax = b$, the vector b is also sparse. It is readily seen that if $b_1 = b_2 = \dots = b_k = 0$, $b_{k+1} \neq 0$, then in the forward substitution step

$$Lc = b \tag{7.9.1}$$

$c_1 = c_2 = \dots = c_k = 0$. This suggests that we may be able to save computation by taking advantage of sparsity in b. If the matrix L is being saved for use with other right-hand sides and they also have k leading zeros, then we may also save storage since the first k columns of L (shaded in Figure 7.9.1) are not needed in the forward substitution. However, if equation $k+1$ is placed first in the ordering, we may need to store all of the resulting L, and perform all of the forward substitution. This suggests that sparsity in b should influence the ordering of A. Note that with only one right-hand side b, the forward substitution can be done along with the factorization and none of L needs to be saved.

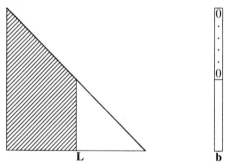

Figure 7.9.1. Storage saved in L from sparsity in b.

Next, we show a similar property for x. The combined effect of forward substitution and back-substitution usually leads to x being a dense vector. This happens if c_n is nonzero and U has an entry in each off-diagonal row except the last, see Section 12.6. This will be the case structurally for irreducible matrices A (see Exercise 7.4). However, we may be interested in only a subset of x. In this case, if the components of interest in x are numbered last, we have to carry the back-substitution computation only far enough to compute all x_i of interest. This also saves computation and storage in U as shown in Figure 7.9.2, where the shaded portion of U is not required. Again the ordering of x to achieve this saving influences the ordering of A.

Looking at Figures 7.9.1 and 7.9.2, we could make the required triangles (unshaded regions in L and U) as small as possible by constraining the sparse ordering on A as follows. All equations (rows of A) corresponding to $b_i = 0$ must be numbered before equations corresponding to $b_i \neq 0$. And

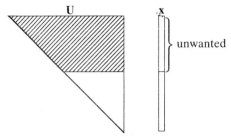

Figure 7.9.2. Savings in **U** from computing part of **x**.

all variables (columns of **A**) corresponding to x_i of interest must be numbered after the columns in **A** corresponding to unwanted variables. Erisman (1973) reported that he obtained a five-fold reduction of storage on practical problems by reusing those locations that would otherwise hold entries in the shaded part of **L\U**.

To preserve symmetry for the symmetrically-structured case requires partitioning **x** and **b** as shown in Figure 7.9.3, where the shaded portion corresponds both to unwanted variables and to zero right-hand side components (other unwanted variables and zero right-hand side components are not exploited).

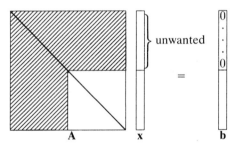

Figure 7.9.3. The structurally symmetric case.

This strategy has two shortcomings. First, the objective of making the order of the unshaded triangles as small as possible is not usually best. Recall that the amount of computation associated with forward substitution and back-substitution is equal to the number of entries in **L** and **U** respectively. Constraining the pivotal sequence, and possibly increasing the factorization costs and the overall density of the factors, may more than offset the resulting saving in forward substitution and back-substitution. Second, the severe constraint on the pivot sequence may lead to an unstable factorization. The second difficulty may be overcome by continuing the restricted pivot choice only as long as sufficiently large pivots are available. The first difficulty must be addressed further.

Tinney, Powell, and Peterson (1973) suggest monitoring the number of entries in the active part of the matrix. As the pivot choice becomes more restricted in the course of Gaussian elimination, the number of entries in the remaining submatrix may increase sharply. They recommend a two-pass algorithm, the first of which monitors the number of entries in the submatrix. If this increases in the neighbourhood of the restriction, they return to where the increase began and release the restriction. This effectively increases the order of the stored parts of the factors in exchange for improving their sparsity.

This concept is illustrated by the example in Figure 7.9.4. If we keep to the constraints indicated, one step of Gaussian elimination with the (1,1) pivot would be performed, leading to complete fill-in. Releasing the constraints means that d is computed and the leading zero on the right-hand side is ignored, but there is no fill-in and overall computation is much reduced.

$$\begin{bmatrix} \times\times\times\times\times\times\times\times\times \\ \times\times \\ \times\ \ \times \\ \times\ \ \ \ \times \\ \times\ \ \ \ \ \ \times \\ \times\ \ \ \ \ \ \ \ \times \\ \times\ \ \ \ \ \ \ \ \ \ \times \\ \times\ \ \ \ \ \ \ \ \ \ \ \ \times \\ \times\ \ \ \ \ \ \ \ \ \ \ \ \ \ \times \end{bmatrix} \begin{bmatrix} d \\ \times \\ \times \\ \times \\ \times \\ \times \\ \times \\ \times \\ \times \end{bmatrix} = \begin{bmatrix} 0 \\ \times \\ \times \\ \times \\ \times \\ \times \\ \times \\ \times \\ \times \end{bmatrix}$$

Figure 7.9.4. Constrained pivot sequence illustration. We do not want the value of d.

Another approach to this problem is to increase the counts in the rows corresponding to nonzeros in $\mathbf{b}$ and in the columns corresponding to wanted components of $\mathbf{x}$. This is one interpretation of a suggestion of Hachtel (1976).

A dramatic example of the saving from taking advantage of both sparsity in the right-hand side and the requirement for only a partial solution is provided by Duff and Reid (1976), who examine the use of the augmented form

$$\begin{pmatrix} \mathbf{I} & \mathbf{A} \\ \mathbf{A}^T & \mathbf{0} \end{pmatrix} \begin{pmatrix} \mathbf{r} \\ \mathbf{x} \end{pmatrix} = \begin{pmatrix} \mathbf{b} \\ \mathbf{0} \end{pmatrix} \tag{7.9.2}$$

in the solution of the problem

$$\min_{\mathbf{x}} \|\mathbf{b} - \mathbf{A}\mathbf{x}\|_2, \tag{7.9.3}$$

where the matrix $\mathbf{A}$ is of dimension $m \times n$, $m \geqslant n$. The matrix of equation (7.9.2) is symmetric and we would normally expect that a solution scheme that took advantage of this fact would have much lower storage requirements for the factors than an unsymmetric code. By treating the matrix as a general matrix, however, we can bias the pivot selection so that early pivots are chosen from rows $m+1$ to $m+n$ (thereby exploiting sparsity in the right-hand side) and late pivots are chosen from columns $m+1$ to $m+n$ (thereby exploiting the fact that equation (7.9.2) need only be solved for $\mathbf{x}$). If $\mathbf{r}$ is required it can be obtained from the computation

$$\mathbf{r} = \mathbf{b} - \mathbf{Ax}. \qquad (7.9.4)$$

By making this judicious choice of pivots and discarding the portions of the factors which are not required, Duff and Reid (1976) obtain the results shown in Table 7.9.1 which indicate the benefits (although the factorization costs were broadly comparable).

Number of rows	200	219	331
Number of columns	199	85	104
Number of nonzeros	702	438	662
Using symmetry	3748	2046	3103
Ignoring symmetry	2581	1441	2109

Table 7.9.1. Number of multiplications in SOLVE phase of some least-squares problems.

7.10 Variability-type ordering

Suppose a problem requires the factorization of a sequence of matrices $\mathbf{A}^l$, all with the same sparsity structure but with some of the entries changing in numerical value. In principle, any operation involving quantities unchanged since the last factorization need not be repeated. This led Hachtel, Brayton, and Gustavson (1971) to introduce the concept of **variability type** (see also Hachtel 1972) to label the way each entry changes in the sequence $\mathbf{A}^l$, $l = 1, 2,...$. Before discussing an ordering strategy for this case we give examples of the way such problems arise.

In the solution of the system of nonlinear algebraic equations $\mathbf{g(x)}=\mathbf{0}$ by Newton's method we generate the sequence

$$\mathbf{J(x}^i)\Delta\mathbf{x}^i = -\mathbf{g(x}^i) \qquad (7.10.1a)$$
$$\mathbf{x}^{i+1} = \mathbf{x}^i + \Delta\mathbf{x}^i \qquad (7.10.1b)$$

from some appropriate starting value $\mathbf{x}^0$, where $\mathbf{J(x}^i)$ is the Jacobian matrix evaluated at the current estimate $\mathbf{x}^i$. If some of the variables in $\mathbf{x}$ enter linearly in some of the equations in $\mathbf{g(x)} = \mathbf{0}$, this will lead to constants in $\mathbf{J(x}^i)$. We aim for the presence of such constants to lead to much of the factorization of $\mathbf{J(x}^i)$ remaining fixed between iterations, allowing work to be saved.

When solving a system of nonlinear differential equations by an implicit method, a set of equations (7.10.1a) has to be solved at each time step. Some of the coefficients in $\mathbf{J}$ may vary with time but not depend on $\mathbf{x}^i$. This leads to three variability types: constants, t-variables and $\mathbf{x}$-variables. At the beginning of a time step (other than the first) only quantities depending on t-variables or $\mathbf{x}$-variables need be recalculated. Within the time step only quantities depending on $\mathbf{x}$-variables need be recalculated when a fresh Jacobian is wanted.

Hachtel *et al.* (1971) considered the design problem for electrical circuits. This involves the implicit solution of a set of ordinary differential equations for various values of the set of design parameters $\mathbf{p}$. They therefore worked with another variability type, that associated only with the design parameters $\mathbf{p}$. There are four variability types: constants, $\mathbf{p}$-variables, t-variables and $\mathbf{x}$-variables. The code contains three nested loops: the $\mathbf{p}$-loop, where the design parameters are changed; the t-loop, corresponding to stepping across the time range in solving the differential equations; and the $\mathbf{x}$-loop, corresponding to the solution of the nonlinear algebraic equations at each time step. We seek ordering strategies that aim to reduce work in the innermost loop at the expense of outer loops in the hope of reducing the overall work. Hachtel *et al.* (1971) actually included two more variability types, corresponding to the constants ± 1. Their use permits multiplications to be avoided during the Gaussian elimination, though we do not discuss this variation.

In general we assume that each matrix entry a_{ij} has variability type v_{ij}, which depends on the depth of nesting of the innermost loop in which it changes. Zeros are taken to have variability type zero and it is assumed that the variability type increases with the depth of nesting. We also associate variability type $v_{ij}^{(k)}$ with the intermediate quantities $a_{ij}^{(k)}$ computed in Gaussian elimination. Recall that these are computed from the formula

$$a_{ij}^{(k+1)} = a_{ij}^{(k)} - a_{ik}^{(k)} a_{kj}^{(k)} / a_{kk}^{(k)}. \tag{7.10.2}$$

If $a_{ik}^{(k)}$ and $a_{kj}^{(k)}$ are entries, we take $v_{ij}^{(k+1)}$ to be given by the formula

$$v_{ij}^{(k+1)} = \max(v_{ij}^{(k)}, v_{ik}^{(k)}, v_{kj}^{(k)}, v_{kk}^{(k)}), \tag{7.10.3}$$

because then $v_{ij}^{(k)}$ labels the depth of nesting at which $a_{ij}^{(k)}$ will need to be recomputed. The aim is to seek pivot choices that keep the variability type low, in order to avoid computations in inner loops.

We need a function that takes account of variability type to replace the Markowitz count. Hachtel *et al.* use

$$\sum_{i>k} \sum_{j>k} \omega(\max(v_{ik}^{(k)}, v_{kj}^{(k)}, v_{kk}^{(k)})) \tag{7.10.4}$$

as the cost of the pivot $a_{kk}^{(k)}$ in (7.10.2), where $\omega(v)$ is a weight function associated with the variability type and satisfies the equations

$$\omega(0) = 0 \; ; \; \omega(v) > 0, \; v > 0. \qquad (7.10.5)$$

Because $v(0) = 0$, the sum (7.10.4) ignores zeros in the pivot row and column and, if $\omega(v) = 1$ for $v > 0$, then it reduces to the Markowitz count.

Little evidence is available on the effectiveness of different choices for the weights $\omega(v)$. Dembart and Erisman (1973, p. 648) found that widely-spaced weights give a lot of fill-ins of type 1, which later give rise to much fill-in of higher type, and they recommend $\omega(v)$, $v > 0$, to be nearly equal.

In spite of the fact that variability-type ordering has been successful for special problems, its implementation is very difficult. It is unlikely that it will ever be incorporated in a general-purpose sparse matrix package, though it may be part of a special-purpose package (for example, in circuit design).

7.11 The symmetric indefinite case

For full symmetric matrices that are not positive definite (or are not known a priori to be so), stable symmetric decompositions are available with the help of symmetric permutations and the use of 2×2 pivots as well as 1×1 pivots (see Section 4.7). Duff, Reid, Munksgaard, and Nielsen (1979) considered the sparse case and proposed an extension of the Markowitz strategy. For the Markowitz count of a 2×2 pivot (to be compared with (7.2.1) for 1×1 pivots) they propose the square of the number of entries in the two pivot rows that are not in the pivot itself. This is a readily evaluated upper bound for the possible fill-in, generalizing one interpretation of the Markowitz cost. Since a 2×2 pivot results in two eliminations being performed together, they compared the Markowitz cost of the best 2×2 pivot with double that of the best 1×1 pivot. Their experimental results showed broad similarity between fill-in with and without 2×2 pivots, though the additional freedom to choose 2×2 pivots sometimes gave a slight advantage. More recently Duff and Reid (1983) have incorporated 2×2 pivoting within a multifrontal framework (see Sections 10.5 and 10.8) to produce a code for solving indefinite systems which retains full efficiency if the system is in fact positive definite.

7.12 Solution methods based on orthogonalization

Most of this book is concerned with solving linear sets of equations using variants of Gaussian elimination. In this section we consider approaches based on orthogonalization (see Section 4.4).

In the sparse case, the factorization

$$\mathbf{A} = \mathbf{QU} \qquad (7.12.1)$$

is even less attractive than in the dense case since an orthogonal rotation causes fill-in to both the pivot and the non-pivot row, unlike Gaussian elimination where only the non-pivot row is affected. Thus the factors $\mathbf{Q}$ and $\mathbf{U}$ are usually much denser than the $\mathbf{LU}$ factors produced by Gaussian elimination to the extent that the straightforward application of the factorization (7.12.1) to the solution of the equation $\mathbf{Ax} = \mathbf{b}$ is impractical for most problems.

It is, however, possible to solve the system $\mathbf{Ax} = \mathbf{b}$ without storing $\mathbf{Q}$ by performing the operations on the right-hand side at the same time as $\mathbf{A}$ is reduced to upper triangular form. Premultiplication of both sides of equation (7.12.1) by their transposes shows that the matrix $\mathbf{U}$ is just the Choleski factor of the normal matrix $\mathbf{A}^T\mathbf{A}$ (apart from possible sign changes in the rows of $\mathbf{U}$). George and Heath (1980) exploit this fact by using the efficient analysis and symbolic factorization techniques available for symmetric matrices (see Section 10.9) to obtain the structure of $\mathbf{U}$ and a column ordering for $\mathbf{A}$. This enables the orthogonal factorization (7.12.1) to proceed row by row using a static data structure for $\mathbf{U}$. The pattern of $\mathbf{U}$ is independent of the row ordering of $\mathbf{A}$, but the amount of work required to construct it is not. Row ordering strategies are discussed further by George and Ng (1983). Coleman, Edenbrandt, and Gilbert (1986) have shown that the pattern of $\mathbf{U}$ generated by George and Heath's approach does not overestimate the actual pattern in the factorization (7.12.1) if $\mathbf{A}$ is irreducible, that is if it cannot be permuted to block triangular form. In the reducible case, it can happen that numerical cancellation occurs for any set of values for the entries of $\mathbf{A}$ so that a severe overestimate of the pattern is obtained.

Order	113	199	541
Nonzeros	655	701	4285
Total storage in K-words			
LU (MA28)	3.9	6.5	40
QU	5.5	5.2	32
Time (IBM 3033 secs)			
LU (MA28)	0.07	0.11	1.22
QU	0.30	0.25	5.31

Table 7.12.1. Comparison of $\mathbf{QU}$ and $\mathbf{LU}$ factorizations (storing only $\mathbf{U}$ in $\mathbf{QU}$).

Although the $\mathbf{QU}$ factorization is normally used for solving overdetermined systems in the least-squares sense, it can be used when $\mathbf{A}$ is square and we show in Table 7.12.1 some results of Heath (private communication 1983, see also Duff 1984d), comparing his implementation of $\mathbf{QU}$

factorization (George and Heath 1980) with Harwell code MA28 (Duff 1977c) which uses **LU** factorization.

The results of Table 7.12.1 indicate that the storage for the U of the **QU** factorization is similar to that for **LU**, although it is significantly more expensive to compute. Additionally, if one wishes to solve $\mathbf{Ax} = \mathbf{b}$ for further right-hand sides, the normal equations

$$\mathbf{A}^T \mathbf{Ax} = \mathbf{A}^T \mathbf{b} \tag{7.12.2}$$

must be used so that a multiplication by $\mathbf{A}^T$ (**A** must be kept) is required in addition to forward substitution through $\mathbf{U}^T$ and back-substitution through **U**.

On balance the **QU** factorization (7.12.1) does not, in spite of its good stability properties, look attractive for solving general unsymmetric systems, although it is a competitive algorithm for least-squares problems. An implementation of this factorization is in an appendix to SPARSPAK called SPARSPAK–B (George and Ng 1984).

Exercises

Exercise 7.1 Construct a sparsity pattern such that expression (7.2.1) is greater than expression (7.2.2) for the first step of Gaussian elimination.

Exercise 7.2 Show that if the graph of a symmetric matrix is a tree, the minimum degree ordering introduces no fill-in.

Exercise 7.3 Show that leading zeros in **b** will also be present in the solution of $\mathbf{Lc} = \mathbf{b}$, where **L** is a nonsingular lower triangular matrix.

Exercise 7.4 If **A** is irreducible, show that the active submatrix of $\mathbf{A}^{(2)}$ is structurally irreducible. Deduce that the active part of every $\mathbf{A}^{(k)}$ is structurally irreducible. If **A** has the triangular factorization **LU**, show that every row of **U**, except the last, has an off-diagonal entry.

Exercise 7.5 Suppose we wish to solve $\mathbf{Ax}^i = \mathbf{b}^i$ for a sequence of vectors $\mathbf{b}^i$ each having the same set of numerical values in its leading positions. Show how this property may be exploited as in the case where there are leading zeros in **b**.

Exercise 7.6 Give an example where no fill-in occurs when performing Gaussian elimination but fill-in is significant when Givens' rotations are used to effect the **QU** factorization (7.12.1).

Exercise 7.7 Give an example where the pattern of **U** generated by George and Heath's technique (Section 7.12) severely overestimates the actual pattern for the factorization (7.12.1).

8 Ordering sparse matrices to special forms

We consider a different approach to reordering equations to preserve sparsity. Some alternative objective functions and heuristic algorithms to achieve these objectives are described. We compare the resulting a priori orderings with the local strategies of Chapter 7.

8.1 Introduction

In this chapter we discuss a different approach to the ordering problem from that of Chapter 7. We replace the objective of minimizing fill-in or minimizing the number of operations by the objective of permuting the matrix to some particular form and so confining the fill-in. Although this often results in more fill-in, it is not necessarily the case since the local strategies do not minimize the fill-in globally. A case where we get less fill-in overall is discussed in Section 8.7. The main advantage of the global approach is that the solution algorithm is usually simpler and is often amenable to the efficient use of vector and parallel processors.

8.2 Desirable forms

There are numerous matrix forms which have been proposed as objectives for ordering strategies. Forms with a particularly long history are band and variable-band (also called skyline, profile, and envelope) matrices such as those shown in Figure 8.2.1.

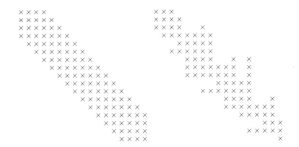

Figure 8.2.1. Band and variable-band matrices.

If no interchanges are performed, these forms are preserved during Gaussian elimination. Algorithms for obtaining these forms are discussed in Section 8.4. They are, in fact, based on finding permutations to block tridiagonal form, illustrated in Figure 8.2.2. If the blocks are small and numerous, such a matrix is banded with a small bandwidth.

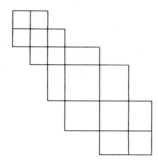

Figure 8.2.2. A block tridiagonal matrix.

We have already discussed the block triangular form (Chapter 6). Its merit is that it permits us to work with a sequence of smaller problems; in fact there is no fill-in except within the diagonal blocks and we retain the off-diagonal blocks in their original form. Similar advantages are available with the block tridiagonal form; we will show that it suffices to retain only the diagonal blocks of the factorized form, together with the original off-diagonal blocks. We discuss this form and a generalization that shares this property in Section 8.5.

Sometimes one of these desirable forms can be obtained for a submatrix. With the submatrix in the leading position we have a bordered form. Prominent examples are the doubly-bordered block diagonal form (Figure 8.2.3) and the bordered block triangular form (Figure 8.2.4) of which a simple case is the bordered triangular form. An important example of a doubly-bordered block diagonal matrix, that obtained by a one-way dissection ordering, is discussed in Section 8.6, and its generalization to the case where the diagonal blocks are themselves in this form is discussed in Section 8.7. Finally in Sections 8.8 and 8.9 we discuss an algorithm for the bordered block triangular form, including the case where the diagonal blocks are themselves in this form. It is due to Hellerman and Rarick (1971, 1972) and is used in linear programming.

A number of other forms have been proposed and are considered (for example by Tewarson 1973), but to our knowledge these others have not been so widely used. For a survey of algorithms for these forms, we refer the reader to Duff (1977a).

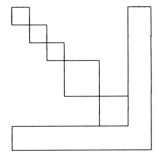

Figure 8.2.3. Doubly-bordered block diagonal form.

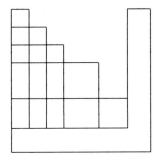

Figure 8.2.4. Bordered block triangular form.

8.3 Band and variable-band matrices

We first consider the band and variable-band forms which were illustrated
in Figure 8.2.1. We say that a symmetrically structured matrix **A** has
bandwidth $2m+1$ and **semibandwidth** m if m is the smallest integer such
that $a_{ij}=0$ whenever $|i-j|>m$. When treating it as a band matrix, all
matrix entries a_{ij} with $|i-j| \leqslant m$ (including zeros) are stored explicitly. For
the unsymmetric case, we define the lower (upper) semibandwidth as the
smallest integer m_l (m_u) such that $a_{ij}=0$ whenever $i-j>m_l$ $(j-i>m_u)$. In
this case the bandwidth is m_l+m_u+1.

Gaussian elimination without interchanges preserves the band structure.
Hence band matrix methods provide an easy way to exploit zeros in a
matrix.

More flexibility is allowed in the variable-band form. In the
symmetrically structured case, we store for each row every coefficient
between the first entry in the row and the diagonal. The total number of
coefficients stored is called the **profile**. In the unsymmetric case we also
store for each column every coefficient between the first entry in the

column and the diagonal. If Gaussian elimination without interchanges is applied, no fill-in is created ahead of the first entry in any row or ahead of the first entry in any column (Exercise 8.1). This shows that the form of a variable-band matrix is preserved. In applications it often happens that worthwhile savings are obtained by exploiting bandwidth variability. A simple example, arising from the triangulation of a square plate, is shown in Figure 8.3.1 and the lower triangular half of the corresponding matrix **A** is shown in Figure 8.3.2. Here the maximum semibandwidth of 4 is attained in only 6 of the rows, whereas pagewise ordering gives a band matrix with semibandwidth 4. The extra organizational overheads when using these schemes are quite slight and both forms are well-suited to vector machines (see Section 10.14).

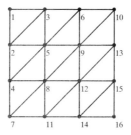

Figure 8.3.1. The triangulation of a square plate.

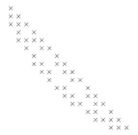

Figure 8.3.2. The pattern of the lower triangular part of the matrix of Figure 8.3.1.

Any zeros within the band or variable-band form are stored explicitly because this part of the matrix usually fills in totally. In fact George and Liu (1975) have shown that for a symmetric matrix, if all rows after the first have a nonzero ahead of the diagonal, then they all fill in totally between this nonzero and the diagonal. This generalizes to any unsymmetric matrix with a nonzero ahead of the diagonal in every row and every column (except the first). The proof is straightforward and we leave it as an exercise for the reader (Exercise 8.2).

Unfortunately, if interchanges are included for the sake of numerical stability, these properties no longer hold. If only row interchanges are used then the variable (or fixed) band form is preserved in the lower triangular part, but it widens in the upper triangular part. For a fixed bandwidth matrix, the upper triangular semibandwidth at most doubles (Exercise 8.3) and many codes have been written which allocate storage for exactly double and hold explicit zeros at the heads of columns whose lengths do not increase as much. It would be possible to write code that dynamically allocated more space for the columns of a variable-band matrix whenever necessary, but we do not know of one.

In the next section we describe algorithms for finding orderings for small bandwidth. The data structures for actually carrying out the factorization will be discussed in Section 10.3.

8.4 Ordering for small bandwidth

It is often natural to use an ordering that gives a small bandwidth. For instance the matrix whose graph is shown in Figure 8.4.1, with the node ordering shown there, has semibandwidth 3. In general, however, an automatic ordering algorithm is clearly desirable. Such algorithms for the symmetric case are the subject of this section.

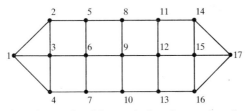

Figure 8.4.1. A graph with an obviously good node order.

Symmetric permutations of a symmetric matrix correspond to relabellings of the nodes of the associated graph (see Section 1.2) and it is easier to describe algorithms in terms of relabelling graphs. We will follow usual practice and do this. Most algorithms divide the nodes into **level sets**, S_i, with S_1 consisting of a single node. The next, S_2, consists of all the neighbours of this node. The set S_3 consists of all neighbours of nodes in S_2 that are not in S_1 or S_2. The general set S_i consists of all the neighbours of the nodes of S_{i-1} that are not in S_{i-2} or S_{i-1}. Once these sets have been constructed, an ordering that takes the nodes of S_1 followed by those of S_2, etc., corresponds to a permuted matrix which is block tridiagonal with diagonal blocks corresponding to the sets S_i. For example, the ordering of the Figure 8.4.1 graph could have come from level sets (1), (2,3,4), (5,6,7),

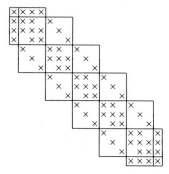

Figure 8.4.2. The matrix corresponding to the Figure 8.4.1 graph.

(8,9,10), (11,12,13), (14,15,16), (17) and the corresponding matrix, shown in Figure 8.4.2, is block tridiagonal with diagonal blocks having orders 1, 3, 3, 3, 3, 3, and 1.

A widely-used algorithm of this kind was proposed by Cuthill and McKee (1969). They ordered within each block S_i by taking first those nodes that are neighbours of the first node in S_{i-1}, then those that are neighbours of the second node in S_{i-1}, and so on. The several other orderings that have been proposed since do not appear to offer any significant advantages.

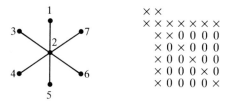

Figure 8.4.3. A graph and its associated matrix, ordered by Cuthill-McKee.

Rather surprisingly, George (1971) found that reversing the Cuthill-McKee order often yields a worthwhile improvement, not in the bandwidth but in the total storage required within the variable-band form (the profile) and in the number of arithmetic operations required for variable-band factorization (see also Liu and Sherman 1976). The resulting algorithm is called the Reverse Cuthill-McKee (RCM) algorithm. A simple example of this is shown in Figures 8.4.3 and 8.4.4. The Cuthill-McKee order (with level sets (1), (2), (3-7)) is shown in Figure 8.4.3 and there are 20 zeros within the variable-band form (shown explicitly in the Figure 8.4.3 matrix).

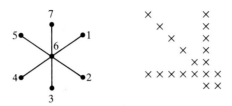

Figure 8.4.4. As Figure 8.4.3, but with order reversed.

On reversing the order (Figure 8.4.4) all these zeros move outside the form and will not need storage.

The reason for the improvement on reversing the Cuthill-McKee order is explained by Jennings (1977). Each zero marked explicitly in the lower triangular part of the matrix in Figure 8.4.3 has a nonzero ahead of it in its row and so fills in when Gaussian elimination is applied. It does not, however, have a nonzero below it in its column and so, on reversing the order, it has no nonzero ahead of it in its row and therefore does not fill in. The opposite effect does not occur because the Cuthill-McKee order gives leading nonzeros which are in columns that form a monotonic increasing sequence. This is straightforward to prove and we leave it as an exercise for the reader (Exercise 8.4).

Any column in which such a gain takes place (a row in the reversed order) corresponds to a node in S_i which has no neighbours in S_{i+1} or whose only neighbours in S_{i+1} are themselves neighbours of earlier nodes in S_i. In our simple example, the gains were all in the last level set whose nodes cannot have neighbours in the next level set. Another simple example is shown in Figure 8.4.5. In this case node 3 in S_2 has only one neighbour, node 5, in S_3 and this is a neighbour of node 2.

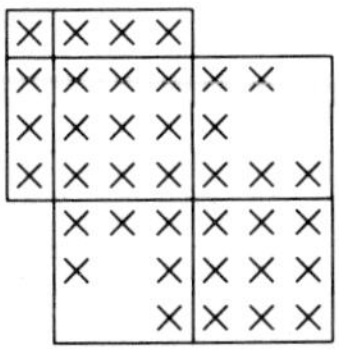

Figure 8.4.5. Another example showing the advantage of reversing Cuthill-McKee ordering.

An important practical case is in finite-element problems involving elements with interior and edge nodes. For example, in Figure 8.4.6 we have a finite-element problem with four quadratic triangular elements and four biquadratic rectangular elements. The graph of the matrix is as in Figure 8.4.6 except that within each element every node is connected to every other node. The Cuthill-McKee order is shown and has level sets (1), (2-9), (10-19), (20-29), (30-33). On reversing the order there are gains associated with nodes 2, 3, 7, 10, 11, 12, 16, 17,... . Notice that they are all interior or edge nodes.

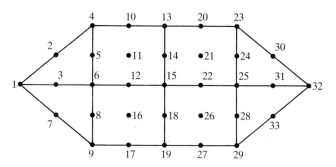

Figure 8.4.6. A finite-element problem.

Figure 8.4.4 also illustrates the interesting property that, if a symmetric matrix has a graph that is a tree (has no cycles), no fill-in is produced with the Reverse Cuthill-McKee ordering. The proof is straightforward and is left as an exercise for the reader (Exercise 8.5). The minimum degree algorithm also produces no fill-in in this case (see Section 7.3). Unfortunately, the variable-band format may contain embedded zeros which in this case do not fill in. This is illustrated in Figure 8.4.7.

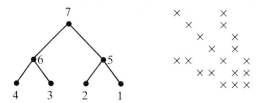

Figure 8.4.7. A tree graph and its associated matrix, ordered by Reverse Cuthill-McKee. There are zeros within the variable-band format.

So far we have not discussed the choice of starting node. It is clear that for small bandwidth we should aim for the level sets to be small and numerous. We obtained this for the Figure 8.4.1 graph by starting at node 1. If instead we had started at node 10 we would have had level sets (10), (7,9,13), (4,6,8,12,16), (1,3,5,11,15,17), (2,14), that is only 5 sets instead of 7 and the semibandwidth would have been 6 instead of 3. For a problem associated with a rectangular or cigar-shaped region it is obviously desirable to start at an end rather than in the middle of a side, but a good choice is not always so apparent. Cuthill and McKee (1969) suggested trying several variables with small numbers of neighbours. Collins (1973) tried starting from every variable, but abandoned each attempt immediately after he encountered a bandwidth greater than the smallest so far obtained.

Gibbs, Poole, and Stockmeyer (1976) suggest that each variable in the final level set S_k should be tried as a starting node. If one of these nodes produces more than k level sets, then this variable replaces the starting node and the new level sets are examined, continuing in this way until no further increase in the number of level sets is obtained. The node in the final starting set and the node in the terminating set from which the restart is made are called **pseudoperipheral** nodes because each is a node on the 'edge' of the graph. The final ordering used by Gibbs *et al.* is based on a combination of the two final sets of level sets. They report that, in all their test examples, the number of level sets was actually maximized. A simple example of the application of this procedure is given as Exercise 8.6. The reader will find it instructive to see how quickly the algorithm recovers from a poor starting node. A good implementation of the Gibbs-Poole-Stockmeyer algorithm is given by Lewis (1982). One of his performance enhancements is to begin looking in the last level set with a node of lowest degree; this reduces the number of starting points tested.

We compare variable-band methods with the minimum degree ordering, but defer the comparison until Section 8.7 in order to include the nested dissection ordering also.

8.5 Refined quotient trees

The algorithm described in the previous section produces a block tridiagonal matrix, each block corresponding to the nodes in a level set. George (1977b) pointed out that the factorization of such a matrix **A** can be written in the form

$$\begin{bmatrix} A_{11} & A_{12} & & & & \\ A_{21} & A_{22} & A_{23} & & & \\ & A_{32} & A_{33} & \cdot & & \\ & & \cdot & \cdot & \cdot & \\ & & & \cdot & \cdot & \cdot \\ & & & & \cdot & A_{NN} \end{bmatrix} = \begin{bmatrix} I & & & & \\ A_{21}D_1^{-1} & I & & & \\ & A_{32}D_2^{-1} & I & & \\ & & \cdot & \cdot & \\ & & & \cdot & \cdot \\ & & & & \cdot & I \end{bmatrix} \begin{bmatrix} D_1 & A_{12} & & & \\ & D_2 & A_{23} & & \\ & & D_3 & A_{34} & \\ & & & \cdot & \cdot \\ & & & & \cdot & \cdot \\ & & & & & D_N \end{bmatrix} \qquad (8.5.1)$$

where

$$\mathbf{D}_1 = \mathbf{A}_{11} \qquad\qquad (8.5.2a)$$

$$\mathbf{D}_i = \mathbf{A}_{ii} - \mathbf{A}_{i,i-1}\mathbf{D}_{i-1}^{-1}\mathbf{A}_{i-1,i}, \, i = 2, 3,..., N. \qquad (8.5.2b)$$

To use this form to solve a set of equations $\mathbf{Ax} = \mathbf{b}$ requires the forward substitution steps

$$\mathbf{c}_1 = \mathbf{b}_1, \tag{8.5.3a}$$

$$\mathbf{c}_i = \mathbf{b}_i - \mathbf{A}_{i,i-1}\mathbf{D}_{i-1}^{-1}\mathbf{c}_{i-1}, i = 2, 3,..., N, \tag{8.5.3b}$$

followed by the back-substitution steps

$$\mathbf{x}_N = \mathbf{D}_N^{-1}\mathbf{c}_N, \tag{8.5.4a}$$

$$\mathbf{x}_i = \mathbf{D}_i^{-1}(\mathbf{c}_i - \mathbf{A}_{i,i+1}\mathbf{c}_{i+1}), i = N-1,..., 1. \tag{8.5.4b}$$

It therefore suffices to keep the off-diagonal blocks of $\mathbf{A}$ unmodified and the diagonal blocks $\mathbf{D}_i$ in factorized form

$$\mathbf{D}_i = \mathbf{L}_i\mathbf{U}_i, i = 1, 2,... . \tag{8.5.5}$$

This is a generalization of the implicit factorization considered in Section 3.12. The matrices $\mathbf{L}_i$ and $\mathbf{U}_i$ are just the diagonal blocks of the ordinary LU factorization of $\mathbf{A}$, as may be seen (Exercise 8.7) by substituting $\mathbf{L}_i\mathbf{U}_i$ for $\mathbf{D}_i$ in equation (8.5.1). The fact that all fill-in is avoided in off-diagonal blocks therefore means that storage is saved overall. The operation count for forming the factorization is the same and there may or may not be a saving in operation count when using the factorization to solve a set of equations.

The same storage gain is available for any block matrix whose pattern is symmetric and which has at most one off-diagonal block in the upper triangular part of each block row (and in the lower triangular part of each block column). Of course a block tridiagonal matrix is a special case of this form. If block Gaussian elimination is applied to such a matrix, each block pivot row and column contains a single nonzero block so the pivotal step modifies no off-diagonal blocks and at most one diagonal block. The generalized forms of equations (8.5.2 – 8.5.4) are

$$\mathbf{D}_1 = \mathbf{A}_{11}, \tag{8.5.6a}$$

$$\mathbf{D}_i = \mathbf{A}_{ii} - \sum_{j<i}\mathbf{A}_{ij}\mathbf{D}_j^{-1}\mathbf{A}_{ji}, i = 2, 3,..., N, \tag{8.5.6b}$$

$$\mathbf{c}_1 = \mathbf{b}_1, \tag{8.5.7a}$$

$$\mathbf{c}_i = \mathbf{b}_i - \sum_{j<i}\mathbf{A}_{ij}\mathbf{D}_j^{-1}\mathbf{c}_j, i = 2, 3,..., N, \tag{8.5.7b}$$

and

$$\mathbf{x}_N = \mathbf{D}_N^{-1}\mathbf{c}_N, \tag{8.5.8a}$$

$$\mathbf{x}_i = \mathbf{D}_i^{-1}(\mathbf{c}_i - \sum_{j>i}\mathbf{A}_{ij}\mathbf{c}_j), i = N-1,..., 1. \tag{8.5.8b}$$

We get this effect whenever the graph of the block form is a tree and the ordering is monotonic (that is each node is ordered ahead of its father). In the block tridiagonal case, the tree is a simple chain. For example, the problem shown in Figure 8.4.6 has five level sets and the graph of its block form is shown in Figure 8.5.1. Graphs obtained by grouping nodes together in this way are called **quotient graphs**. Each node i of such a quotient graph corresponds to a set I of nodes in the original graph, and an edge in the quotient graph connects node i to node j if the original graph has one or more edges connecting a node of the set I to a node of the set J.

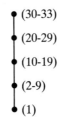

Figure 8.5.1. The quotient graph (chain) corresponding to the level sets of Figure 8.4.6.

George and Liu (1978a) proposed an algorithm for automatically constructing a **refined quotient tree** from the Reverse Cuthill-McKee quotient chain. By breaking the blocks into smaller sub-blocks, it reduces the fill-in within the diagonal blocks. We first illustrate the idea on the problem shown in Figure 8.5.2.

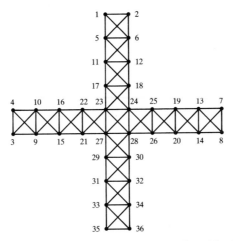

Figure 8.5.2. A plus-shaped problem, ordered by Reverse Cuthill-McKee.

The problem has been ordered by Reverse Cuthill-McKee and the level sets are shown in Figure 8.5.3. By breaking three of the level sets in parts according to the arm of the problem to which they correspond we get the quotient tree shown in the second part of Figure 8.5.3. The block matrix corresponding to this quotient tree is shown in Figure 8.5.4. The block tridiagonal form corresponding to the quotient chain is shown by heavy lines. Notice that the refined quotient tree form has 46 2×2 blocks of zeros within the block tridiagonal form that remain zero during Gaussian elimination. 20 of these 46 are also preserved by working with a variable-band form (those before the first nonzero in a row or in a column) but the remaining 26 are within the variable band. The implicit factorization (8.5.1) on the original block tridiagonal form preserves 28 of these blocks of zeros, but wastes those within the diagonal blocks since these are normally stored as full matrices.

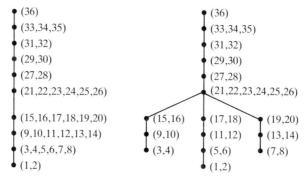

Figure 8.5.3. The level set chain and the refined quotient tree for the Figure 8.5.2 problem.

Although their description is somewhat different, the algorithm of George and Liu (1978a) is essentially a two-pass algorithm with the steps

(i) in each level set, group together those nodes for which there is a path from any one to any other one through nodes in the level set, and

(ii) whenever a group at one level is connected to two groups at the next higher level (that is nearer the root), amalgamate the two groups at the higher level.

The first step yields a quotient graph with no connections between nodes at the same level. For our example, it actually gives the quotient tree of Figure 8.5.3, but in general this gives a quotient graph containing nodes with two 'fathers' and where this happens step (ii) remedies it by amalgamating the two fathers into one.

Note that, unless the quotient tree is a chain, the procedure gives blocks

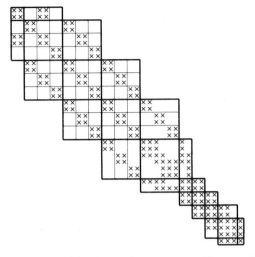

Figure 8.5.4. The block matrix corresponding to the tree of Figure 8.5.3.

that are smaller than those of the straightforward variable-band block tridiagonal algorithm because the level sets are divided between different arms of the tree. Another advantage is that the arms represent independent computations that could be handled as independent processes.

8.6 One-way dissection

The ideas discussed in the last section will not help if the graph is without appendages such as the arms of the cross of Figure 8.5.2. This led George (1980) to propose a very simple ordering that leads to a quotient tree having just two levels (a father node and its sons) and a matrix of doubly-bordered block diagonal form (Figure 8.2.3). For example on the regular grid of Figure 8.6.1, we might take blocks corresponding to the sets of columns (1,2), (4,5), (7,8), (10,11), (3,6,9) to give the tree of Figure 8.6.2 and the matrix of Figure 8.6.3.

With this special form, each block pivotal step of Gaussian elimination modifies the last diagonal block and equations (8.5.6 – 8.5.8) simplify to the following:

$$\mathbf{D}_i = \mathbf{A}_{ii}, \ i=1,2,\dots,N-1. \tag{8.6.1a}$$

$$\mathbf{D}_N = \mathbf{A}_{NN} - \sum_{j=1}^{N-1} \mathbf{A}_{Nj}\mathbf{D}_j^{-1}\mathbf{A}_{jN} \tag{8.6.1b}$$

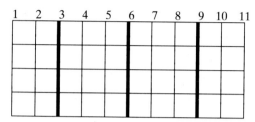

Figure 8.6.1. A regular 5×11 grid problem, one-way dissected.

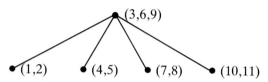

Figure 8.6.2. A tree arising from one-way dissection of the problem in Figure 8.6.1.

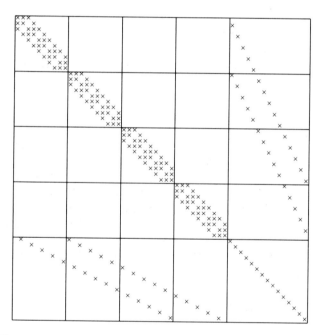

Figure 8.6.3. The matrix arising from one-way dissection of the problem in Figure 8.6.1.

$$\mathbf{c}_i = \mathbf{b}_i, \, i = 1, 2,\ldots, N-1 \tag{8.6.2a}$$

$$\mathbf{c}_N = \mathbf{b}_N - \sum_{j=1}^{N-1} \mathbf{A}_{Nj} \mathbf{D}_j^{-1} \mathbf{c}_j \tag{8.6.2b}$$

$$\mathbf{x}_N = \mathbf{D}_N^{-1} \mathbf{c}_N \tag{8.6.3a}$$

$$\mathbf{x}_i = \mathbf{D}_i^{-1} (\mathbf{c}_i - \mathbf{A}_{iN} \mathbf{c}_N). \tag{8.6.3b}$$

The blocks $\mathbf{A}_{ii}$, $i = 1, 2,\ldots, N-1$, are factorized without modification so their band structure can be exploited. For instance in our example they have order 10 and semibandwidth 2 if sensibly ordered. As in the last section, the off-diagonal blocks are used unmodified, so their structure can be fully exploited.

For automatic one-way dissection of a general problem, George (1980) again commences by generating a good level structure, as in Section 8.4. He next computes the average number of nodes at each level, say m, and takes points from each of the level sets

$$j = \lfloor i\delta + 0.5 \rfloor, \, i=1,2,\ldots \tag{8.6.4}$$

with spacing

$$\delta = \sqrt{\frac{3m+13}{2}}, \tag{8.6.5}$$

where $\lfloor \; \rfloor$ denotes the integer part of a number. The formula (8.6.5) for the spacing was chosen by George on the basis of numerical experiments and analysis of regular grids, with the aim of keeping storage requirements low. It would suffice to place all the points in the level sets (8.6.4) in the last (separator) block and the points in the intervening groups of level sets into the other blocks. However it can happen that a point in one level set has no neighbours in the next level set (though it must have one or more neighbours in the previous level set). There is no need for any such point to be placed in the separator set. For example, points 10-19 of Figure 8.4.6 lie in a level set, but (13,14,15,18,19) suffices as a separator set. A point not needed in the separator set should be grouped with the points of the previous level. Once the separators have been chosen, the ordering within each diagonal block can be chosen by a fresh application of the Reverse Cuthill-McKee algorithm to the submatrix alone.

8.7 Nested dissection

The term 'nested dissection' was introduced by George (1973), following a suggestion of G. Birkhoff. Its roots are in finite-element substructuring and it is closely related to the tearing methods discussed in Chapter 11. Here we present the basic algorithm and relate it to the other orderings of this and the previous chapter.

The central concept for nested dissection, as for one-way dissection (Section 8.6), is the removal of a set of nodes from the graph of a symmetric matrix that leaves the remaining graph in two or more disconnected parts. For example, the removal of nodes 11 to 15 of the graph of Figure 8.7.1 leaves two subgraphs with no connections between each other. In nested dissection the parts are themselves further divided by the removal of sets of nodes, with the dissection nested to any depth.

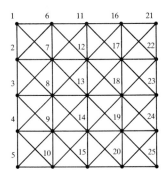

Figure 8.7.1. Graph for 25-node example.

If the variables of each subgraph are grouped together, we obtain the block form

$$\begin{pmatrix} \mathbf{A}_{11} & & \mathbf{A}_{13} \\ & \mathbf{A}_{22} & \mathbf{A}_{23} \\ \mathbf{A}_{31} & \mathbf{A}_{32} & \mathbf{A}_{33} \end{pmatrix}. \tag{8.7.1}$$

The blocks $\mathbf{A}_{11}$ and $\mathbf{A}_{22}$ may themselves be ordered to such a form by using the dissection sets (3,8) and (18,23). Grouping the variables like this at every level defines a nested dissection order.

A major feature of the algorithm is that its performance can be analysed for some model problems. George (1973) showed that on a regular $q \times q$ grid such as that of Figure 8.7.1 ($q=5$) the fill-in introduced can be identified precisely and it is $O(q^2\log_2 q)$. Further, the operation count is $O(q^3)$. Hoffman, Martin, and Rose (1973) show that the algorithm is optimal for such grids in the sense that all other orderings require at least $O(q^3)$ operations. Comparable figures for the banded approach are $O(q^3)$ and $O(q^4)$ for storage and work respectively and for one-way dissection are $O(q^{5/2})$ and $O(q^{7/2})$ respectively. Actual gains for practical values of q, though worthwhile, are not enormous because in the exact formula the multiplying factors for the leading terms disfavour nested dissection. George's (1973) figures show, for instance, that if $q = 32$, the number of operations is reduced to about 38% and the number of storage locations needed (excluding overheads) to about 52% of the corresponding figures for band elimination using pagewise ordering. Experiments comparing one-way dissection with nested dissection and pagewise ordering are described by George (1977a).

Nested dissection gives a more scattered sparsity pattern than is obtained with the banded approach, so that a more complicated data structure (with greater overheads) is required to take advantage of the gain. The data structure used is similar to that for the minimum degree algorithm (see Section 10.9) and is discussed in detail by George and Liu (1981).

There are no known formulae for the order of fill-in or operation count for the Markowitz ordering, even for the very regular pattern of the matrix whose graph is shown in Figure 8.7.1. The difficulty with this matrix and the Markowitz ordering is that the effect of tie-breaking is critical (see Section 7.7). Experimental results comparing the nested dissection and minimum degree orderings are shown in Table 8.7.1. We use the tie-breaking strategy of choosing the node of minimum degree which is first in the original order. These results indicate that the minimum degree algorithm is broadly comparable with the nested dissection algorithm. Table 7.7.1 shows that poor tie-breaking choices can lead to much worse results.

q	Nested dissection	Minimum degree
4	100	111
6	269	321
8	572	621
11	1 283	1 421
16	3 336	3 954
21	6 654	7 381
26	11 177	12 794
32	18 748	23 142
64	100 256	129 046

Table 8.7.1. Number of multiplications required for matrix factorization using minimum degree and nested dissection.

This points to the fundamental reason for the success of the nested dissection strategy. It is a truly global ordering in the sense that decisions made at the very first stage take the entire matrix into account.

So far we have not discussed how suitable dissection sets can be chosen. When a complicated design is assembled from simpler substructures, it makes sense to exploit these natural substructures. The resulting ordering is likely to be good simply because, when each variable is eliminated, only the other variables of its substructure are involved. In general, however, an automatic procedure is needed. George and Liu (1978b, 1979a) use the algorithm of Gibbs, Poole, and Stockmeyer (1976) to find a set of level sets and take the middle level set as a separator. They then repeat on the halves, and so on. On the regular square problem they found results that were very nearly as good as those of the theoretical dissection of George (1973).

George and Liu (1978b) and George (1980) used problems arising from a sequence of refinements of a triangulation of an L-shaped region to compare four algorithms:

(i) variable-band solution following Reverse Cuthill-McKee ordering (see Sections 8.3 and 8.4),

(ii) minimum degree ordering (see Section 7.3) using the Yale Sparse Matrix Package (Eisenstat, Gursky, Schultz, and Sherman, 1982),

(iii) one-way dissection (see Section 8.6), and

(iv) nested dissection (this section).

We show some of their results in Table 8.7.2. We also include some results of Lewis (1983) for the largest problem when run on the CRAY-1. It is clear that one-way dissection is outstandingly successful on storage, needing about half that of the other methods for larger problems. However, it is possible to implement the minimum degree algorithm without storing all the entries in the factors (Eisenstat, Schultz, and Sherman 1979) which would probably require even less storage but at some computing expense. Apart from the slower factorization time for one-way dissection, the factorization and solution times show remarkably small differences. Of course, none of these problems was particularly large.

To obtain some indication of the behaviour on a bigger problem, Reid (1981) experimented with one supplied by Manteuffel (1980) who describes this problem and gives a diagram. It is a cylindrical shell with some irregularities and is built from 784 20-node bricks to give a total of 17 928 variables. Manteuffel estimated that a band technique would involve a semibandwidth of 1300, storage for 23×10^6 entries and 15×10^9 arithmetic operations. The SPARSPAK code (Chu, George, Liu, and Ng 1984) is unable to handle such a big problem, but we can make estimates from its analysis of the equivalent problem with one degree of freedom per node. This indicates that the Reverse Cuthill-McKee algorithm would reduce the storage required by about the factor 2 and that nested dissection would further reduce it by about the factor 2.4. Reid (1981) found that the minimum degree algorithm on this problem needs storage for 4.8×10^6 entries and involves 1.15×10^9 arithmetic operations. Here the performance of minimum degree and nested dissection orderings is not significantly different.

Computer	IBM	IBM	IBM	IBM	IBM	CRAY
Order	265	577	1009	1561	2233	2233
Nonzeros	1753	3889	6865	10681	15337	15337
Ordering time (seconds)						
Band(RCM)	0.07	0.16	0.27	0.42	0.62	0.09
Min. deg.	0.39	1.0	1.55	2.6	3.5	–
One-way	0.12	0.27	0.48	0.70	1.05	0.15
Nested	0.20	0.49	0.95	1.6	2.4	0.35
Allocation time (seconds)						
Band(RCM)	0.01	0.03	0.05	0.07	0.10	0.02
Min. deg.	0.06	0.13	0.24	0.39	0.55	–
One-way	0.07	0.17	0.28	0.43	0.62	0.07
Nested	0.06	0.14	0.26	0.39	0.58	0.07
Total storage (thousands of words)						
Band(RCM)	4.8	14	30	56	93	93
Min. deg.	6.7	17	34	54	85	–
One-way	4.2	10	19	32	49	49
Nested	7.0	18	34	56	84	84
Factorization work (millions of mults and divs)						
Band(RCM)	0.03	0.13	0.37	0.87	1.74	1.74
Min. deg.	0.03	0.10	0.31	0.55	1.19	–
One-way	0.04	0.18	0.45	0.97	1.77	1.77
Nested	0.03	0.13	0.31	0.63	1.11	1.11
Factorization time (seconds)						
Band(RCM)	0.33	1.3	3.4	7.5	14.4	0.73
Min. deg.	0.36	1.3	3.4	5.9	12.1	–
One-way	0.65	2.4	5.5	11.1	19.6	1.29
Nested	0.46	1.6	3.8	7.3	12.8	0.85
Solution work (thousands of mults and divs)						
Band(RCM)	7.5	23	53	100	169	169
Min. deg.	6.5	18	39	65	107	–
One-way	6.9	19	37	64	99	99
Nested	7.2	20	40	68	106	106
Solution time (seconds)						
Band(RCM)	0.07	0.21	0.46	0.83	1.40	0.04
Min. deg.	0.07	0.18	0.38	0.60	0.98	–
One-way	0.09	0.22	0.41	0.68	1.04	0.06
Nested	0.07	0.19	0.37	0.63	0.94	0.08

Table 8.7.2. Comparison of methods on sequence of triangulations of L-shaped region (George and Liu 1981; Lewis 1983) on IBM 4341 and CRAY-1.

8.8   The Boeing version of the Hellerman-Rarick algorithm

Algorithms described earlier in this chapter, though applicable to unsymmetric problems, have primarily been used with symmetric or near-symmetric matrices. In the remainder of this chapter we describe an algorithm which is primarily used for very unsymmetric problems. As for the earlier algorithms, we aim to permute the matrix so that its pattern has a special form.

Hellerman and Rarick (1971) introduced an algorithm which they called the preassigned pivot procedure (P^3) and later (Hellerman and Rarick 1972) suggested an initial step of permuting to block lower triangular form to be followed by the application of the earlier (P^3) algorithm to each diagonal block. The later algorithm is called the partitioned preassigned pivot procedure (P^4). We have already described (Chapter 6) a two-stage algorithm for permuting to block triangular form, and we believe that this is the most efficient algorithm available for practical problems. We therefore describe only the treatment of each diagonal block of the block triangular form. Each such block is irreducible (cannot be permuted to block triangular form), so it suffices to limit our description to the case where the original matrix is irreducible.

In this section we consider a variant of the P^3 algorithm due to Erisman, Grimes, Lewis, and Poole (1985), because it is a simplification and is therefore easier to describe. They call it P^5 (precautionary partitioned preassigned pivot procedure). It produces a bordered block triangular form (Figure 8.2.4). All the diagonal blocks of the block triangular submatrix are full and the algorithm tries to make the border thin. The original P^3 algorithm also produces a bordered block triangular form, but tries harder to make the border thin and the diagonal blocks may themselves be bordered block triangular matrices. We describe this in the next section.

At a typical intermediate stage the permuted matrix has the form illustrated in Figure 8.8.1. The leading $p \times p$ submatrix is block lower triangular with full diagonal blocks. The diagonal entries are **assigned pivots**. The q columns in the border are called **spike columns** by Hellerman and Rarick for reasons that will be apparent in the next section. Each has a leading full block in rows corresponding to a block of the block triangular submatrix. Each of these spike columns extends at least as far up the matrix as its predecessors. The submatrix of rows 1 to p and columns $p+1$ to $n-q$ is zero. We call the submatrix of rows $p+1$ to n and columns $p+1$ to $n-q$ the **active submatrix** since it is within this submatrix that further permutations take place. We commence with $p = q = 0$ and the whole matrix active and end with $p+q = n$ and no columns left in the active submatrix.

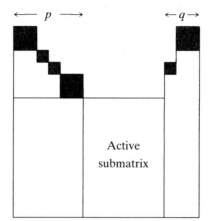

Figure 8.8.1. The permuted matrix at a typical intermediate stage of the P^5 algorithm.

In the first major step of the algorithm, m columns are chosen, where m is the minimum number of entries in a row. Some or all of them make up the leading block column and the rest make up the end of the border. A column with most entries in rows with count m (we defer the resolution of ties for the present) is chosen first. After removing this column from the active submatrix, we will have the greatest possible number of rows with the minimum row count of $m-1$. Next a column with most entries in rows with count $m-1$ is chosen. The process continues similarly until m columns have been chosen. If the last column chosen has s singletons (rows with one entry), the rows containing these singletons are permuted to the front and assigned as pivotal rows. The last s columns selected ($s < m$, since the matrix is irreducible) are permuted to the front and assigned as pivotal columns. The remaining $m-s$ columns are permuted to the back and become the end of the border.

```
       1 2 3 4 5 6
    1  × × ×   ×
    2  ×     ×   ×
    3  × × × ×
    4  ×     ×   ×
    5    × × × × ×
    6  × × ×   ×
```

Figure 8.8.2. A 6×6 matrix pattern.

In the simple example shown in Figure 8.8.2, the minimum row count m is 3 and column 1, being a column with most entries in rows having 3 entries, is chosen first. We revise the row counts to exclude this column and find that column 4 has most entries (2) in rows with the new minimum row count of 2, so this is chosen next. Now column 6 has singletons in rows 2 and 4. Therefore we permute rows 2 and 4 and columns 6 and 4 forward to the pivotal block, and permute column 1 to the back to become a border spike. The resulting matrix is shown in Figure 8.8.3. The active submatrix is now rows 3 to 6 and columns 3 to 5 of the permuted matrix.

$$
\begin{array}{c}
\;6\;4\;2\;5\;3\;1 \\
\begin{array}{c}
4 \\ 2 \\ 3 \\ 1 \\ 5 \\ 6
\end{array}
\left.\begin{array}{ccccccc}
\times & \times & & & & \times \\
\times & \times & & & & \times \\
& \times & \times & & \times & \times \\
& & \times & \times & \times & \times \\
\times & \times & \times & \times & \times & \\
& & \times & \times & \times & \times
\end{array}\right.
\end{array}
$$

Figure 8.8.3. The matrix of Figure 8.8.2 after the first major stage.

After the first major step, the matrix will always have the form illustrated in Figure 8.8.4. There is a leading block of order s, there are $m-s$ spikes at the end of the matrix and the first s rows are otherwise zero. Since every row in the original matrix had at least m entries, the $s\times s$ pivotal block and the first s rows of the border are full.

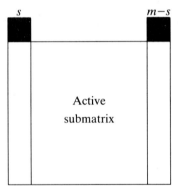

Figure 8.8.4. The general form after selection of the first set of spikes.

The algorithm now treats the active submatrix in exactly the same way, continuing until it produces the form illustrated in Figure 8.8.1. The active submatrix is rectangular, so it is possible that all the columns selected are assigned as pivotal. The simplest case is when $m=1$, in which case a row singleton is moved to the leading position and assigned immediately as a pivot.

In describing the algorithm, we purposely omitted to say which column is chosen if several have the maximum number of entries in rows with minimum row count. In this case, Hellerman and Rarick aim to reduce the number of spikes in later stages by choosing the candidate that has greatest column count unless they have a single entry in a row of minimum count. In the latter case, account is first taken of the number of entries in rows of second least row count, then a remaining column with greatest count is taken.

This completes our description of the algorithm. For ease of reference, we summarize it in pseudo-algol in Figure 8.8.5. It is convenient to describe each column as being permuted to the end of the active submatrix as it is chosen. If it has s singletons then it and the appropriate number of its successors are moved forward into the pivotal block.

Count the number of entries in the rows and columns;
Initialize the active submatrix to be the whole $n \times n$ matrix;
$j := 1$; **while** $j \leqslant n$ **do**
begin
    find the minimum row count m;
    for $m1 := m$ **step** -1 **until** 1 **do**
    begin
        find the set T of columns with maximum number s of entries in rows with count of $m1$;
        if there is more than one column in T and $s = 1$ **then** find the second least row count m' of rows with entries in columns of T and reduce T to those columns with maximum number of entries in rows with count m';
        choose a column $\bar{j}$ of T with greatest column count;
        exchange column $\bar{j}$ with the last column of the active submatrix;
        remove column $\bar{j}$ from the active submatrix and revise the row counts to correspond;
    end;
    $j := j + m$;
    assign_pivots
end;

procedure assign_pivots
begin
    permute the rows so that the s rows that have just had count 1 are the leading rows of the active submatrix;
    for $i := 1$ **step** 1 **until** $\min(s,m)$ **do** move the first column in the border ahead of the active submatrix;
    $i = \min(s,m)$;
    remove the leading i rows from the active submatrix;
end;

Figure 8.8.5. A summary of the P^5 version of the Hellerman-Rarick algorithm in pseudo-algol.

Erisman *et al.* (1985) experimented with the use of the implicit factorization (8.5.1) of the block matrix

$$\begin{pmatrix} \mathbf{A}_{11} & \mathbf{A}_{12} \\ \mathbf{A}_{21} & \mathbf{A}_{22} \end{pmatrix}, \tag{8.8.1}$$

where $\mathbf{A}_{11}$ is the block lower triangular submatrix of the P^5 ordering and the rest is the border, thereby confining the fill-in to the block $\mathbf{A}_{22}$. Their test comparisons with the algorithm of Markowitz showed the two to be broadly comparable, with neither consistently better than the other. Unfortunately they did not have any proposals for safeguards against numerical instability, so the algorithm cannot be regarded as a serious challenger to that of Markowitz.

8.9 The Hellerman-Rarick ordering

The aim of Hellerman and Rarick was not so much to produce a bordered block triangular matrix as to produce a spiked matrix of the form illustrated in Figure 8.9.1. By a **spike**, we mean the part of a spike column that lies on and above the diagonal. Any pair of spikes have the property that the set of rows for the first is either contained in or disjoint from the set of rows for the second. This property corresponds to being able to regard each spike as the border of a block lower triangular form in a properly nested set, as illustrated in Figure 8.9.2.

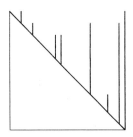

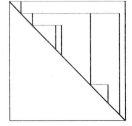

Figure 8.9.1. A spiked matrix.

Figure 8.9.2. The matrix of Figure 8.9.1, regarded as a nested bordered block triangular form.

If Gaussian elimination without interchanges is applied to such a sparse matrix, fill-ins can take place only within spike columns. Indeed the fill-in may be confined to the spikes themselves if the matrix is reduced to a diagonal matrix by a sequence of elementary row operations that create zeros successively in the lower triangular part of row k then in the upper triangular part of column k, for $k = 2, 3, \ldots, n$. Hellerman and Rarick aim to minimize both the number of spikes and the extent to which they project above the diagonal.

```
× |   | × ×        ×      × ×        ×   ×   ×
×|×  |× ×        × ×    × ×        × × ×    ×
×|×  |× ×        × × ×    ×        × × ×    ×
×|× ×|  ×        × ×    × ×        × ×    × ×
×|× ×|× ×        × × × × ×        × × × × ×
```

Figure 8.9.3. A simple example showing extra spikes being moved forward.

The total number of spikes produced by Hellerman and Rarick is the same as for the P^5 variant described in the last section, but they try harder to move spikes forward. If an active submatrix has minimum row count m and the last of the m columns chosen has more than m singletons, then some of the previously assigned spikes may be moved forward to become pivotal, which shortens them. A trivial illustration is shown in Figure 8.9.3. On the left, the active submatrix is in rows 2 to 5 and columns 2 and 3, and column 2 contains two singletons. Erisman et al. (1985) would accept entry (2,2) as a pivot, then find the singleton in column 3 and assign this to give the form shown in the middle of Figure 8.9.3. Hellerman and Rarick, on the other hand, after finding 2 singletons in column 2 would bring the spike forward to give the last form in Figure 8.9.3.

```
   ×          ×    × ×
     × × ×         × ×
     × × ×         × ×
     × × ×         × ×
     × × ×         × ×
   × × × × × × × ×
   × × × × × × × ×
   × × × × × × × ×
```

Figure 8.9.4. The 8×8 example of Erisman et al.

If extra spikes are always brought forward in this way, we may place a zero in a pivotal position, as is the case for the matrix of Erisman et al. shown in Figure 8.9.4, where column 5 is a spike column that has been moved forward. In this particular case, not only is the fifth pivot zero, but it remains zero during Gaussian elimination. Hellerman and Rarick do not make clear how they treat such a situation. Erisman et al. make the interpretation that if a column has s singletons, $s-m$ previously assigned spikes are always moved forward, and found that this variant of the algorithm failed quite often. Since the algorithm is widely used in linear programming codes, it seems more likely that the authors intended spikes to be brought forward only if nonzero pivots can be assigned, perhaps with the help of permutations of the singleton rows. The P^5 approach of never assigning previous spikes seems over-cautious. The pseudo-algol procedure assign_pivots in Figure 8.9.5 summarizes this part of the algorithm and replaces the procedure of the same name in Figure 8.8.5. It uses our interpretation of the algorithm, rather than that of Erisman et al. (1985).

```
procedure assign_pivots
begin
        for i := 1 step 1 until s do
        begin
                move the first spike column in the border ahead of
                the active submatrix;
                if it is possible to permute the rows so that the i
                diagonal coefficients ahead of the active submatrix
                are entries
                then do so
                else move the column ahead of the active matrix back
                to the border and goto quit;
        end
        i := s+1;
        quit: remove the leading i−1 rows from the active
        submatrix;
end;
```

Figure 8.9.5. The alternative procedure assign_pivots, that
converts the pseudo-algol program of Figure 8.8.5 to the
Hellerman-Rarick algorithm.

These extra movements can mean that the leading square submatrix is
no longer a block lower triangular matrix with full diagonal blocks. Instead
it is a bordered block lower triangular matrix whose diagonal blocks are
themselves bordered block triangular matrices, nested to any depth. When
an extra spike is moved forward we add a border to a block triangular
matrix because spikes always start in the same row as a block.

For actual factorization of the matrix, further permutations may be
needed, since the diagonal blocks can be singular or nearly so, even with
the P^5 algorithm which avoids structural singularity. For instance the
Figure 8.9.6 example (Westerberg, private communication 1974) is already
permuted to the form the algorithm (both versions) leaves. Even though
the matrix is structurally nonsingular, pivoting in the order given will yield
a seventh pivot that is zero (that is the leading 7×7 submatrix is structurally
singular). Thus the normal implementation of the Hellerman and Rarick
algorithm would fail. Note, however, that if we consider the matrix of
Figure 8.9.6 to have a border of order 3 and regard the leading block
diagonal matrix of order 6 as A_{11} in the matrix (8.8.1), then the Schur
complement $A_{22} - A_{21}A_{11}^{-1}A_{12}$ is not structurally singular and any method
which forms this and allows pivoting when solving for the border variables
will succeed (provided A_{11} is not numerically singular).

Hellerman and Rarick (1971) suggested permuting the spikes whenever
necessary, and indeed interchanging columns 7 and 9 cures the structural
singularity of the Figure 8.9.6 example. In general, however, such
interchanges alter the form of the matrix even to the extent that the spikes
no longer give a properly nested bordered block triangular form unless

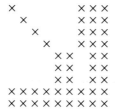

Figure 8.9.6. Example with structurally singular leading
7×7 submatrix.

some of them are artificially elongated. For numerical stability it is
necessary to avoid all small pivots, even in non-spike columns. Saunders
(1972) suggests column interchanges on these grounds too. This may
further worsen the structure since additional spikes are introduced, but in
practice the problem is not serious if a mild relative pivot tolerance is used.
The structure is not altered by row or column interchanges within inner
blocks (for instance we did not alter the structure of the Figure 8.9.6
matrix) so it may be worthwhile to try to use such interchanges first.

Exercises

8.1 Verify that if Gaussian elimination without interchanges is applied to a
variable-band matrix, this structure is preserved.

8.2 Show by induction that if every row of a matrix except the first has a nonzero to
the left of the diagonal and every column except the first has a nonzero above the
diagonal, then Gaussian elimination without interchanges fills in the rows and
columns completely between the first nonzero and the diagonal.

8.3 Show that if Gaussian elimination with row interchanges is applied to a band
matrix of bandwidth $2m+1$, there is no fill-in outside the band in the lower
triangular part, but there may be fill-in in the upper triangular part in positions (i,j)
with $j-i \leq 2m$.

8.4 Let A be a symmetric matrix that has been ordered by the Cuthill-McKee
algorithm. Show that the leading nonzeros in the rows are in columns that form a
monotonic increasing sequence.

8.5 If a symmetric matrix has a graph that is a tree, show that no fill-in is produced
with the Reverse Cuthill-McKee ordering.

8.6 Construct level sets for the following cigar-shaped graph starting from node 10, then generate more level sets according to the Gibbs-Poole-Stockmeyer algorithm.

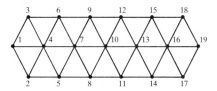

8.7 By substituting $L_i U_i$ for D_i in equation (8.5.1), show that L_i and U_i are the diagonal blocks of the ordinary **LU** factorization of **A**.

8.8 Consider Gauss-Jordan elimination applied to a Hellerman-Rarick spiked matrix, with row operations applied to produce zeros in row i, then column i, for $i = 2, 3,..., n$. Show that this changes entries and produces fill-in in only the upper triangular parts of the spikes.

9 Implementing Gaussian elimination: ANALYSE with numerical values

We describe the implementation of the ordering and solution algorithms of Chapter 7, emphasizing the use of a dynamic data structure for the pivoting strategy of Markowitz (1957). We also study the use of static data structures for numerical factorization and solution. We examine hybrid techniques that combine sparse and full matrix manipulation, methods based on code generation, and the use of drop tolerances.

9.1 Introduction

We have shown in the two previous chapters that a significant reduction in floating-point computation results from the use of a good pivotal ordering in Gaussian elimination. In this chapter and the next we discuss implementation strategies for achieving this goal without unreasonable overheads.

As in Chapter 5, we consider the three phases

ANALYSE, where the sparsity pattern is analysed to find an ordering that preserves sparsity;

FACTORIZE, where numerical factors are computed using information from ANALYSE; and

SOLVE, where the factors are used to solve a set of equations.

Implementations can be divided broadly into two categories according to whether or not ANALYSE works on actual numerical values. This chapter is concerned with the case where it does and the next chapter with the case where it does not. Having numerical values during ANALYSE allows numerical stability tests to be performed and leads to matrix factors as a by-product, so that a separate FACTORIZE phase is not required until another matrix with the same pattern needs to be processed. Because a numerical factorization is performed at the same time as the analysis, the first phase is sometimes termed ANALYSE-FACTORIZE.

We showed in Chapter 6 how a general unsymmetric system could be permuted to block triangular form. This permutation is often considered to be part of the ANALYSE phase. We do not consider it further in this chapter because we assume that, if desired, such a step has already been performed and that we are now working with one of the irreducible blocks on the diagonal of the block triangular form. This assumption does not preclude the use of the techniques of this chapter on a reducible matrix, although when the matrix is reducible the initial formation of the block triangular form is likely to be more efficient.

Once a matrix has been analysed, the positions of all the fill-ins for another matrix having the same pattern will be known provided the same pivotal sequence can be employed. Thus our position in this chapter is that dynamic data structures are needed for ANALYSE but not for FACTORIZE or SOLVE. This, together with the work involved in the actual pivot choice, leads to ANALYSE times that are typically three to seven times greater than FACTORIZE times, see Table 5.6.1.

We emphasize the pivotal strategy of Markowitz and its variations because this has been very successful (see Chapter 7). Our description of Markowitz ANALYSE in Section 9.2 includes an emphasis on the avoidance of 'traps' that can lead to unnecessarily slow execution. The FACTORIZE and SOLVE phases, both of which use static data structures created by ANALYSE (whether or not it uses Markowitz pivoting), are considered in Sections 9.3 and 9.4. In Section 9.5 we consider the benefits of switching to code for dense matrices towards the end of the calculation and in Sections 9.6 and 9.7 we discuss the generation of directly executable or interpretative code for FACTORIZE and SOLVE. This avoids or reduces looping control and permits fast execution on scalar machines, but at the expense of significant storage for the code. In Section 9.8 we discuss performing interchanges within FACTORIZE, and in Section 9.9 we consider discarding numerically small entries, which increases the sparsity of the factors at the expense of having to iterate in SOLVE.

The unifying theme of this chapter is a willingness to tolerate an expensive ANALYSE for the sake of good sparsity preservation and rapid subsequent FACTORIZE and SOLVE. For matrices that are very unsymmetric there may be no good alternative, but for symmetric or nearly symmetric matrices the rapid ANALYSE of the next chapter, based on the sparsity pattern alone, is attractive.

9.2 Markowitz ANALYSE

We begin our detailed discussion of the implementation of the ANALYSE phase for unsymmetric matrices by recalling some of the remarks of Chapter 7. In this section we will consider the implementation of the ordering strategy of Markowitz (1957) with threshold pivoting for numerical stability. That is, we will choose a_{ij} as pivot if it minimizes the quantity (the **Markowitz count**)

$$(r_i - 1)(c_j - 1) \qquad (9.2.1)$$

over all the entries of the reduced matrix that satisfy the inequality

$$|a_{ij}^{(k)}| \geq u \max_{l \geq k} |a_{il}^{(k)}| \qquad (9.2.2a)$$

or the inequality

$$|a_{ij}^{(k)}| \geq u \max_{l \geq k} |a_{lj}^{(k)}| , \qquad (9.2.2b)$$

where r_i is the number of entries in row i of the reduced matrix (the **row count**), c_j is the number of entries in column j (the **column count**), and u is a preset and fixed parameter in the range $0 < u \leq 1$. Most of our discussion applies also to algorithms that use other formulae involving r_i and c_j.

This pivotal strategy might be implemented by looking at each entry in turn but would then be prohibitively expensive. For example, suppose a matrix of order 10 000 has 40 000 entries and the chosen pivot has counts $r_i = c_j = 3$, then we will have performed 40 000 tests for a step that involves only 10 floating-point operations. It is, however, practical to store all the counts r_i and c_j, and indeed some storage schemes (see Section 2.11) need them anyway. These counts were used by Curtis and Reid (1971a) who proposed that the rows and columns should be searched in order of increasing counts. They terminate the search as soon as an entry is found that satisfies inequality (9.2.2a) and has a Markowitz count (9.2.1) that is as low as any unsearched entry could possibly have. Although (see Section 7.2) it is not necessarily true that the chosen pivot will be in a row with minimum row count or a column with minimum column count, we can expect the row and column counts of the pivot to be near their minima.

It is important to have an efficient way to access the rows and columns in order of their counts. A pocket sort (Knuth 1973, pp. 170-178) permits n numbers, all in the range 1 to n, to be sorted in $O(n)$ operations (Exercise 9.1), but even this would be an intolerable burden if performed at every step. Returning to the example at the beginning of the last paragraph, it would need about 20 000 operations for a step involving 10 floating-point operations. Therefore the sort should be performed once at the beginning and a data structure that allows rapid updating of the sorted information at each step should be used. One possibility involves storing all rows (and columns) with the same number of entries in a set of doubly-linked lists (chains) with an array of pointers to the heads of the lists. We illustrate this scheme in Table 9.2.1 for the matrix of order 6 shown in Figure 9.2.1.

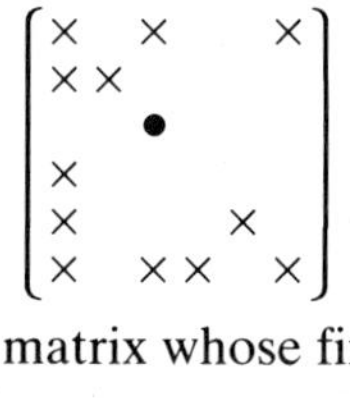

$$\begin{bmatrix} \times & \times & & \times & & \\ \times & \times & & & & \\ & & \bullet & & & \\ \times & & & & & \\ \times & & & & \times & \\ \times & \times & \times & & \times & \end{bmatrix}$$

Figure 9.2.1. A sparse matrix whose first pivot is shown as ●.

It is possible to reduce the pointer storage from six to five vectors of length the matrix order, n, by using only one header-pointer array and

Subscripts	1	2	3	4	5	6
Row counts	3	2	1	1	2	4
Forward links	0	0	4	0	2	0
Backward links	0	5	0	3	0	0
Header pointers	3	5	1	6	0	0
Column counts	5	1	3	1	1	2
Forward links	0	4	0	0	2	0
Backward links	0	5	0	2	0	0
Header pointers	5	6	3	0	1	0

Table 9.2.1. Linked-list storage of row and column counts
for the matrix of Figure 9.2.1.

linking each chain of rows with the corresponding chain of columns (see
Exercise 9.2). Notice that since all the entries are integers in the range 1 to
n, short integers can be used when available without unduly restricting the
size of problem that can be handled. For example, with 16-bit integers, n is
restricted to 32 767.

It is very easy to use the data structure illustrated in Table 9.2.1 to
discover which columns have a specified number of entries. In our
example, for the singleton columns we find the chain 5, 2, 4 by beginning
with column header pointer 1 (value 5) and continuing through the column
forward links until a zero is found (marking the end of the chain).

It is equally easy to update the structure after an elimination. The pivot
row and column are removed from their chains, which is simple to do
because of the forward and backward links. Any row or column which has
had its number of entries altered is similarly removed from its chain and is
then placed at the head of the chain corresponding to its new number of
entries. The Fortran code necessary to do this is the object of Exercise 9.3.
If the entry (3,3) of the matrix of Figure 9.2.1 is chosen as pivot, the
resulting data structure for the row and column counts of the reduced
matrix is as shown in Table 9.2.2. Note that the locations previously used to
hold the links for the pivotal row and column are now completely free and
so could be used to hold other information (for example the step at which
the pivot was used).

The use of the data structure of Table 9.2.1 in the selection of pivots is
also straightforward. For the rest of this section we assume that the matrix
entries are held by rows (see Section 2.10 for a discussion of how such an
ordering may be constructed), although all the arguments are equally valid
if storage is by columns (usually with 'row' replaced by 'column' and
vice-versa). This means that the stability test is better done by rows, that is
it is better to use inequality (9.2.2a). Clearly any row singleton can be

Subscripts	1	2	3	4	5	6
Row counts	2	2	–	1	2	3
Forward links	5	0	–	0	2	0
Backward links	0	5	–	0	1	0
Header pointers	4	1	6	0	0	–
Column counts	5	1	–	1	1	2
Forward links	0	4	–	0	2	0
Backward links	0	5	–	2	0	0
Header pointers	5	6	0	0	1	–

Table 9.2.2. Storage of row and column counts after the pivotal step shown in Figure 9.2.1.

chosen immediately (it is impossible to obtain a Markowitz count better than zero) and, being the only entry in its row, such a singleton satisfies inequality (9.2.2a) for any valid value of u. A column singleton also has lowest Markowitz count but if we require (9.2.2a) to be satisfied we must first compare its magnitude against those of other entries in its row. Therefore, rows and columns are scanned in order of increasing numbers of entries, taking rows before columns with the same number of entries because of the greater ease with which their stability checks can be made. The first entry encountered which satisfies (9.2.2a) and has lowest Markowitz count found so far is kept as a potential pivot and our remaining problem with pivot selection lies in terminating the search.

The main problem with the termination is that (see Exercise 9.4) the scan does not guarantee that entries are encountered in order of increasing Markowitz count so the search cannot be stopped as soon as an entry satisfying (9.2.2a) is found. We are, however, searching the rows and columns in order of row/column counts and so at any stage we have a bound on the count of all unsearched entries. Thus, if we are about to search rows with r_i entries, the minimum possible later Markowitz count is $(r_i-1)^2$ since all columns with less than r_i entries have already been searched. Since we search rows before columns, the minimum count possible among unsearched entries when we are searching a column with c_j entries is $(c_j-1)c_j$. As soon as the count of our potential pivot is less than or equal to this minimum possible count, we terminate the search.

It might be thought that this search is likely to be very long but our experience (Duff 1979) is that usually the pivot is chosen after looking at only a few rows and columns. Typical figures are shown in Table 9.2.3. It is possible, however, for the search to be long (see Exercise 9.5). Zlatev (1980) has proposed searching the matrix by rows in order of increasing

Order of matrix	147	57	292
Number of nonzeros	2449	281	2208
Average number of rows and columns searched	8.2	2.8	3.1

Table 9.2.3. Examples of the amount of searching during pivot selection.

row counts and restricting the search to a specified number of rows (recommending three). Although this (for similar reasons to the 'min col. in min row' strategy of Section 7.4) will often yield worse fill-in than a strict implementation of the Markowitz criterion, it will not invariably do so and will always require a low search time. For example, on a five-diagonal matrix of order 2000 with bandwidth 42 and with diagonal entries 4 and off-diagonal entries -1, a strict Markowitz ordering required 51.0 seconds on an IBM 3081K and gave factors with 91 487 entries while a restricted Markowitz ordering required 25.9 seconds and gave factors with 90 797 entries. It is easy, however, to 'hedge one's bets' and the current (1984) version of the Harwell code MA28 allows the user to choose between the two pivoting strategies; by default strict Markowitz is used, partly for compatibility with the original code and partly because of our usually good experience with this strategy (see Table 9.2.3).

Another minor variant of these pivoting schemes, again suggested by Zlatev (1980), is to replace choosing the first pivot candidate with lowest Markowitz count by choosing the one with lowest count that satisfies the stability test by the best margin. This gives a little added stability in some instances for only a slight complication to the pivot selection process. It has again been adopted as an option in MA28.

After selecting the pivot, the elimination operations must be performed on the reduced matrix. For storage by rows, the structure of the matrix is scanned by columns to find out which rows are involved in this elimination step. Then appropriate multiples of the pivot row are subtracted from each of these rows in turn using the first algorithm of Section 2.4 for the addition of sparse vectors. The first algorithm of Section 2.4 is preferable to the second since the pivotal row remains the same throughout the elimination step and this can be exploited by a minor variation of the code of Figure 2.4.1. Fill-in is accommodated as illustrated in Section 2.4. The column structure is adjusted in a similar manner.

The pivotal row is a row of U and must be stored. The multiples of it that are added to other rows are entries of a column of L and they, too, must be stored if it is desired to solve for further right-hand side vectors or if the right-hand side vector is not available at the time of factorization. There are two convenient possibilities for storing the entries of a column of L:−

(i) they may be stored as a sparse vector, or

(ii) each entry may be held with the corresponding row of **U**, provided it is distinguished from the row in some way.

The first choice eventually leads to **L** being stored by columns and the second leads to **L** being stored by rows. Harwell's MA28 uses the second choice, keeping the elements of **L** ahead of the elements of **U** in each row.

Provided no advantage is taken of 'accidental zeros', that is provided each zero produced by exact cancellation is stored as an entry with the value zero, the resulting structure for **L\U** provides a suitable framework for FACTORIZE. If the same pivotal sequence is applied to a matrix with the same initial pattern, there will be room for all the fill-ins that result. The nonzeros of the matrix must be loaded into the corresponding entries in the structure of **L\U** and the fill-in positions must be set initially to zero. This is discussed further in the next section.

In the implementation of ANALYSE, there are many possibilities for saving storage. The most important is that after a row (column) is pivotal it need not be accessed again so that the corresponding entries of most of the work arrays (see, for example, Table 9.2.2) are free for other purposes. This is convenient for storing permutations and gives useful workspace on termination of the main loop. The space thus freed can often be simply reused but sometimes it is necessary to flag these portions of arrays to identify their status.

Although we have examined the implementation of the Markowitz ANALYSE in some detail, there are further programming and data structure considerations which are outside the scope of this book. The reader is referred to Duff (1977c), Duff and Reid (1982), or George and Liu (1979b) for a more detailed discussion.

9.3 FACTORIZE

In the FACTORIZE phase we assume that the pivotal sequence and the data structure of **L\U** (including all fill-in locations) are known from ANALYSE. The first step in FACTORIZE is to map the new set of numerical values for **A** into the appropriate positions in **L\U**, with zero numerical values embedded where the fill-ins will occur. We will consider the case where **L\U** is stored by rows, with each row of **L** preceding the corresponding row of **U**. An array of length n contains pointers to the starts of the rows. For FACTORIZE we also need another array of length n which points to the diagonal positions in all the rows. Each corresponds to the position of the first entry in a row of **U**.

It is now an easy matter to use the Doolittle implementation of Gaussian elimination (Figure 3.7.2) where each row of **A** in sequence is transformed to a row of **L\U**. Gustavson (1972) attributes this critical step to

W. F. Tinney's group from the Bonneville Power Administration and to Chang (1969). It makes use of a vector of length n as discussed in Chapter 2. Since the information in this vector need be preserved only for the transformations on a single row, we call this a temporary vector in the rest of this section.

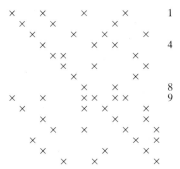

Figure 9.3.1. Stage 9 of the reduction of a 15×15 matrix. The active rows are 1, 4, 8, and 9.

Assume that the decomposition has been performed on the first $k-1$ rows. The status of the partially reduced matrix is illustrated in Figure 9.3.1 (where $k = 9$). The entries to the left of the diagonal (that is, the entries of row k of L) determine which rows of U (termed the active rows) will be used to transform row k. These are displayed in Figure 9.3.1 in full form, but are actually accessed using the packed form. For each active row, the extra integer array of pointers to the diagonal entries allows us to access the U portion of the row directly.

The first step is to take the packed form of row k and expand it into the temporary vector of length n. In turn, each multiplier is calculated and stored, and this multiple of the corresponding packed row of U is added to the temporary vector.

Since all fill-in has been anticipated, no new entries are created in row k of L\U. Observe that it is crucial that the column indices are ordered in the L part; for example, if row 4 in Figure 9.3.1 were used to modify row 9 before row 1, the multiplier (9,4) would be wrong. On the other hand, the order of the off-diagonal entries in the rows of U is irrelevant.

After all active rows have been processed, the revised row k is returned to its packed form and processing of row $k+1$ is commenced.

Figure 9.3.2 displays Fortran code to perform the sequence of operations which have just been described. In this figure, ILPTR (K) is the location of the first column index of row k of L, while IUPTR (K) is the location of the first column index of row k of U (the diagonal position in the k-th row of L\U). The column indices themselves are stored in JCN.

```
C Place row k in array W
      DO 10 KK = ILPTR(K), ILPTR(K+1)-1
        W(JCN(KK)) = VAL(KK)
   10 CONTINUE
      DO 30 KK = ILPTR(K), IUPTR(K)-1
C Determine the next active row,
C compute multiplier, store it in L
        J = JCN(KK)
        ALPHA = -W(J)/VAL(IUPTR(J))
        W(J) = ALPHA
C Add in multiple of row j of U
        DO 20 JJ = IUPTR(J)+1, ILPTR(J+1)-1
          W(JCN(JJ)) = W(JCN(JJ))+ALPHA*VAL(JJ)
   20     CONTINUE
   30 CONTINUE
C Now set row k of L\U back into packed form
      DO 40 KK = ILPTR(K), ILPTR(K+1)-1
        VAL(KK) = W(JCN(KK))
   40 CONTINUE
```

Figure 9.3.2. Code for converting a row of **A** to a row of **L\U**.

Observe that the innermost loop (discussed in Section 5.7) is of the form of the 'DO 20' loop in Figure 9.3.2.

It is evident that we have made extensive use of the prior knowledge of the pattern of **L\U**. It is possible to use a similar scheme given only the pivotal sequence with the pattern of **L** and **U** being generated as the elimination proceeds (see Exercise 9.7).

Because the pivotal sequence that we are now using in FACTORIZE was chosen from a matrix with different numerical values, it may now be unstable. We can check stability by looking directly at the generated entries, or we can use the bound of Erisman and Reid (1974) or that of Barlow (1986), see Section 4.5, both of which avoid any overheads in the innermost loop but sometimes give a severe overestimate for the growth. If the factorization is found to be unstable, it is usual to apply ANALYSE to the new matrix. There are two other alternatives. Stewart (1974) has suggested simply changing very small pivots and then correcting later with the Woodbury formula (see Section 11.4). Stewart's suggestion leads to overheads in each subsequent SOLVE and so should not be used when many SOLVE calls are expected. We are not familiar with any codes that have implemented Stewart's suggestion. The second alternative is to allow some change in the sparsity pattern during FACTORIZE. This is discussed in Section 9.8.

9.4 SOLVE

We now consider the use of the factors generated by the algorithm of Section 9.3 for the solution of a set of linear equations by forward substitution and back-substitution. This is the simplest and fastest phase of the computation, although it is important to do it efficiently since there are several applications where many solutions (tens or hundreds) are performed for each FACTORIZE, for example when using extrapolation methods to solve stiff systems of ordinary differential equations (Duff and Nowak 1987). We also consider taking advantage of sparsity in the right-hand side vector.

Other considerations may hold if the factors are stored differently. Examples are the frontal and multifrontal schemes of Sections 10.5 to 10.8, where the appropriate variations of SOLVE are discussed.

The amount of arithmetic needed to solve a triangular system (with a dense right-hand side) is clearly proportional to the number of entries in the triangular matrix, each being used in a single computation. So long as we can get ready access to the entries by rows or by columns, the SOLVE phase is easy to implement. We show in Figure 9.4.1 code for the solution of a lower triangular system where the factors are stored as in Section 9.3. The matrix is assumed to be unit triangular and the computation is equivalent to subtracting a sparse inner product from each component of $\mathbf{x}$ in turn. Observe that this is simpler than the inner product of two packed vectors (see Section 2.3) since $\mathbf{x}$ is in full form. This formulation has the advantage of writing to each component of $\mathbf{x}$ just once, but it is not possible to take advantage of any zeros in the right-hand side since any test would have to be performed within the inner loop and, on many computing systems, would be as expensive as the actual numerical calculation.

```
        DO 20 K = 2,N
C Scan row k of L taking inner product
C with vector x
            PROD = 0.0
            DO 10 KK = ILPTR(K), IUPTR(K)-1
                PROD = PROD + VAL(KK)*X(JCN(KK))
   10       CONTINUE
            X(K) = X(K) - PROD
   20   CONTINUE
```

Figure 9.4.1. Code for solving a system with unit lower triangular matrix stored by rows.

If access to $\mathbf{L}$ is by columns, however, then the inner loop consists of adding a multiple of a sparse column to the right-hand side and the whole inner loop may be avoided for zero components (see Section 7.9). This is implemented by multiplying the packed form of the sparse vector by the appropriate scalar and then using the indices to add the results to the proper positions in the solution vector. The only defect with this scheme is that a component of the right-hand side is updated for every multiplication in the inner loop. The code for this is quite simple (see Exercise 9.8).

In the back-substitution phase the solution will be full unless the matrix is reducible, see Section 7.9. Hence in general there is no sparsity in the solution vector to exploit in this phase, but there are occasions when only a partial solution is wanted. In this case a substantial saving can be made if U is held by rows and the required components are all near the bottom of the solution vector.

Therefore, to take advantage of both zeros in the right-hand side and partial solutions, it is advantageous to hold L by columns and U by rows. In the symmetric case, if we store L by columns, we implicitly store $U = L^T$ by rows.

Note that, although the storage needed for the factors of a symmetric matrix is about half that needed if the symmetry is ignored, the work needed to solve the system by forward and back-substitution is about the same since the factor is used twice.

So far in this chapter, we have not considered permutations because this would have obscured the main points of our discussion. However, in most cases permutations will be required and we examine their role here since their implementation can have a significant effect on the efficiency of the solution. We consider the solution of the equation

$$L y = P b, \qquad (9.4.1)$$

followed by the solution of the equation

$$U z = y, \qquad (9.4.2)$$

and the calculation

$$x = Q z, \qquad (9.4.3)$$

where P and Q are permutation matrices, held as ordering vectors (see Section 2.14). The solution proceeds as follows:

(a) permute b to $P b$ and store it in the full-length vector w,

(b) solve $L y = w$ and $U z = y$, and

(c) permute z to $Q z$ and store the result in x.

Although it appears that five arrays (for b, w, y, z, and x) are required, it is normal to hold w, y, and z in the same array. Therefore three arrays are needed if b is to be kept and only two are needed if the solution vector x overwrites b.

If we hold the permutations as sequences of interchanges (see Exercise 9.9), it is possible to use a single array for the solution process since the sequence of interchanges can be used to reorder b and z in place.

The scheme of the last two paragraphs requires that permuted indices are stored for L and U. An alternative is to store the original indices (see

Exercise 9.10). In this case, the working vector is not permuted initially but is permuted between the forward substitution and the back-substitution.

Whatever scheme is used internally by SOLVE, it is of course important that the user of the software is not involved in the level of detail that we have discussed in this chapter. A user should have to supply only the vector **b** and should receive the solution **x**, rather than a permutation of it.

9.5 Switching to full form

From our considerations so far, it is clear that complicated data structures and programming are needed to take advantage of sparsity. On most computer architectures an inner loop involving indirect addressing of the form X(JCN(KK)), common in sparse codes (see, for example, Figure 9.4.1), will run two to three times slower than the corresponding loop with direct addressing (as used for full matrices). With some vector processors, for example the CRAY-1 or CYBER 205, the difference can be much more marked since loops containing indirect addressing do not vectorize. Interestingly, some recent vector computers (for example, the CRAY X-MP and the Fujitsu FACOM VP) have reduced this degradation by providing hardware instructions for indirect addressing. For very sparse systems the overheads are clearly insignificant compared with the saving from exploiting sparsity, but equally clearly it would not be sensible to solve a full system with a sparse code. However, this is exactly what happens towards the end of Gaussian elimination when fill-in causes the reduced submatrices to become increasingly full. We illustrate this in Table 9.5.1, where we show examples of the densities of reduced submatrices of various orders.

Order	1176	292	130	541
Nonzeros	18552	2208	1282	4285
Order of reduced submatrix				
200	0.32	0.04	–	0.13
150	0.44	0.08	–	0.26
100	0.68	0.18	0.10	0.42
50	1.0	0.50	0.23	0.75
20	1.0	1.0	0.72	1.0
10	1.0	1.0	1.0	1.0

Table 9.5.1 Examples of the densities of the reduced submatrices in Gaussian elimination. A full matrix has density 1.0.

Towards the end of the elimination we are thus dealing with matrices which would be more efficiently factorized by full matrix code. This suggests using a hybrid approach in which a switch is made from the use of sparse code to full code at a point determined by the order and density of the reduced submatrix and by the computing environment.

Dembart and Erisman (1973) discussed such hybrid codes and indicated the possible savings. The first code to employ such a switch was included in the SLMATH package of IBM (1976); it switched at 100 per cent density in the ANALYSE entry only. Recent modifications to the Harwell code MA28 employ the switch in all phases and the results of Duff (1984a) show that significant gains can be obtained when using the CRAY-1, with overall reductions in time of up to factors of about four, two, and two on the ANALYSE, FACTORIZE, and SOLVE phases, respectively.

The best point for the switch to full form depends both on the particular computer architecture and the relative importance of storage and time. The data of Duff (1984a) in Tables 9.5.2, 9.5.3, and 9.5.4 illustrate some of these issues.

Order	1176	292	130	541
Nonzeros	18552	2208	1282	4285
Density for switch				
Never	4.14	0.46	0.16	1.21
1.0	4.28	0.45	0.16	1.18
0.8	2.77	0.45	0.16	1.16
0.6	2.87	0.42	0.15	1.14
0.4	3.37	0.43	0.15	1.18

Table 9.5.2. Times (IBM 3033 seconds) for ANALYSE when the switch to full form is made at different densities.

Table 9.5.2 shows times on an IBM 3033 for the ANALYSE step in MA28 for the problems of varying densities given in Table 9.5.1. Table 9.5.3 shows the corresponding times for the same problems on a CRAY-1. Because of the more pronounced degradation in performance for indirect addressing on the CRAY-1, it is not surprising that there are gains at much lower densities on this computer. On a vector machine that has hardware for gather and scatter operations, we expect the relative performance to be similar to that of the IBM 3033.

There is a potential storage price to pay for switching to full form at lower densities. Table 9.5.4 shows total storage (both integer and real) for the LU factors. Storage for other arrays of length the order of the system,

Order	1176	292	130	541
Nonzeros	18552	2208	1282	4285
Density for switch				
Never	2.59	0.28	0.11	0.75
1.00	2.61	0.28	0.10	0.74
0.80	1.54	0.25	0.10	0.72
0.60	1.43	0.23	0.10	0.70
0.40	1.06	0.20	0.10	0.66
0.20	0.81	0.17	0.09	0.59
0.05	1.03	0.25	0.09	0.81

Table 9.5.3. Times (CRAY-1 seconds) for ANALYSE when the switch to full form is made at different densities.

Order	1176	292	130	541
Nonzeros	18552	2208	1282	4285
Density for switch				
Never	42 354	10 430	2 662	25 142
1.00	42 293	10 264	2 496	24 469
0.80	37 255	9 865	2 421	24 249
0.60	39 655	9 872	2 481	24 301
0.40	46 874	10 745	2 780	22 301
0.20	58 158	14 258	4 396	36 029
0.05	91 430	36 298	16 903	81 976

Table 9.5.4. Storage requirements on the CRAY-1 when the switch to full form is made at different densities.

for example permutations, is the same for all densities and is not included. If the switch is made when the matrix is dense, no extra storage is needed for the reals and storage is saved for integers. Data is shown only for the CRAY-1; we would get less saving from integer storage on the IBM if half-word integers were in use.

If a symmetric and positive-definite system is solved with an ANALYSE like that of Section 9.2, there is scope for substantial savings by switching to full code since no further pivoting will be needed after the switch is made. A multifrontal approach (see Section 10.8) or a similar technique always works with full submatrices, so needs no special switch. Such a method is advantageous on a computer with high overheads for indirect addressing, but we are not able to draw general conclusions.

9.6 Loop-free code

One of the early approaches to dealing with overheads in sparse matrix computation involved using the ANALYSE step to develop a computer program specifically tailored to the given sparse matrix. Gustavson, Liniger, and Willoughby (1970) discuss the approach GNSO (GeNerate and SOlve) whereby the sparsity pattern is analysed to determine precisely the numerical computations required to factorize and solve a system of equations with the given sparsity pattern. Their program then produced a Fortran program as output.

```
 × × ×            × × ×            1 2 3
   ×                ×                4
 ×   ×            × × ×            5 7 6

 pattern          pattern          positions of
 of A             of L\U           entries
```

Figure 9.6.1. Generation of loop-free code.

```
VAL (5)  =  VAL (5) /  VAL (1)
VAL (7)  =  -VAL (2) *  VAL (5)
VAL (6)  =  VAL (6) -  VAL (3) *VAL (5)
VAL (7)  =  VAL (7) /  VAL (4)
```

Figure 9.6.2. Loop-free code for factorizing the Figure 9.6.1 matrix.

A simple illustration of this idea is given in Figures 9.6.1 and 9.6.2. Similar code may be generated for SOLVE (Exercise 9.11). The features of the generated program are that it is a sequence of statements with no loops or branches (this is why it is called **loop-free** code), so it has no looping overheads, but it can be very long. Compilers tend to have difficulty with such code, so one of the early changes was to generate machine code directly, rather than through Fortran.

Even then the program gets so long for problems of even modest order that the approach becomes impractical. Willoughby (1971) reports on code of length 1.3 Megabytes for a matrix of order 1024 with 15 000 nonzeros, indicating that auxiliary storage, with the associated penalties in performance, was required at that time. Erisman (1972) reported a larger problem where the generated code overflowed onto a second disk. Since loop-free codes work only with symbolic information at the ANALYSE stage, there are stability questions as well.

On the positive side, this approach has allowed the only successful implementation of variability-type ordering (Section 7.10) of which we are aware. Calahan (1982) has recently revisited this approach to try to adapt it to vector computing for some classes of problems.

9.7 Interpretative code

A rather natural extension of the loop-free code idea is to abstract the basic operations from loop-free code and build them into data for a simple Fortran program. Each different computational construct is identified by an operation code which, together with the indices, become data for the program.

```
      K = 1
    5 GO TO (10, 20, 30, 40), INC(K)
C Do operations of type 1
   10 VAL(INC(K+1)) = VAL(INC(K+1))/VAL(INC(K+2))
      K = K+3
      GO TO 5
C Do operations of type 2
   20 VAL(INC(K+1)) = -VAL(INC(K+2))*VAL(INC(K+3))
      K = K+4
      GO TO 5
C Do operations of type 3
   30 VAL(INC(K+1)) = VAL(INC(K+1))
      *              - VAL(INC(K+2))*VAL(INC(K+3))
      K = K+4
      GO TO 5
C Terminate (type 4)
   40 ....
```

Figure 9.7.1. Illustrative Fortran code driven by generated data.

$$INC = 1\ 5\ 1\ \mathbf{2}\ 7\ 2\ 5\ \mathbf{3}\ 6\ 3\ 5\ 1\ 7\ \mathbf{4}\ 4$$

Figure 9.7.2. Data for the program in Figure 9.7.1, with the operator codes shown in bold.

Figures 9.7.1 and 9.7.2 contain an illustrative program and the appropriate data to perform the factorization of the sample problem of Section 9.6. Again, a similar sequence of operation codes may be generated for SOLVE (see Exercise 9.12).

This idea was presented by Chang (1969) and was the basis of the electrical network frequency domain analysis program described by Erisman (1972). Some variations of the approach are discussed by Dembart and Erisman (1973), including the use of procedure calls to replace blocks of code that occur frequently and hybrid techniques that use loop-free code only when the matrix is very sparse.

We would expect reduced storage compared with loop-free code because only the indices, rather than the entire statement, need to be stored. On the other hand, we also expect somewhat slower execution because of the need to interpret the code.

We will illustrate these comments by a comparison of the code GNSOIN of Gustavson, which is a later version of GNSO (Gustavson *et al.* 1970) for generating loop-free code, the code TRGB of Bending and Hutchison (1973), which generates interpretative code and is used in chemical engineering calculations, and the Harwell code MA28, which uses conventional looping. We show the results for just one matrix in Table 9.7.1 because the code produced by GNSOIN on larger or denser matrices was too long for our Fortran compiler. The results support our remarks, namely that the loop-free code is very fast but requires a large amount of storage while the interpretative code provides a compromise.

	FACTORIZE (IBM 3033 seconds)	SOLVE (IBM 3033 seconds)	Storage (K-bytes)
Loop-free (GNSOIN)	0.006	0.002	85
Interpretative (TRGB)	0.025	–	20
Looping (MA28)	0.051	0.008	6

Table 9.7.1. Comparison of different approaches on a matrix of order 199 with 701 nonzeros. GNSOIN times do not include compilation and TRGB does not have a separate SOLVE entry.

More recently the sparse solver from the TRAFFIC program, described by Erisman (1972) and based on interpretative code, was replaced by a complex arithmetic adaptation of SPARSPAK (Chu, George, Liu, and Ng 1984). The result was a simple and portable program replacing a very complex and machine-dependent program. Comparable computational speeds were achieved with greatly reduced storage requirements. This illustrates the tremendous progress that has been made with general-purpose codes over the years and underscores our comments in Chapter 5 that these general-purpose codes provide a good starting point for new applications. Loop-free and interpretative codes remain of interest for special-purpose adaptation. The investigation of Calahan (1982) suggests that they may also be useful on advanced architectures.

9.8 A posteriori ordering for stability

So far in this chapter we have assumed that FACTORIZE uses a pivotal sequence chosen during ANALYSE, and our principal suggestion for the case where the sequence is unstable for the new numerical values was to run ANALYSE afresh on the new matrix. Since this is an expensive procedure, we now explore the possibility of making minor changes to the ordering to ensure numerical stability. We call such an approach an **a posteriori ordering** for stability. Other examples of this approach are discussed in Chapter 10.

Given a good ordering for sparsity, there is no loss of generality in assuming that the matrix is already permuted accordingly. Thus we assume that solving the system

$$\mathbf{Ax} = \mathbf{b} \tag{9.8.1}$$

by pivoting straight down the diagonal is satisfactory for preserving sparsity. A posteriori pivoting control consists of testing, at each stage, the size of the modified diagonal element. For example, we may use the usual threshold criterion

$$|a_{kk}^{(k)}| \geq u \max_{i>k} |a_{ik}^{(k)}| \tag{9.8.2}$$

and, if necessary, perform a row interchange to ensure its satisfaction. Unfortunately, even a few such interchanges may wreck the sparsity of the factors.

The extreme case of this situation is seen where the numerical values are ignored completely in the initial pivot selection, and then a full partial pivoting strategy is enforced, that is $u = 1$ in inequality (9.8.2). This was the strategy used by the code NSPIV of Sherman (1978). For the case where significant numerical pivoting is required, such as the fourth problem in Table 9.8.1, very poor performance results. Sherman has since written a version called NSPFAC that allows $u < 1$ and is a significant improvement over NSPIV although it can still perform poorly relative to MA28 for problems requiring significant numerical pivoting.

Order	147	199	822	541
Nonzeros	2441	701	4790	4285
Entries in factors				
MA28 (L\U)	5095	1478	6653	13660
NSPIV (U)	4836	1622	8345	51358
Time (IBM 370/168 secs) for one-off				
MA28	1.28	0.23	1.53	4.09
NSPIV	1.60	0.24	2.04	40.30

Table 9.8.1. Comparison between a Markowitz code, MA28, and an a posteriori code, NSPIV. Note that the storage figures for NSPIV include U only since L is not kept.

There are some compensations with a posteriori codes. They are much shorter than more general codes, can be implemented with much simpler data structures, and, unless there is much more fill-in, can be competitive for 'one-off' cases. This latter point is illustrated in Table 9.8.1.

9.9 The use of drop tolerances to preserve sparsity

One of the main problems associated with the use of direct methods for solving sets of sparse linear equations lies in the increase in the number of entries due to fill-in. This increase is particularly marked when solving very regular sparse problems such as those obtained from discretizing partial differential equations. Indeed, the storage of the matrix factorization often limits the size of problem which can be solved. A simple way to extend this limit is to remove from the sparsity pattern (and from any subsequent calculations) any entry which is less than some prespecified absolute or relative tolerance, usually called a **drop tolerance.** Thus, if during the process of Gaussian elimination, an intermediate value satisfies the inequality

$$|a_{ij}^{(k)}| < TOL_a \qquad (9.9.1)$$

or the inequality

$$|a_{ij}^{(k)}| < TOL_r \times \max_l |a_{lj}^{(k)}|, \qquad (9.9.2)$$

for a preset non-negative value of TOL_a or TOL_r, then $a_{ij}^{(k)}$ is dropped from the sparsity pattern and subsequent consideration. For drop tolerances to be useful, they must be set high enough to reduce substantially the number of entries in the factors.

When using drop tolerances the factorization obtained, $\bar{L}\bar{U}$ say, will be such that

$$A + E = \bar{L}\bar{U}, \qquad (9.9.3)$$

where the size of the entries of E may be significant relative to those of A. Hence the solution of

$$\bar{L}\bar{U}x = b \qquad (9.9.4)$$

may differ substantially from the actual solution to the original system, even if the problem is well-conditioned.

One way of using the **incomplete factorization** $\bar{L}\bar{U}$ to solve

$$Ax = b \qquad (9.9.5)$$

is to use it as an acceleration or 'preconditioning' to some iterative method. One of the simplest iterative schemes is that of iterative refinement (also called defect correction), defined for $k = 0, 1, 2,...$ by the equations

$$r^{(k)} = b - Ax^{(k)}, \qquad (9.9.6a)$$

$$\bar{L}\bar{U}\,\Delta x^{(k)} = r^{(k)}, \qquad (9.9.6b)$$

and

$$x^{(k+1)} = x^{(k)} + \Delta x^{(k)}, \qquad (9.9.6c)$$

where the starting iterate $x^{(0)}$ is usually taken as 0 so that $x^{(1)}$ is the solution of equation (9.9.4). Iterative refinement was discussed in Section

4.11 in connection with allowing for roundoff errors in **LU** factorization. The iteration (9.9.6) will converge so long as the spectral radius of the matrix

$$\mathbf{I} - (\bar{\mathbf{L}}\bar{\mathbf{U}})^{-1}\mathbf{A} \qquad\qquad (9.9.7)$$

is less than 1.0 (see Exercise 9.13) so that the error matrix **E** in equation (9.9.3) can have quite large entries. The rate of convergence is, however, dependent on the value of this spectral radius and can be quite slow. We illustrate the performance of this technique in Table 9.9.1 (Duff 1985), where we have used the Harwell code MA28.

Value of drop tolerance, TOL_a	0.	10^{-4}	10^{-2}	0.05	0.1
Storage for factors	48 055	42 701	38 911	28 325	25 130
Time for factorization	2.18	1.72	1.41	0.53	0.38
Number of iterations	1	2	3	8	–
Time for solution	0.066	0.094	0.121	0.235	–

Table 9.9.1. Iterative refinement with drop tolerances on a five-diagonal system of order 1000 with semibandwidth 100. Times are in seconds on an IBM 3081K. The case in the last column did not converge.

Drop tolerances were incorporated in the sparse matrix code SSLEST (Zlatev, Barker, and Thomsen 1978) and were later used with iterative refinement by the codes Y12M (see Østerby and Zlatev 1983) and a 1983 version of the Harwell code MA28. Although there has been some discussion of the use of iterative methods which are more rapidly convergent than iterative refinement, for example Chebyshev acceleration, we are not aware of any codes for the solution of general systems which use these more powerful methods.

In the case of symmetric positive-definite matrices, however, the use of drop tolerances is much better established. Here the iterative technique employed is normally conjugate gradients and, with reasonable care in forming the preconditioning matrix, convergence can be guaranteed. Munksgaard (1980) maintains stability during the factorization by testing the inequality

$$a_{kk}^{(k)} \geq u \max_{j>k} |a_{kj}^{(k)}|, \qquad\qquad (9.9.8)$$

where u usually has the value 0.01. If this inequality is not satisfied, he sets $a_{kk}^{(k)}$ to $\max_{j>k} |a_{kj}^{(k)}|$. Jennings and Malik (1977) preserve positive definiteness by adding appropriate quantities to the diagonal entries $a_{ii}^{(k)}$ and $a_{jj}^{(k)}$

whenever $a_{ij}^{(k)}$ is dropped (see Exercises 9.14 and 9.15). We illustrate the performance of this approach in Table 9.9.2 with some of the results given by Munksgaard (1980). Munksgaard has implemented his scheme in the Harwell code MA31 (used for the results in Table 9.9.2), which also has an option for dynamically adjusting the drop tolerance to obtain a partial factorization which uses no more storage than that allocated by the user.

Value of drop tolerance, TOL_r	0. (Direct method)	10^{-2}	1.0	∞ (Normal CG)
Storage for factors	8254	5215	4340	3185
Time for one-off	1.00	0.71	0.79	1.31
Number of iterations	–	11	23	92
Time for subsequent solutions	0.042	0.322	0.554	1.20

Table 9.9.2. Use of drop tolerances with conjugate gradients. Matrix is of order 406 with 1904 nonzeros from a discretization of an L-shaped region. Times are in seconds on an IBM 3033.

In this section we have been concerned with dropping entries less than a given absolute or relative tolerance. It is also possible to drop any fill-in occurring outside a particular sparsity pattern (which can be that of the original matrix). Partial factorizations using this criterion, when used as a preconditioning for the method of conjugate gradients, give rise to the ICCG (Incomplete Choleski Conjugate Gradient) methods of Meijerink and van der Vorst (1977) which have proved extremely successful in the solution of positive-definite systems from partial differential equation discretizations. Further consideration of these techniques lies outside the scope of this book and the reader may consult Golub and Meurant (1983) or Hageman and Young (1981) for a more detailed discussion.

Exercises

9.1 Indicate, by means of a Fortran program, how the rows of a matrix can be ordered in $O(n)$ time so that they are in order of ascending row counts, given that the row counts are already available.

9.2 Show how the pointer storage in Table 9.2.1 can be reduced from $6n$ to $5n$ integers by holding only one header array.

9.3 Give Fortran code for removing a row from a chain of rows having NZ entries and inserting it into a chain of rows having NZ1 entries.

9.4 Give an example where scanning rows and columns in the order given by the algorithm in Section 9.2 does not access entries in order of increasing Markowitz count.

9.5 Give an example where the search strategy of Section 9.2 must search more than half the matrix before termination.

9.6 Modify the code of Figure 2.4.1 so that W is left holding the vector **y**.

9.7 Construct code for performing the operations of Figure 9.3.2 in the case where the structure of L\U is not known in advance. Assume that the entries in the rows of A are stored in order and that the L\U factorization has proceeded to row $k-1$.

9.8 Write Fortran code to solve a lower triangular system where the matrix is stored by columns. Include statements to take advantage of zeros in the right-hand side.

9.9 Express the permutation $(1,2,3,4,5) \to (3,5,1,2,4)$ as a sequence of interchanges and show that the effect of applying them to a vector is the same. Write Fortran code that uses interchanges to reorder a vector in place.

9.10 Write code to solve $\mathbf{Lx} = \mathbf{Pb}$ when $\mathbf{L}$ is unit lower triangular and $\mathbf{P}^T\mathbf{L}$ is held by columns.

9.11 Write loop-free code for SOLVE for the example in Figure 9.6.1.

9.12 Using the operation codes in Figure 9.7.1, write data for the interpretative code for the SOLVE phase for the example in Figure 9.6.1.

9.13 Show that convergence of the method outlined in (9.9.6) depends on the spectrum of the matrix (9.9.7).

9.14 Find a small quantity that when added to the diagonals $a_{ii}^{(k)}$ and $a_{jj}^{(k)}$ ensures positive definiteness, regardless of the values of the other entries, when entry $a_{ij}^{(k)}$ is dropped during the partial factorization of a symmetric positive-definite matrix.

9.15 It may be verified that the matrix

$$\mathbf{A} = \begin{pmatrix} 1 & 0.71 & \sigma \\ 0.71 & 1 & 0.71 \\ \sigma & 0.71 & 1 \end{pmatrix}$$

is positive definite for σ in the range $0.0164 < \sigma < 1$. Thus the two matrices

$$\mathbf{A}_1 = \begin{pmatrix} 1 & 0.71 & 0.02 \\ 0.71 & 1 & 0.71 \\ 0.02 & 0.71 & 1 \end{pmatrix} \text{ and } \mathbf{A}_2 = \begin{pmatrix} 1 & 0.71 & 0 \\ 0.71 & 1 & 0.71 \\ 0 & 0.71 & 1 \end{pmatrix}$$

are 'neighbours', but $\mathbf{A}_2$ is indefinite. If we were to use the cautious drop tolerance approach of Exercise 9.14 with threshold 0.025, what resulting sparse matrix would be used rather than $\mathbf{A}_2$?

10 Implementing Gaussian elimination with symbolic ANALYSE

We discuss in more detail implementation techniques for sparse Gaussian elimination, based on an analysis of the sparsity pattern without regard to numerical values. This includes a discussion of data structures for pivot selection, the use of cliques, and the use of both dynamic and static data structures. We examine the division of the solution into the distinct phases of reordering, symbolic factorization, numerical factorization, and solution, indicating the high efficiency with which these steps can now be performed. We study methods for band matrices and generalizations which include variable-band (profile) techniques and frontal methods. The further generalization to multifrontal techniques is also discussed in some depth.

10.1 Introduction

The unifying theme of this chapter is that an initial pivotal sequence is chosen from the sparsity pattern alone and is used unmodified or only slightly modified in the actual factorization. When it is applicable, this approach can exhibit a significant advantage over the methods of Chapter 9. Here ANALYSE will usually be less expensive than FACTORIZE, rather than three to seven times more expensive. Additionally, ANALYSE can usually be performed in place, so its success is not dependent on predicting the fill-in. At the completion of ANALYSE we can compute the work and storage requirements for FACTORIZE provided no numerical pivoting is performed. We show how the ordering can be modified to accommodate numerical pivoting during FACTORIZE, thereby producing an a posteriori ordering for stability (see Section 9.8).

The simplest example of this approach is the use of band and variable-band orderings (see Sections 8.3 and 8.4) for symmetric and positive-definite systems. Here diagonal pivots are unconditionally stable so the ordering can be chosen once and for all on sparsity grounds and used for any matrix whose nonzeros fit the pattern. The same is true for unsymmetric matrices that are diagonally dominant. Band-matrix forms can accommodate row pivoting by a straightforward enlargement of the bandwidth, but no such extension of the variable-band form is available. We discuss band and variable-band methods in Sections 10.2, 10.3, and 10.4.

The frontal method is a variation of the variable-band technique that was developed for finite-element problems though it is not restricted to them. It, too, is most straightforward for symmetric and positive-definite systems, but a relatively minor modification permits numerical pivoting, thereby allowing general systems to be handled. Such pivoting may lead to additional fill-in but does not necessarily do so. The frontal method is described in Sections 10.5 and 10.6.

The generalization to more than one front (Sections 10.7 and 10.8) accommodates any pivotal ordering for symmetric patterns. In fact, the use of multiple fronts permits very efficient implementations of minimum degree orderings (see Section 10.9).

The frontal approaches have the merits of permitting numerical pivoting and involving full-matrix code in the innermost loop, but at some expense in data movement. An alternative is to set up static data structures once the initial pivotal sequence has been chosen. This is discussed in Sections 10.10 and 10.11 for the case when numerical pivoting is not needed and in Section 10.12 for the case when it is needed.

In Section 10.13, we explain how the pivotal orderings of nested dissection, refined quotient tree, and one-way dissection fit into this overall pattern. In Section 10.14, we collect some remarks on the effects of new computer architectures, namely those employing vector or parallel processing.

10.2 Band methods

Fixed bandwidth methods are very straightforward to implement. We merely hold the matrix by rows or columns in a normal rectangular array. Because Fortran arrays are stored by columns, it is generally best to take the array to have as many rows as the bandwidth and as many columns as the order, rather than vice-versa. In this way the rows (or columns) are contiguous in storage, which may be important for efficiency on some computers. The alternative of having as many rows as the order and columns as the bandwidth is also used and allows access to diagonals in contiguous storage and rows (or columns) with uniform stride. These storage schemes are illustrated for a tridiagonal matrix in Figures 10.2.1 and 10.2.2. We may use a similar storage pattern for the computed $\mathbf{L}\backslash\mathbf{U}$ factorization. If the matrix is symmetric, the superdiagonal part need not be stored.

For a symmetric and positive-definite matrix, no interchanges are needed for numerical stability. Without interchanges the symmetry is preserved so an $m+1$ by n array suffices for a matrix of bandwidth $2m+1$ and order n. A choice is available (see Section 3.8) between a Choleski factorization

$$\mathbf{A} = \bar{\mathbf{L}}\bar{\mathbf{L}}^T \qquad\qquad (10.2.1)$$

$$\begin{array}{l} a_1 \; c_1 \\ b_1 \; a_2 \; c_2 \\ \quad b_2 \; a_3 \; c_3 \\ \qquad b_3 \; a_4 \; c_4 \\ \qquad\quad b_4 \; a_5 \; c_5 \\ \qquad\qquad b_5 \; a_6 \; c_6 \\ \qquad\qquad\quad b_6 \; a_7 \end{array} \quad \Rightarrow \quad \begin{array}{l} 0 \;\; c_1 \; c_2 \; c_3 \; c_4 \; c_5 \; c_6 \\ a_1 \; a_2 \; a_3 \; a_4 \; a_5 \; a_6 \; a_7 \\ b_1 \; b_2 \; b_3 \; b_4 \; b_5 \; b_6 \; 0 \end{array}$$

Figure 10.2.1. Storing a band matrix by columns.

$$\begin{array}{l} 0 \;\; b_1 \; b_2 \; b_3 \; b_4 \; b_5 \; b_6 \\ a_1 \; a_2 \; a_3 \; a_4 \; a_5 \; a_6 \; a_7 \\ c_1 \; c_2 \; c_3 \; c_4 \; c_5 \; c_6 \; 0 \end{array} \qquad \begin{array}{l} 0 \; a_1 \; c_1 \\ b_1 \; a_2 \; c_2 \\ b_2 \; a_3 \; c_3 \\ b_3 \; a_4 \; c_4 \\ b_4 \; a_5 \; c_5 \\ b_5 \; a_6 \; c_6 \\ b_6 \; a_7 \; 0 \end{array}$$

Figure 10.2.2. Storing the band matrix of Figure 10.2.1 by rows and by diagonals.

and the symmetric decomposition

$$\mathbf{A} = \mathbf{LDL}^T. \tag{10.2.2}$$

We prefer the decomposition (10.2.2) because there is no need to calculate square roots and because the algorithm does not necessarily break down in the indefinite case, although it is potentially unstable.

Another choice lies in the order in which the elimination operations are performed. In Gaussian elimination by rows (see Figure 3.8.1), step k of the factorization involves the following minor steps:

$$d_{kk} = a_{kk}^{(k)}, \tag{10.2.3a}$$

and, for $i = k+1,\ldots, \min (n, k+m)$,

$$u_{ki} = a_{ik}^{(k)}, \tag{10.2.3b}$$

$$l_{ik} = a_{ik}^{(k)}/d_{kk}, \tag{10.2.3c}$$

and

$$a_{ij}^{(k+1)} = a_{ij}^{(k)} - l_{ik} u_{kj}, \; j = k+1,\ldots, i. \tag{10.2.3d}$$

Step (10.2.3b) is included because the pivotal row must be stored temporarily for use in step (10.2.3d), although $\mathbf{U} = \mathbf{DL}^T$ as a whole is not stored. The most computationally intensive step is (10.2.3d) but the components l_{ik}, $i = k+1,\ldots$, and u_{kj}, $j = k+1,\ldots$, will be evenly or contiguously spaced in storage (depending on how the matrix is stored), so that it is easy to arrange for a vector computer to execute step (10.2.3d) rapidly.

In the Doolittle and Crout versions (Figures 3.8.2 and 3.8.3), the entries in the factors are found directly by the operations $d_{11} = a_{11}$ and, for $k = 2, 3, \ldots, n$,

$$l_{kj} = \left(a_{kj} - \sum_{i=m_k}^{j-1} l_{ki} d_{ii} l_{ji}\right)/d_{jj}, \qquad j = m_k, \ldots, k-1 \qquad (10.2.4)$$

and

$$d_{kk} = a_{kk} - \sum_{i=m_k}^{k-1} l_{ki} d_{ii} l_{ki}, \qquad (10.2.5)$$

where $m_k = \max(1, k-m)$. If we are willing to tolerate an extra multiplication in the innermost loop, these formulae may be used directly. However, it is easy to avoid the extra multiplication in the Doolittle variant by keeping

$$a_{kj}^{(j)} = l_{kj} d_{jj}, \qquad j = m_k, m_k+1, \ldots, k-1, \qquad (10.2.6)$$

in a temporary vector. At the k-th major step we have the minor steps

$$a_{kj}^{(j)} = a_{kj} - \sum_{i=m_k}^{j-1} a_{ki}^{(i)} l_{ji} \qquad (10.2.7a)$$

and

$$l_{kj} = a_{kj}^{(j)}/d_{jj} \qquad (10.2.7b)$$

for $j = m_k, m_k+1, \ldots, k-1$, and

$$d_{kk} = a_{kk} - \sum_{i=m_k}^{k-1} a_{ki}^{(i)} l_{ki}. \qquad (10.2.7c)$$

The SOLVE phase involves solving the triangular sets of equations

$$\mathbf{Lc} = \mathbf{b} \qquad (10.2.8)$$

and

$$\mathbf{L}^T\mathbf{x} = \mathbf{D}^{-1}\mathbf{c} \qquad (10.2.9)$$

There are two alternative computational sequences for the forward substitution (10.2.8):

$$c_k = b_k - \sum_{j=m_k}^{k-1} l_{kj} c_j, \qquad k = 1, 2, \ldots, n, \qquad (10.2.10)$$

where $m_k = \max(1, k-m)$ and, for $k = 1, 2, \ldots, n$,

$$c_k = b_k^{(k)} \tag{10.2.11a}$$

and

$$b_i^{(k+1)} = b_i^{(k)} - l_{ik} c_k, \quad i = k+1,\dots, \min(n,k+m), \tag{10.2.11b}$$

with $\mathbf{b}^{(1)} = \mathbf{b}$. The first sequence may be preferable if $\mathbf{L}$ is stored by rows and the second form may be preferable if $\mathbf{L}$ is stored by columns. Similar considerations apply to the back-substitution (10.2.9).

In the unsymmetric case, interchanges are not needed if the matrix is diagonally dominant and in this case the band form is preserved. If contiguously stored vectors are wanted for the inner loop of FACTORIZE, Gaussian elimination by rows is suitable for storage by rows and row Doolittle is suitable for storage by columns.

At the risk of instability, we may continue to use these methods for matrices that are symmetric but not definite or are unsymmetric but not diagonally dominant. It is not difficult to check for instability by monitoring the size of each number computed. Alternatively we may use row interchanges, thereby destroying symmetry in the symmetric case and increasing the stored bandwidth of $\mathbf{U}$ from $m+1$ to $2m+1$ in both instances.

Since these methods at no time require access to more than $m+1$ rows at once, they will automatically work well in a paged virtual storage environment provided the rows or columns (rather than the diagonals) are stored contiguously. Similarly they are well-suited to working out of main storage. We summarize the storage and computational requirements in Table 10.2.1.

	Symmetric and positive definite	Diagonally dominant	General
Storage	$(m+1)n$	$(2m+1)n$	$(3m+1)n$
Active main storage	$\frac{1}{2}m^2$	m^2	$2m^2$
Number of multiplications	$\frac{1}{2}m^2n$	m^2n	$2m^2n$

Table 10.2.1. Leading terms in storage and operation counts for factorization of band matrices of bandwidth $2m+1$ and order n (valid for $m \ll n$).

10.3 Variable-band (profile) methods

As we explained in Section 8.3, a variable-band form (Jennings 1966), also called skyline (Felippa 1975), is preserved if no interchanges are performed. The important case is when the matrix is symmetric and positive definite. The lower triangular part may be stored by rows almost as conveniently as if the bandwidth were fixed. One possibility is to hold a set of pointers to the positions of the diagonals. For example, the matrix whose lower triangular part is shown in Figure 10.3.1 might be stored in the Fortran arrays shown in Table 10.3.1.

$$\begin{bmatrix} 1.7 & & & & \\ 0.6 & 3.2 & & & \\ & & 4.5 & & \\ & 2.3 & 4.7 & 5.7 & \\ & & & 3.1 & 6.9 \end{bmatrix}$$

Figure 10.3.1. The lower triangular part of a symmetric variable-band matrix.

Subscript	1	2	3	4	5	6	7	8	9
VAL	1.7	0.6	3.2	4.5	2.3	4.7	5.7	3.1	6.9
ILPTR	1	3	4	7	9				

Table 10.3.1. Fortran arrays holding the Figure 10.3.1 matrix.

The computational variants (10.2.3) and (10.2.7) for the (fixed) band form both generalize to this case, but the row Doolittle variant (10.2.7) is more straightforward and is the one normally used. For example, it is used by SPARSPAK (Chu, George, Liu, and Ng 1984). If rows j and k have their first entries in columns m_j and m_k, respectively, then the inner product in (10.2.7a) commences at $i = \max(m_j, m_k)$ so does not necessarily involve all the components of the recently computed vector $a_{ki}^{(i)}$, $i = m_k, \dots, j-1$. The pointers to diagonal entries readily allow us to calculate the row starts m_k when needed. For instance, in Table 10.3.1 row 4 starts at position ILPTR(3)+1 = 5 and ends at position ILPTR(4) = 7; the diagonal is in column 4 so the first entry is in column $4 - (7-5) = 2$. Apart from this, the Doolittle code is no more complicated than in the (fixed) band case and the inner loop is still an inner product between contiguously stored vectors.

The problem with Gaussian elimination by rows (10.2.3) is that there may be some sparsity within the pivot row and column (see, for example, column 2 in Figure 10.3.1). If the first entry in row i is in a column after column k, we have $l_{ik} = 0$ and it is easy enough to skip steps (10.2.3c) and (10.2.3d), but an explicit zero must be stored in u_{ki} and used in later steps (10.2.3d). We will be doing some multiplications involving explicit zeros in step (10.2.3d), something we normally try to avoid in sparse matrix computations. However this may be sensible on vector architectures that can handle this operation better than the inner products involved in (10.2.7).

If the maximum semibandwidth is m, at no time will access to more than m rows be needed at once. Thus, as for the (fixed) band case, it is straightforward to arrange for out-of-core working. This facility is provided, for example, by Harwell's code MA15 (Reid 1972).

10.4 Special methods for banded systems

Banded systems occur very frequently, sometimes as a subproblem within a larger computation. For example, methods for partial differential equations often solve many independent one-dimensional subproblems, each of which has a banded set of equations. The simplicity of such systems and the fact that they arise frequently has given rise to many special methods. We discuss some of them in this section, considering in particular those techniques that are efficient on vector and parallel architectures.

The most straightforward strategy in a parallel computing environment is to solve the one-dimensional problems simultaneously on separate processors. Similarly, vectorization techniques may be applied over the systems by considering each set of corresponding variables as a vector. Unfortunately the systems sometimes have to be solved one by one successively, so we now consider the case of a single system.

The simplest and most common banded system is tridiagonal, for example,

$$
\begin{aligned}
a_{11}x_1 + a_{12}x_2 &= b_1 \\
a_{21}x_1 + a_{22}x_2 + a_{23}x_3 &= b_2 \\
a_{32}x_2 + a_{33}x_3 + a_{34}x_4 &= b_3 \qquad (10.4.1)\\
a_{43}x_3 + a_{44}x_4 + a_{45}x_5 &= b_4
\end{aligned}
$$

$$\cdot \qquad \cdot \qquad \cdot \qquad \cdot$$

If we eliminate variables $x_1, x_3, x_5, \ldots$, the resulting set of equations will be of the form

$$a_{22}^{(2)} x_2 + a_{24}^{(2)} x_4 \qquad\qquad = b_2^{(2)}$$

$$a_{42}^{(2)} x_2 + a_{44}^{(2)} x_4 + a_{46}^{(2)} x_6 \qquad = b_4^{(2)} \qquad\qquad (10.4.2)$$

which is again a tridiagonal system, but with half as many variables as before. This reduction can be applied repeatedly to yield a method called odd-even (or cyclic) reduction (Buneman 1969). It is equivalent to using nested dissection on the tridiagonal system.

Although cyclic reduction requires about twice as many floating-point operations as straightforward elimination and creates fill-in (for example $a_{42}^{(2)}$, $a_{24}^{(2)}$, and $a_{46}^{(2)}$ in (10.4.2)), it is useful for two main reasons.

First, if we rewrite the system ordering odd-numbered variables first, we have a system of the form

$$\begin{pmatrix} \mathbf{D}_1 & \mathbf{B}_1 \\ \mathbf{B}_2 & \mathbf{D}_2 \end{pmatrix} \begin{pmatrix} \mathbf{x}_{odd} \\ \mathbf{x}_{even} \end{pmatrix} = \begin{pmatrix} \mathbf{b}_{odd} \\ \mathbf{b}_{even} \end{pmatrix} \qquad\qquad (10.4.3)$$

where $\mathbf{D}_1$ and $\mathbf{D}_2$ are diagonal matrices, $\mathbf{B}_1$ is lower bidiagonal, and $\mathbf{B}_2$ is upper bidiagonal. It is apparent that, unlike the standard solution scheme which is inherently recursive, the cyclic reduction scheme vectorizes because within each block the equations are decoupled. During the first stage the vectors have length $n/2$ if the matrix has order n.

Second, if the matrix has the block tridiagonal form

$$\begin{bmatrix} \mathbf{A} & -\mathbf{B} & & & & & \\ -\mathbf{B}^T & \mathbf{A} & -\mathbf{B} & & & & \\ & -\mathbf{B}^T & \mathbf{A} & -\mathbf{B} & & & \\ & & \cdot & \cdot & \cdot & & \\ & & & \cdot & \cdot & \cdot & \\ & & & & \cdot & \cdot & -\mathbf{B} \\ & & & & & -\mathbf{B}^T & \mathbf{A} \end{bmatrix}, \qquad (10.4.4)$$

which often occurs when a partial differential equation on a rectangular region is discretized, the same operations are applied to the matrix (but not the right-hand side) in each block row. The reduced submatrix has the form

$$\begin{bmatrix} \mathbf{A}^{(2)} & -\mathbf{B}^{(2)} & & & \\ -\mathbf{B}^{(2)\,T} & \mathbf{A}^{(2)} & -\mathbf{B}^{(2)} & & \\ & & \cdot & \cdot & \cdot \\ & & & \cdot & \cdot & \cdot \\ & & & & \cdot & \cdot \end{bmatrix} \qquad (10.4.5)$$

where

$$\mathbf{A}^{(2)} = \mathbf{A} - 2\mathbf{B}^T\mathbf{A}^{-1}\mathbf{B} \qquad (10.4.6a)$$

and

$$\mathbf{B}^{(2)} = \mathbf{B}\mathbf{A}^{-1}\mathbf{B}. \qquad (10.4.6b)$$

Clearly $\mathbf{A}^{(2)}$ and $\mathbf{B}^{(2)}$ need be evaluated only once and similar economy is possible during successive stages. In the special case where $\mathbf{B} = \mathbf{I}$, the reduction may be performed by premultiplying the even-numbered block rows by $\mathbf{A}$ and then adding each odd-numbered block row to its neighbours to yield a system of the form (10.4.5) with

$$\mathbf{A}^{(2)} = \mathbf{A}^2 - 2\mathbf{I} \qquad (10.4.7a)$$

and

$$\mathbf{B}^{(2)} = \mathbf{I}. \qquad (10.4.7b)$$

If this method were applied successively, the matrices $\mathbf{A}^{(k)}$ would become steadily more dense. Such dense matrices may be avoided by observing that the diagonal matrix at the r-th reduction step, $\mathbf{A}^{(r)}$, can be obtained by solving the recurrence

$$\mathbf{A}^{(r)} = \left(\mathbf{A}^{(r-1)}\right)^2 - 2\mathbf{I} \qquad (10.4.8)$$

to obtain

$$\mathbf{A}^{(r)} = \prod_{v=1}^{2^{r-1}} \left(\mathbf{A} - \alpha_v^{(r)}\mathbf{I}\right) \qquad (10.4.9a)$$

where

$$\alpha_v^{(r)} = 2\cos((2v-1)\pi/2^r), \qquad (10.4.9b)$$

giving a sparse representation for $\mathbf{A}^{(r)}$. Each matrix $\mathbf{A} - \alpha_v^{(r)}\mathbf{I}$ is sparse (in fact, is usually tridiagonal) so can easily be factorized, which allows systems with matrix $\mathbf{A}^{(r)}$ to be solved. For further details, we refer the reader to Swarztrauber (1977). The method is unstable in the form we have just described, but modifications suggested by Buneman (1969) produce a stable variant.

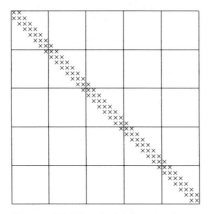

Figure 10.4.1. A tridiagonal matrix of order 35 partitioned into 5 blocks of order 7.

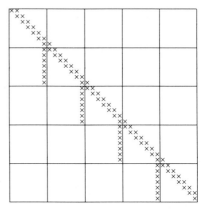

Figure 10.4.2. The Figure 10.4.1 matrix after step (i) of Wang's algorithm.

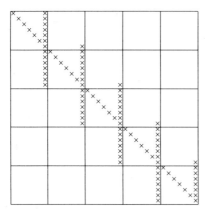

Figure 10.4.3. The Figure 10.4.1 matrix after step (ii) of Wang's algorithm.

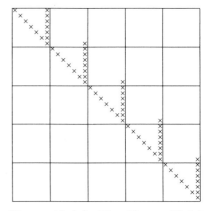

Figure 10.4.4. The Figure 10.4.1 matrix after step (iii) of Wang's algorithm.

An alternative algorithm for a tridiagonal set of equations was proposed by Wang (1981). He partitions the matrix, perhaps after augmenting it (Exercise 10.1), into p blocks of size k. We illustrate the algorithm for a matrix of order 35 in Figures 10.4.1 to 10.4.4. Wang applies the following steps

(i) For each block, add a multiple of the first row to the next row to eliminate the subdiagonal entry (note that these operations can be performed in parallel or as a vector operation). Do likewise for rows $2, 3,..., k-1$ of the blocks. These operations fill the last column of each subdiagonal block, as illustrated in Figure 10.4.2.

(ii) For each block, add a multiple of row $k-1$ to the previous row to eliminate the superdiagonal entry (note that these operations can again be performed in parallel or as a vector operation). Do likewise for rows $k-2, k-3,..., 1$ of the blocks. These operations fill the last column of each diagonal block and the last coefficient of the last column of each superdiagonal block, as illustrated in Figure 10.4.3.

(iii) Add a multiple of the last row of the first block to the rows of the next block to eliminate entries in the subdiagonal block (note that these operations can again be performed in parallel or as a vector operation). Do likewise for blocks $2, 3,..., p-1$. These operations produce no fill-ins and reduce the subdiagonal blocks to zero, as illustrated in Figure 10.4.4.

(iv) Add a multiple of the last row of the last block to the previous k rows to eliminate superdiagonal entries (note that these operations can again be performed in parallel or as a vector operation). Do likewise for blocks $p-1, p-2,..., 1$. These operations produce no fill-ins and reduce the superdiagonal blocks to zero, to produce a diagonal matrix.

(v) Solve the diagonal system as a parallel or vector operation.

Where the number of processors is modest in comparison with the matrix order, the scheme ensures that they are all fully used at every stage of the algorithm, and the total number of arithmetic operations is very similar to that required for odd-even reduction. If a number of small systems are to be solved simultaneously, they may be placed end-to-end to yield a large system for which adequate parallelism is available. As for odd-even reduction, the algorithm can be unstable and it is hard to see how pivoting can be introduced if it is needed. The algorithm is analogous to one-way dissection (Section 8.6), although the pivotal sequence is different.

An unsatisfactory feature of this algorithm is that the parallelism in steps (iii) and (iv) is different from that in steps (i) and (ii) because it is within rather than across the blocks, giving k-fold instead of p-fold parallelism. This will not matter on a vector computer provided neither p nor k is too small to exploit the vector hardware effectively. However on a parallel computer, the communication or synchronization costs may be substantial. The following alternatives in steps (iii) and (iv) involve more sequential operations, but keep the direction of parallelism.

(iii') Add a multiple of the last row of the first block to the last row of the next block to eliminate the subdiagonal entry. Do likewise for blocks $2, 3, \ldots, p-1$. Add a multiple of the last row of each block to the first row of the next block to eliminate the subdiagonal entries (note that these operations can again be performed in parallel or as a vector operation). Do likewise for rows $2, 3, \ldots, k-1$.

(iv') Add a multiple of the last row of the last block to the last row of the previous block to eliminate the superdiagonal entry. Do likewise for blocks $p-1, p-2, \ldots, 2$. Add a multiple of the last row of each block to the first row of the same block to eliminate the superdiagonal entries (note that these operations can again be performed in parallel or as a vector operation). Do likewise for rows $2, 3, \ldots, k-1$.

Partitioning schemes can also be used to solve banded systems other than tridiagonal. Dongarra and Johnsson (1987) discuss such an algorithm and compare it with other techniques for the parallel solution of banded systems.

Another approach that has been adopted (Kung and Leiserson 1979) is to design a special array of microprocessors (a systolic array) for solving banded systems of equations. The actual operations are as for ordinary Gaussian elimination, but the result of each operation is passed immediately to a neighbouring microprocessor for execution of the next step. It is hard to see how any interchanges can be incorporated into this approach, but it looks attractive for cases that do not need them.

10.5 Frontal methods for finite-element problems

Frontal methods have their origin in the solution of finite-element problems in structural mechanics, but they are not restricted to this application nor need the matrix be symmetric and positive definite. It is, however, easiest to describe the method in terms of its use in this application, and we do this in the first instance. In a finite-element problem the matrix is a sum

$$\mathbf{A} = \sum_{l} \mathbf{A}^{[l]}, \qquad (10.5.1)$$

where each $\mathbf{A}^{[l]}$ has entries only in the principal submatrix corresponding to the variables in element l and represents the contributions from this element. It is normal to hold each $\mathbf{A}^{[l]}$ in packed form as a small full matrix together with a list of the variables that are associated with element l, which identifies where the entries belong in $\mathbf{A}$ (see Section 2.15). The formation of the sum (10.5.1) is called **assembly** and involves the elementary operation

$$a_{ij} := a_{ij} + a_{ij}^{[l]}. \qquad (10.5.2)$$

We have used the algol symbol ':=' here to avoid confusion with the superscript notation of (10.2.3). We call an entry **fully summed** when all contributions of the form (10.5.2) have been summed.

It is evident that the basic operation of Gaussian elimination

$$a_{ij}^{(k+1)} = a_{ij}^{(k)} - a_{ik}^{(k)} a_{kk}^{(k)\,-1} a_{kj}^{(k)} \qquad (10.5.3)$$

may be performed before all the assemblies (10.5.2) are complete, provided only that the terms in the triple product in (10.5.3) are fully summed (otherwise we will be subtracting the wrong quantity). Each variable can be eliminated as soon as its row and column is fully summed, that is after its last occurrence in a matrix $\mathbf{A}^{[l]}$ (a preliminary pass through the data may be needed to determine when this happens). If this is done, the elimination operations will be confined to the submatrix of rows and columns corresponding to variables that have not yet been eliminated but are involved in one or more of the elements that have been assembled (we call these the **active** variables). This permits all intermediate working to be performed in a full matrix whose size increases when a variable appears for the first time and decreases when one is eliminated. The pivotal order is determined from the order of the assembly. If the elements are ordered systematically from one end of the region to the other, the active variables form a **front** that moves along it. For this reason the full matrix in which all arithmetic is performed is called the **frontal matrix** and the technique is called the **frontal method**.

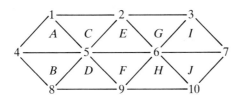

Figure 10.5.1 A simple triangulated region.

It is perhaps easier to envisage this by examining the process of frontal elimination on a small example. We show such a problem in Figure 10.5.1, where there are three variables associated with each triangle (one at each vertex). To keep the example simple, we suppose that the element matrices are not symmetric but any diagonal entry can be used as a pivot without fear of instability. We perform the assemblies from left to right (in the order shown lexicographically in Figure 10.5.1). After the assembly of the first two elements, the matrix has the form shown in Figure 10.5.2 if the variables 4, 8, 1, 5 are ordered first. The remaining triangles will make no further contribution to the row and column corresponding to variable 4, which is why we permuted it to the leading position. We may immediately eliminate variable 4 to yield the matrix shown in Figure 10.5.3. Next we

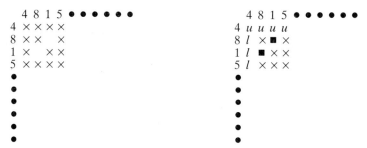

Figure 10.5.2. The matrix after the assembly of the first two triangles.

Figure 10.5.3. The matrix after the elimination of variable 4. Entries of **L** and **U** are shown as *l* and *u*, and fill-ins are shown as ■.

add the contribution from triangle *C* to give the pattern shown in Figure 10.5.4. Now the row and column of variable 1 are fully summed, so we perform a symmetric permutation to bring it to the pivotal position (here 2, 2) and perform the elimination step to give the pattern shown in Figure 10.5.5. The next element is *D* with variables 5, 8, and 9, and now variable 8 becomes fully assembled and may be eliminated. The procedure continues similarly. The set of active variables (the front) at successive stages is (4,8,1,5), (8,1,5), (8,1,5,2), (8,5,2), (8,5,2,9), (5,2,9), (5,2,9,6),... .

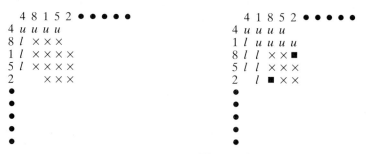

Figure 10.5.4. The matrix after the assembly of triangle *C*. Entries of **L** and **U** are shown as *l* and *u*.

Figure 10.5.5. The matrix after the elimination of variable 1. Entries of **L** and **U** are shown as *l* and *u*, and fill-ins are shown as ■.

In general, the partially processed matrix has the form illustrated in Figure 10.5.6. We show the situation after a set of eliminations and just prior to another assembly. The fully-summed rows and columns in blocks (1,1), (1,2) and (2,1) contain the corresponding rows and columns of **L** and **U**. They will not be needed until the SOLVE stage, so may be stored on auxiliary storage as packed vectors. Blocks (3,1) and (1,3) are zero because the eliminated variables are fully summed. Block (2,2) is the frontal matrix, normally held in main storage. Blocks (2,3), (3,2), and (3,3) as yet have no contributions, so require no storage. Thus only the frontal matrix needs to be in main memory and full matrix storage is suitable.

	Eliminated	Frontal	Remainder
Fully summed	$L_{11} \backslash U_{11}$	U_{12}	0
Frontal	L_{21}	$F^{(k)}$	
Remainder	0		

Figure 10.5.6. Matrix of partially processed finite-element problem, just before another assembly.

As further assemblies are performed, permutations are needed to maintain the form shown in Figure 10.5.6. Variables involved for the first time move into the front and frontal variables involved for the last time (identified by a preliminary pass of the data) are eliminated and move out. The frontal matrix thus varies in size and variables do not necessarily leave it in the same sequence as they enter it. It is obviously necessary to be able to accommodate the largest frontal matrix that occurs.

Returning to the finite-element matrix (10.5.1), suppose there is a variable associated with element 1 but with no others. Then the frontal method will begin with frontal matrix $A^{[1]}$ and will immediately eliminate this variable. The amount of arithmetic needed will depend solely on the size of $A^{[1]}$ itself and not on the size of the overall problem. A variable associated with a later element and no others will enter the front with the assembly of its element and will be immediately eliminated, but now the amount of arithmetic will depend on the current front size. However, since the variable is fully summed within its element, there is no reason why the operations (10.5.3) associated with its elimination should not be done in temporary storage before its assembly into the frontal matrix; then assembly consists of placing the pivot row and column with the eliminated rows and columns and adding the remaining reduced submatrix (including the modifications caused by the elimination operations) into the appropriate positions in the frontal matrix. This elimination within the element is known as **static condensation** and means that the true cost of adding internal variables to finite elements is very slight indeed. For instance the algebraic cost of working with 9-variable elements as shown in Figure 10.5.7 is virtually identical with that of working with the usual 8-variable element (Figure 10.5.8). Adding this simple technique to Harwell's frontal code MA32 (Duff 1983) substantially improved its performance on appropriate problems.

Figure 10.5.7. 9-node element.　　　　Figure 10.5.8. 8-node element.

Variables may leave the front in a different order from that in which they enter it. Thus a variable associated with a single element enters the front with the assembly of its element and immediately leaves it by elimination; a variable associated with a pair of elements enters the front with the first element and leaves it with the second. This property is important to the efficiency of the method, but does lead to slight programming complications. A preliminary pass through the index lists associated with the matrices $\mathbf{A}^{[l]}$ is needed to determine when each variable appears for the last time (that is when it can be eliminated). There is also some complication in organizing the actual assemblies and eliminations within the frontal array. When the variables enter the front the corresponding rows and columns are normally placed at the end of the array and the eliminations are performed from the middle with the revised matrix closed up to fill the gaps. There is, of course, a need to maintain an index array to indicate the correspondence with the associated variables.

The frontal technique can also be applied to systems which need numerical pivoting. To help visualize this, we show in Figure 10.5.9 the situation after an assembly but prior to another set of eliminations. Blocks (1,1), (1,2), (2,1), and (2,2) correspond to the frontal matrix at this stage. Any entry whose row and column is fully summed (that is, any entry in block (1,1) of Figure 10.5.9) may be used as a pivot. For unsymmetric matrices, we may use the threshold criterion

$$|a_{ij}^{(k)}| \geq u \max_{l} |a_{lj}^{(k)}| \tag{10.5.4}$$

where u is a given threshold, see Section 5.4. The pivot must be chosen from the (1,1) block of Figure 10.5.9, but the maximum is taken over blocks (1,1) and (2,1). Note that if unsymmetric interchanges are included, the index list for the rows in the front will differ from the index list of the

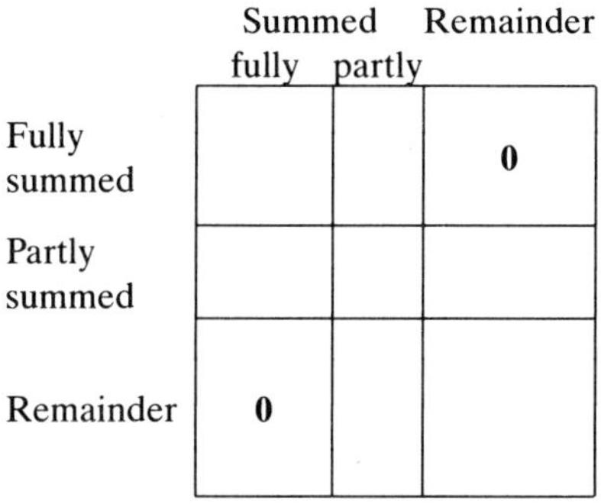

	Summed fully	partly	Remainder
Fully summed			**0**
Partly summed			
Remainder	**0**		

Figure 10.5.9. Reduced submatrix after an assembly and prior to another set of eliminations.

columns and will require separate storage. In the symmetric case, we may maintain symmetry by using either a diagonal pivot that satisfies inequality (10.5.4) or a 2×2 pivot

$$E_k = \begin{pmatrix} a_{ii}^{(k)} & a_{ij}^{(k)} \\ a_{ji}^{(k)} & a_{jj}^{(k)} \end{pmatrix}, \tag{10.5.5}$$

all of whose coefficients are fully summed and that satisfies the inequality

$$\|E_k^{-1}\|_1^{-1} \geq u \max \left(\max_{l \neq i,j} |a_{li}^{(k)}|, \max_{l \neq i,j} |a_{lj}^{(k)}| \right), \tag{10.5.6}$$

see inequality (5.4.7).

It is possible that we cannot choose pivots for the whole of the frontal matrix because of large off-diagonal entries that lie in the fully-summed rows and partly-summed columns or vice-versa (that is, because they lie in block (1,2) or block (2,1) of Figure 10.5.9). In this case we simply leave the variables in the front, continue with the next assembly, and then try again. It is always possible to complete the factorization using this pivotal strategy because each entry eventually lies in a fully-summed row and in a fully-summed column, at which stage it may be used as a pivot. Naturally if many eliminations are delayed, the order of the frontal matrix might be noticeably larger than it would have been without numerical pivoting, but Duff (1984c) has found that in practice such increases are slight. Note that this is an example of a posteriori ordering for stability. We choose an ordering for the variables that keeps the front small without worrying about the possibility of instability, and we ensure the stability during the numerical factorization.

The right-hand side vector may be treated similarly in a frontal fashion. It will have contributions from all the elements and these may be accumulated into a frontal vector, just as the matrix entries are accumulated into the frontal matrix. As eliminations are performed on the frontal matrix, so they are performed on the frontal vector too, and the pivotal component is stored away after each pivotal step. During back-substitution the vector components are recovered and the front is reconstituted in reverse order, so allowing successive components of the solution to be calculated. Alternatively, the right-hand side may be held in a normal full array and treated in a conventional fashion using indirect addressing. This is more straightforward and involves less data movement, especially where the right-hand side is not assembled from the individual elements, for example in an eigenvalue calculation when it arises directly from an earlier step.

10.6 Frontal methods for non-element problems

Our discussion in the previous section was based on the application of the frontal method to finite-element problems, but the technique is not restricted to this application. Indeed, the frontal method applied to a general (fully-summed) problem can be viewed as a generalization of the variable-band solution (Section 10.3).

For non-element problems, the rows may be taken in turn as if they were unsymmetric element matrices. The assembled rows are then always fully summed and a column becomes fully summed whenever the row containing its last nonzero is reached. This situation is illustrated in Figure 10.6.1, where we show the frontal matrix after the assembly of equation 3 of the equation form of the five-point discretization of the Laplacian operator on a 2×4 grid. After this stage no more entries will appear in column 1 so this column is 'fully summed' and can be eliminated.

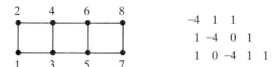

Figure 10.6.1. A 2×4 grid and the first three rows of the corresponding matrix (all entries to the right are zero).

The counterpart to Figure 10.5.6 is Figure 10.6.2. We have ordered the columns so that the complete columns come first, followed by the partially complete columns, followed by those still without entries. After another row is entered, the reduced submatrix has the form shown in Figure 10.6.3, which is the counterpart of Figure 10.5.9. Eliminations may now take place in every fully-summed column and pivots may be chosen from anywhere in the column. Therefore partial pivoting is possible, satisfying inequality (10.5.4) with $u=1$.

Without numerical pivoting, this form of frontal solution is very similar to the variable-band or profile method discussed in Section 10.3. However,

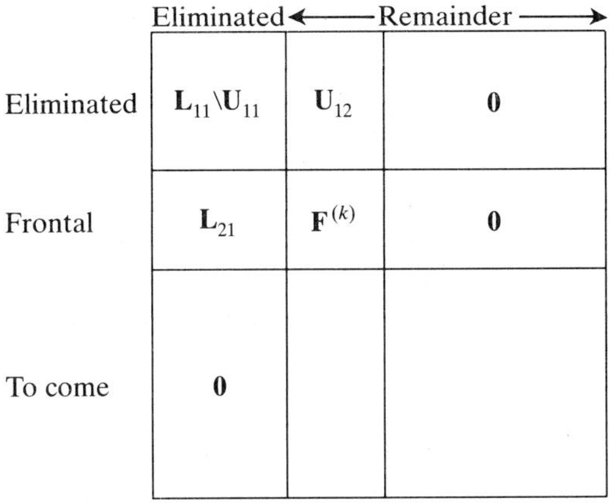

Eliminated ◄——————Remainder ——————►

Eliminated	$\mathbf{L}_{11}\backslash\mathbf{U}_{11}$	$\mathbf{U}_{12}$	$\mathbf{0}$
Frontal	$\mathbf{L}_{21}$	$\mathbf{F}^{(k)}$	$\mathbf{0}$
To come	$\mathbf{0}$		

Figure 10.6.2. Matrix of partially processed non-element problem, just prior to entering a row.

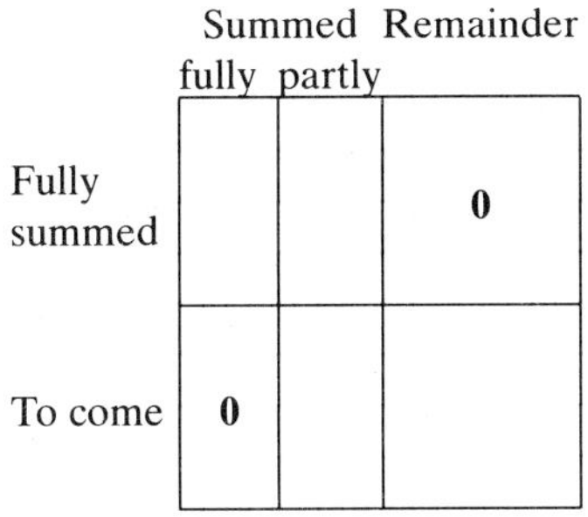

Summed Remainder
fully partly

Fully summed			$\mathbf{0}$
To come	$\mathbf{0}$		

Figure 10.6.3. Reduced submatrix of non-element problem, just after entering a row.

the frontal code does not prescribe the pivotal sequence, which means that the ordering is not quite so critical for good sparsity preservation. The fact that numerical pivoting is straightforward to incorporate gives the method a substantial advantage.

As is the case for band techniques, frontal methods may be applied to any matrix. Of course their success depends on the front size remaining small. In particular they tend to produce more fill-in and require more arithmetic operations than the local methods of Chapter 7. For example any zeros in the frontal matrix $\mathbf{F}^{(k)}$ of Figure 10.6.2 are stored explicitly and there may be many of them.

Two of the main reasons favouring band methods also support frontal techniques. The first is that only the active part of the band or the frontal matrix need be held in main storage. The second is that the arithmetic is performed with full-matrix code without any indirect addressing or complicated adjustment of data structures to accommodate fill-ins (although it is necessary to keep track of the correspondence between the rows and columns of the frontal matrix and those of the problem as a whole). The avoidance of indirect addressing facilitates vectorization on machines capable of it (see Duff 1984e and Section 10.14).

It is important to keep the front small since the work involved in an elimination is quadratic in the size of the front (see Exercise 10.2). It is particularly easy to keep the front small in a grid-based problem if the underlying geometry is that of a long thin region. Automatic methods for bandwidth reduction can also be applied for frontal solution (for example, see Sloan and Randolph 1983). For a finite-element problem, manual ordering of the elements for a small front size is much more natural than ordering the variables for small bandwidth and is likely to be easier simply because there are usually less elements than variables; in general, however, people prefer this to be done automatically for them.

It is interesting that the advantages associated with reversing the Cuthill-McKee ordering (see Section 8.4) for variable-band matrices are automatically obtained in the frontal method since each variable is eliminated immediately after it is fully summed.

Irons (1970) was, to our knowledge, the first to publish a paper on the frontal method. His code was designed for symmetric and positive-definite matrices and was generalized to unsymmetric cases by Hood (1976). Harwell subroutine MA32 (Duff 1981e, 1983, 1984c) incorporates many improvements to Hood's code and we believe it was the first to accept the data by equations as well as elements, which makes it very suitable for finite-difference discretizations of partial differential equations.

10.7 Multifrontal (substructuring) methods for finite-element problems

In this and the next section, we discuss a generalization of the frontal technique that is able to accommodate any ordering scheme based on the structure of the matrix, including minimum degree and nested dissection. Because more than one front is involved, it is called the **multifrontal** technique. As for one front, it is easier to describe in the finite-element case, which is the topic of this section. The general case is discussed in Section 10.8.

The concept of static condensation, discussed in Section 10.5, extends naturally to substructures consisting of groups of elements. Any variables that are internal to a substructure may be eliminated by operations involving only those matrices $A^{[k]}$ that are associated with the substructure. The rows and columns of the resulting reduced matrix are associated with variables on the boundary of the substructure. The reduced matrix has a form just like that of a large element parameterized by the boundary variables. Indeed the term **generated element** (Speelpenning 1978 uses the term 'generalized element') is often used. We are led to considering bracketings of the sum

$$A = \sum_l A^{[l]}$$

(10.7.1)

other than the left-to-right form corresponding to the (single) frontal method,

$$((...((A^{[1]}+A^{[2]}) + A^{[3]}) + A^{[4]}) + ...).$$

(10.7.2)

We illustrate this by means of a small example.

1	2	5	6
3	4	7	8

Figure 10.7.1. A finite-element problem.

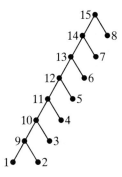

Figure 10.7.2. Assembly tree for frontal elimination on the Figure 10.7.1 example, using summation (10.7.2).

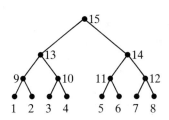

Figure 10.7.3. Assembly tree for the summation (10.7.3).

If we consider the finite-element problem of Figure 10.7.1, the frontal method can be represented by the summation (10.7.2) or the assembly tree shown in Figure 10.7.2. In such an assembly tree each non-leaf node corresponds to an assembly and subsequent eliminations. The alternative summation order

$$((A^{[1]}+A^{[2]}) + (A^{[3]}+A^{[4]})) + ((A^{[5]}+A^{[6]}) + (A^{[7]}+A^{[8]}))$$

(10.7.3)

will give the assembly tree shown in Figure 10.7.3. We may proceed as follows:

(i) eliminate all variables that belong in only one element, storing the resulting pivotal rows and columns and setting the resulting generated element matrices aside temporarily,

(ii) assemble the resulting matrices in pairs and eliminate any variable that is internal to a single pair; store the resulting generated element matrices $A^{[9]}$, $A^{[10]}$, $A^{[11]}$, $A^{[12]}$,

(iii) treat the pairs $(A^{[9]}, A^{[10]})$ and $(A^{[11]}, A^{[12]})$ similarly to produce $A^{[13]}$ and $A^{[14]}$,

(iv) assemble $A^{[13]}$ and $A^{[14]}$ and complete the elimination.

Each stage corresponds to one depth of brackets in expression (10.7.3) or one level in the tree. We are doing the assemblies as indicated by the brackets or the tree and doing each elimination as early as possible. The result is an ordinary **LU** factorization where the rows of **U** and the columns of **L** are generated as usual at each pivotal step. The novel part lies in the different order for the arithmetic operations and the data organization.

For simplicity, we began by considering a very regular example. In general, a tree node may have any number of sons and there will be no particular relationship between the number of nodes at one tree level and the number at another. At each node, the associated operations consist of assembling the element matrices of the sons, eliminating any variables that do not appear elsewhere, and storing the resulting generated element.

At each node, all arithmetic may be performed in a temporary full array labelled to indicate which variables are associated with its rows and columns. In fact, it is exactly like the frontal matrix of the frontal method. Essentially, we have a number of frontal matrices active at once. This is why the method is called the multifrontal method.

The benefit of the multifrontal approach over the (single) frontal approach of Sections 10.5 and 10.6 lies in the added freedom in the way in which the assemblies can be organized. It is thus possible to combine the multifrontal concept with quite general ordering schemes. The tree given in Figure 10.7.3 could be produced by a nested dissection ordering on the finite-element problem in Figure 10.7.1 and our implementation of the minimum degree ordering in Section 10.9 is strongly influenced by the multifrontal concept. It is possible to generate trees in a hybrid fashion where, for example, the user provides a coarse dissection as an assembly tree and the software refines the tree by ordering within the nodes using some heuristic which generates subtrees. This approach was adopted by Reid (1984), who designed his finite-element package to allow the user to provide any number (including none) of substructures; the package then completes the assembly tree automatically using the minimum degree algorithm. For a fuller discussion of the use of substructures, we refer the reader to Noor, Kamel, and Fulton (1977) and Dodds and Lopez (1980).

Given an assembly tree, there is still considerable choice for the order in which operations are performed. For the Figure 10.7.3 example, instead of assembling in the natural order, the order 1, 2, 9, 3, 4, 10, 13, 5, 6, 11, 7, 8, 12, 14, 15 would result in the same arithmetic operations being performed. The only requirement that a reordering must satisfy is that each node is ordered ahead of its father. One possibility, which leads to a simplification in the data structure for numerical factorization, is to order the nodes of the tree by postordering (Aho, Hopcroft, and Ullman 1974), during a depth-first search. Each node is ordered when a single backtrack for it is performed. For example, depth-first search of the tree of Figure 10.7.3 with priority to the left leads to the node order given earlier in this paragraph. The advantage of such an ordering is that the generated elements required at each stage are the most recently generated ones of those so far unused, which means that a stack may be used for temporary storage (see Exercise 10.4). Indeed, the use of postordering gives a nested block structure for the L\U array, as can be seen by comparing Figure D.15 with Figures D.12 to D.14 in Appendix D.

It is sensible to calculate this node order and implied pivot order during ANALYSE. Given the tree and lists of variables associated with the original finite elements (leaf nodes), we proceed as follows:

(1) Perform a depth-first search to establish the node order; for each variable, record which of the original elements with which it is associated is last in this ordering.

(2) Perform a second depth-first search to establish the pivot order. When a leaf-node is reached, any variable appearing only in the element associated with that node can be added immediately to the pivotal sequence. Removing such a variable from the index list corresponds to performing a static condensation. The resulting list is associated with the leaf node. On a backtrack, the index lists of the sons of the node to which we backtrack are merged and the variables are now regarded as associated with the node rather than its sons. Any variable that now occurs only in this new list is added to the pivotal sequence and removed from the list.

Notice that this process is much quicker than actual factorization because:

(1) merging a list of length l with a given list needs only $O(l)$ operations (see Section 2.4) whereas assembling an element of order l requires $O(l^2)$ operations, and

(2) removing a variable from a list of length l needs $O(1)$ operations whereas one step of Gaussian elimination with a front size l requires $O(l^2)$ operations.

10.8 Multifrontal methods for non-element problems

Just as the frontal approach was shown to be applicable to general systems of equations, the same is true of the multifrontal technique. We illustrate this by a simple example. Consider the symmetric matrix whose pattern is shown in Figure 10.8.1 and suppose we pivot down the diagonal in order. At the first step row (and column) 1 is 'assembled' and eliminations are performed using pivot (1,1) to give a reduced matrix of order two with associated row (and column) indices 3 and 4. Note that at this stage no reference has been made to any of the other entries of the original matrix. The reduced matrix can be stored until needed later. Row (and column) 2 is now 'assembled', the (2,2) entry is used as pivot and the reduced matrix of order two, with associated row (and column) indices of 3 and 4, is stored. Before we perform the pivot operations using entry (3,3), the contributions from the first two eliminations (the two stored submatrices of order two) must be 'assembled' with the original entry (3,3). The pivot operation leaves a reduced matrix of order one with row (and column) index 4. The final step sums this matrix with the (4,4) entry of the original matrix. The elimination can be represented by the tree shown in Figure 10.8.2.

```
×   × ×
  × × ×
× × ×
× ×   ×
```

Figure 10.8.1. A non-element problem.

Figure 10.8.2. Elimination tree for the matrix of Figure 10.8.1.

Figure 10.8.3. Elimination tree for the matrix of Figure 10.8.1 after node amalgamation.

The same storage and arithmetic is needed if the (4,4) entry is assembled at the same time as the (3,3) entry, and in this case the two pivotal steps can be performed on the same submatrix. This corresponds to collapsing or amalgamating nodes 3 and 4 in the tree of Figure 10.8.2 to yield the tree of Figure 10.8.3. On typical problems, node amalgamation produces a tree with about half as many nodes as the order of the matrix.

Duff and Reid (1983) considered representing the matrix as an artificial set of element matrices and constructed an algorithm that ensured that the elements were few and large. However, they found that this was less efficient than treating the original matrix directly, which may be regarded as viewing each diagonal entry and each off-diagonal pair of entries as an artificial element (see Section 2.16). Furthermore, only by treating the matrix directly can any given pivotal sequence be encompassed.

As for the frontal method, the multifrontal approach can be extended to incorporate numerical pivoting and asymmetry. In the symmetric case, 1×1 pivots from the diagonal are used only if they satisfy inequality (10.5.4), and are supplemented by 2×2 pivots satisfying inequality (10.5.6). As a result, the matrices stored during factorization may be a little larger than in the corresponding positive-definite case, since eliminations may wait for suitable pivots to be available.

Order	532	1224	183	216
Nonzeros	3474	9613	1069	876
Measure of symmetry	0.74	0.61	0.42	0.0
Entries in factors				
Multifrontal	9714	91136	2775	7058
Unsymmetric pivoting	9297	63313	1271	3677

Table 10.8.1. Effect of asymmetry on diagonal ordering.

An unsymmetric matrix $\mathbf{A}$ may be handled by applying a symmetric ANALYSE (for example, the minimum degree ordering) to the pattern which is the union of the pattern of $\mathbf{A}$ with the pattern of $\mathbf{A}^T$, but allowing off-diagonal pivots during numerical FACTORIZE provided they satisfy inequality (10.5.4). If the pattern of $\mathbf{A}$ is symmetric, or nearly so, this is very successful and the front sizes are usually only a little greater than in the case where any diagonal entry can be used as a pivot. However, this technique will not perform well when $\mathbf{A}$ has a pattern that is far from symmetric. Some examples of Duff (1984f) are shown in Table 10.8.1. We have replaced Duff's measure of asymmetry by the measure of symmetry used in Section 6.11, which is the proportion of off-diagonal nonzeros for which the corresponding entry in the transpose is also nonzero. In these tests, the matrices were made diagonally dominant by adding terms to the diagonal so that the multifrontal code always used the diagonal pivots chosen by ANALYSE. We see here quite clearly that the multifrontal approach is not well suited for unsymmetric systems. Suggestions have been made (Pagallo and Maulino 1983) of ways of extending symbolic analysis to unsymmetric matrices and off-diagonal pivoting by using bipartite graph representations of the matrix. However, we know of no implementations, and an initial ordering is likely to suffer severely from the kind of numerical pivoting considerations which we now discuss.

Diagonal entries	4.0	1.0	0.01	0.001
Nonzeros in factors	8 048	8 344	13 640	14 660
Total storage for factors	13 134	13 440	18 118	19 226
Multiplications	52 838	56 264	140 460	170 380

Table 10.8.2. Effect of numerical pivoting on five-diagonal matrix of order 400.

Because of the results discussed in the previous paragraph, multifrontal techniques for general systems are normally only considered appropriate for matrices which are structurally symmetric (or at least nearly so). Even for this case, unsymmetric pivoting can still cause some problems as we illustrate in the results of Duff (1984f) shown in Table 10.8.2. The example used in this table is a five-diagonal matrix of order 400 whose off-diagonals have value −1 and whose diagonal entries are as shown in the first row of the table. The matrix in the first column (diagonals equal to 4.0) is positive definite. Matrices in succeeding columns are not positive definite and have smaller and smaller diagonal entries, causing an increasing number of off-diagonal pivots to be selected because of the threshold condition (10.5.4). The effect of this change in the original pivot order is quite clear.

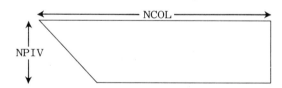

Figure 10.8.4. Block pivot row of **U** with NCOL columns and NPIV rows.

The SOLVE phase of a multifrontal method differs from that discussed in Section 9.4 in only one respect, namely that the factors are stored in blocks corresponding to variables eliminated at the same node of the assembly tree. Such a block pivot row for **U** is shown in Figure 10.8.4. Indexing vectors are held to identify the rows and columns in the block pivot. There are two ways of using factors stored in this form. The first follows the conventional practice of Section 9.4 for back-substitution, where each row is treated separately as a packed vector and indirect addressing is used in the innermost loop. The other possibility involves loading all the active components of the solution vector into a vector of length NCOL, performing all the substitution operations using direct addressing in the innermost loop and then unloading into the main solution vector again. The latter approach will be better if the saving in using direct over indirect addressing more than matches the cost of the loading and unloading. This will be true on most computers if NPIV is large enough, although we would expect the use of direct addressing to be particularly efficacious on vector computers. Of course, similar considerations apply to the columns of **L** during forward substitution.

Recently some other work on multifrontal methods has been done. Liu (1986a) has obtained noticeable savings in the amount of storage needed for the stack by reordering each set of sons of a node, although the amount of storage for the factors and the number of operations is not affected. Lewis (private communication 1985) uses a multifrontal implementation of a nested-dissection ordering in an out-of-core factorization scheme where all that need be held in main storage is information on the factorization pertaining to the current node and all nodes between it and the root. Duff (1986) and Liu (1986b) both examine in some detail the relationship between an elimination tree and the structure of the factors. A multifrontal approach for both Givens and Householder reductions has been considered by Liu (1986c) and George, Liu, and Ng (1988). The adaptation of multifrontal schemes for parallel architectures has been considered by Duff (1986).

Our final comments in this section relate to software. The only codes implementing multifrontal techniques of which we are aware are in the Harwell Subroutine Library, although some of the codes which we discuss in Section 10.9 are very similar to multifrontal codes in the ANALYSE phase. The Harwell code MA27 (Duff and Reid 1982, 1983) is designed for symmetric systems and uses 2×2 pivoting when necessary. The unsymmetric case is handled by the code MA37 (Duff and Reid 1984). Reid (1984) has developed a code, TREESOLVE, for large sets of linear finite-element equations that uses multifrontal techniques without numerical pivoting but allows the stack, and indeed the individual frontal matrices, to reside partly on auxiliary storage so that really huge problems can be solved.

10.9 Minimum degree ordering

The minimum degree algorithm (Tinney scheme 2) was introduced in
Section 7.3 and we will now consider its implementation. It is possible to
proceed as in the unsymmetric case (Section 9.2) and in one step:
determine the ordering, identify the location of the fill-in, set up static data
structures for FACTORIZE, and perform an actual factorization. Early
codes such as Harwell's MA17 (Reid 1972) did this. However, much greater
efficiency can be achieved by separating these functions and not
considering the numerical values until actual factorization. The first step,
determining the minimum degree ordering, is the subject of this section.
Identifying the location of the fill-in and setting up the static data structures
are the subject of the next section. Alternatively, the ordering of this
section may be followed by the multifrontal technique of the previous
section.

To facilitate the pivot selection, we require access to the rows (or
columns) of the matrix. Taking full advantage of symmetry means that only
half of the matrix is stored and access to the whole of a row demands access
to the row and column of the stored half of the matrix. The simplest
alternative is to store the entire matrix, and we make this assumption here.
The storage burden is eased by not having to store the numerical values at
this stage.

The critical factor in the efficient implementation of the minimum
degree ordering is to avoid the explicit storage of the fill-ins. Rather, the
fill-in is treated implicitly by storing the updated submatrices as a sum of
cliques. This clique representation of Gaussian elimination was popul-
arized by Rose (1972) and used in a somewhat different way by George and
Liu (1981).

```
         1 2 3 4 5 6 7
      1 ×    ×   × ×
      2    ×  ×    × ×
      3 ×    ×    × ■ ×
      4    ×  ×    × ×
      5 ×    ×    × ■ ×
      6 × × ■ × ■ × ×
      7    × × × × × ×
```

Figure 10.9.1. The first step of elimination. Fill-ins are
shown as ■.

We will describe the non-finite-element case first, although some
simplifications are possible for finite-element problems (see the last
paragraph of this section). The first pivot is chosen as the diagonal entry in
a row with least entries, say $(k+1)$ entries. Following the operations
involving the first pivot, the reduced matrix will have a full submatrix in the
rows and columns corresponding to the k off-diagonal entries in the pivot

row. The graph associated with this full submatrix is called a pivotal clique. The important thing concerning the storage of cliques (see Section 2.15) is that a clique on k nodes requires only k indices although the corresponding submatrix requires storage for $\frac{1}{2}k(k+1)$ numerical values. Rather than finding out where fill-ins take place and revising the lists of entries in the rows, as we did in the unsymmetric case (see Section 9.2), we keep the original representations of the rows together with a representation of the full submatrix as a clique. To update the count of number of entries in a row that was involved in the pivotal step, we merge the index list of its entries with the index list of the clique. For example, in the case illustrated in Figure 10.9.1, the clique has index list (3,5,6) and the count of entries in the new row 3 is available by merging its index list, (3,5,7), with the clique index list to give the list (3,5,6,7) of length 4. Note that the counts of rows 3, 5, and 6 (corresponding to the clique) change, but those of the rest (2, 4, and 7) do not. The lists may be held as a collection of sparse vectors (Section 2.7) or as a linked list (Section 2.9).

At a typical stage of the elimination, the pivot is chosen from a row with least entries. The set of entries in the pivot row is formed by merging its index list (discounting eliminated columns) with those of all the cliques that involve the pivot. This set, less the index of the pivot, provides the index set of the new clique. Since all the old cliques that involve the pivot are included in the new clique, there is no need to keep them. The number of entries in each altered row is now calculated.

As in the unsymmetric case, it is advisable to store the numbers of entries in the rows as a doubly-linked list. At each pivotal step, those affected are removed from the list, the new counts are calculated, and the rows are inserted in new positions in the list.

The algorithm has a very interesting property. Since each new list is always formed by merging old lists (and removing the pivot index) the total number of list entries does not increase in spite of the possible fill-ins. Hence, provided the original matrix can be stored and provided adequate data structures are in use, there is no possibility of failure through insufficient storage.

Since there is no possibility of failure through lack of storage, we have a tool for determining whether an actual factorization is feasible on a given computer. Although it has not been proved that the time for finding the pivot order is bounded by a limit proportional to the number of entries in the original matrix, typical times for careful implementations exhibit a roughly linear behaviour in n (matrix order) and τ (number of matrix entries), as illustrated by the figures for MA27A in Table 10.9.1.

The expensive step is usually the (logically straightforward) recalculation of the degrees (numbers of entries in the rows) following each elimination. Three devices may be employed to reduce the cost of this step (see Duff and Reid 1983 and Exercise 10.5):

(a) Since each row whose degree we are updating involves all the entries in the pivot row, we may always commence from the index list of the new clique.

(b) It is easy to recognize any clique that lies entirely within the pivotal clique; it may be removed without losing any information about the structure of the reduced system.

(c) It is very worthwhile to recognize rows with identical structure since once identical they will remain so and only one copy is needed and only one degree needs to be calculated; the degree itself provides a convenient hashing function (two rows need not be tested for being identical unless they are found to have the same degree). This is similar to 'mass node elimination' of George and Liu (1981).

(d) Once an entry of the original matrix has been 'overlaid' by a clique, it may be removed without altering the overall pattern.

Order	265	1009	3466
Nonzeros	1753	6865	23896
MA17A (1970)	1.56	29.9	>250
MA17E (1973)	0.68	6.86	62.4
YSMP (1978)	0.24	1.27	6.05
SPARSPAK (1980)	0.27	1.11	4.04
MA27A (1981)	0.15	0.58	2.05

Table 10.9.1. Times in IBM 370/168 seconds for various implementations of minimum degree ordering.

To illustrate the improvements to minimum degree ANALYSE that we have considered in this chapter, we show in Table 10.9.1 specimen times for five codes. The problems are three of the graded-L triangulations of George and Liu (1978b). Codes MA17A and MA17E are from the Harwell Subroutine Library and do not use clique amalgamation; MA17E holds row and column indices explicitly whereas MA17A uses the storage scheme described in Table 2.11.2. YSMP (Yale Sparse Matrix Package), SPARSPAK (University of Waterloo), and MA27A (Harwell Subroutine Library) all use successive refinements of clique amalgamation, which is why they are faster than the first two. Since there is more than one version of each of these codes, the date of the version used in this comparison is given in parentheses. These dates also serve to indicate the advances in the implementation of the minimum degree algorithm over the last few years. Note that Table 10.9.1 should not be interpreted as a comparison of the current versions of these codes.

The finite-element case is simpler in that the problem is given in the form of a set of cliques. The degrees of all the variables should be calculated initially from the clique representation without performing any assemblies. Thereafter the implementation can be just like the non-element case except that the assembled matrix part is always void.

10.10 Generating static data structures for FACTORIZE

Given an ordering (for example, minimum degree from Section 10.9, or nested dissection), we now discuss computing the positions of the fill-ins and setting up a static data structure that accommodates them.

For the minimum degree algorithm of Section 10.9, the structure of each row of $\mathbf{U}$ (or equivalently each column of $\mathbf{L}$) in the factorization

$$\mathbf{PAP}^T = \mathbf{LDL}^T = \mathbf{LU} \tag{10.10.1}$$

of the permuted matrix may be computed trivially as a by-product, since at each stage the index set for the entries in the pivot row/column is computed. However, as we now explain, a straightforward and fast algorithm can compute the structure of $\mathbf{U}$ for any pivot sequence (that is, any permutation matrix $\mathbf{P}$ in (10.10.1)). There are storage advantages in doing this for the minimum degree algorithm and it allows other pivot choices, for example nested dissection, to be accommodated.

A way that this can be done is to use the multifrontal technique of Section 10.8, but without storing or computing any of the numerical values. We start with an index list for each row of the upper triangular part of $\mathbf{PAP}^T$, holding the column indices of the entries in the rows. At the first stage, we generate an element whose index list is that of the first row except that column 1 is omitted. At the k-th stage, we generate an element by merging the index list of row k of $\mathbf{PAP}^T$ with those of any generated elements whose list includes column k and finally omit column k itself (the lists used are discarded). The index lists of the generated elements are nothing other than those of the rows of $\mathbf{U}$ with the diagonal omitted, so separate storage is not needed. In a single pass we can generate the pattern of $\mathbf{U}$ from that of the upper triangle of $\mathbf{PAP}^T$. Notice that each row of $\mathbf{U}$ (each generated element) is used only once, when it is merged with the row corresponding to its first off-diagonal entry. Thus the pattern of $\mathbf{U}$ may be determined in $O(n) + O(\tau)$ operations, where n is the matrix order and τ is the number of entries in $\mathbf{U}$. Furthermore, the amount of storage needed for row i of $\mathbf{U}$ cannot exceed that needed for row i of $\mathbf{PAP}^T$ plus that needed for the previous rows of $\mathbf{U}$ with first entry in column i. Since these previous rows will not be required for determining the patterns of the later rows of $\mathbf{U}$, the algorithm may be performed in place provided the patterns of the rows of $\mathbf{U}$ are stored externally or discarded. Discarding is appropriate in a storage allocation phase. At its conclusion we know precisely the number of entries in each row of $\mathbf{U}$, and a subsequent pass may be made to set up the actual data structure for $\mathbf{U}$.

Sherman (1975) described this algorithm in terms of the structure of the rows of **U** without using our multifrontal description. Of course, the pattern of the rows of **U** is the same as the pattern of the columns of **L**. The algorithm has often been described in terms of **L** (by George and Liu 1981, for example), that is

> while the numerical values in column j $(2 \leq j \leq n)$ of **L** in the factorization (10.10.1) are a linear combination of all previous columns of **L** with an entry in row j, the sparsity pattern of column j is the union of the sparsity patterns of column j of $\mathbf{PAP}^T$ and all previous columns of **L** whose first entry is in row j.

This result is normally proved directly rather than as a corollary of looking at the process from the multifrontal point of view. We leave it as an exercise for the reader (Exercise 10.6) to prove it this way.

To implement the storage determination algorithm, we scan the rows in order, calculating the number of entries in each row of **U** in turn. At stage k we scan a linked list of rows whose first entry is in column k, merge the patterns of these rows with the pattern of row k of $\mathbf{PAP}^T$, and discard each pattern after use. The computed structure for row k of **U** is added to the linked list for its first off-diagonal entry and its number of entries is recorded. The number of arithmetic operations needed for FACTORIZE can be computed trivially from these numbers of entries (Exercise 10.7).

A separate pass can then be made to set up a static data structure for the rows of **U**. When the numerical values from **A** are placed into this data structure, zeros must be placed in fill-in positions and the numerical factorization process can then proceed efficiently (see next section). Since the numerical factorization is the most costly part of the computation, the careful preparatory work reaps its dividends.

Of course, one could combine the storage allocation with the generation of the data structure, but the combined algorithm would not be in place and could fail for lack of storage.

$$
\begin{array}{ccccc}
\times & \times & 0 & \times & \times \\
 & \times & 0 & \times & \times \\
 & & \times & \times & \times \\
 & & & \times & \times \\
 & & & & \times \\
\end{array}
$$

Figure 10.10.1. A matrix pattern.

Some economy of integer storage is possible with the compressed index scheme of Sherman (1975), which we now describe. If we look at the matrix of Figure 10.10.1, our usual storage scheme would hold the off-diagonal entries as a collection of packed vectors (Section 2.7) as shown in Table 10.10.1. The compressed index scheme makes use of the fact that the tail of one row often has the same pattern as the head of the next row

Subscript	1	2	3	4	5	6	7	8
LENROW	3	2	2	1				
IROWST	1	4	6	8	9			
JCN	2	4	5	4	5	4	5	5

Table 10.10.1. Matrix pattern of Figure 10.10.1 stored as a collection of sparse row vectors.

(particularly so for the factors after fill-in has occurred) and so the same indices can be reused. In order to do this, the length of each row (that is the number of off-diagonal entries) must be kept and we show the resulting data structure in Table 10.10.2. A similar gain is obtained automatically when using multifrontal schemes. Referring to Figure 10.8.4, we see that $NCOL*NPIV - NPIV*(NPIV-1)/2$ entries will require only $NPIV+NCOL$ indices.

Subscript	1	2	3	4
LENROW	3	2	2	1
IROWST	1	2	2	3
JCN	2	4	5	

Table 10.10.2. Matrix pattern of Figure 10.10.1 stored as a collection of sparse row vectors, using compressed storage.

We indicate some of the gains obtained on practical problems in Table 10.10.3 where, if compressed storage were not used, the number of indices would be equal to the number of entries in the factors. Indeed, on the regular five-diagonal pattern, the storage for the reals increases as $O(n^{3/2})$ for banded storage (or $O(n \log n)$ at best) while the integer storage for indices can be shown to be $O(n)$ (see Exercise 10.8).

Order	900	130
Nonzeros	7 744	1296
Entries in factors	34 696	1304
Number of indices	11 160	142

Table 10.10.3. Illustration of savings obtained through use of compressed storage scheme.

George and Liu (1980) have shown that if the compressed storage for indices is $r(\mathbf{U})$, symbolic factorization can be performed in time and space that is $O(r(\mathbf{U}))$. Since $r(\mathbf{U})$ can be, as we have seen, substantially less than $|\mathbf{U}|$, the number of entries in $\mathbf{U}$, the symbolic factorization time may grow with the size of problem less rapidly than $|\mathbf{U}|$. This is particularly true for problems arising from discretization of partial differential equations, where $r(\mathbf{U})$ is often $O(n)$.

10.11 Numerical FACTORIZE and SOLVE using static data structures

For the symbolic FACTORIZE we were able to make use of the fact that the only previous rows to contribute to the sparsity pattern of row k are those whose first entry after the diagonal lies in column k. For the numerical FACTORIZE, the situation is not quite so simple since in this case any row with an entry in column k will contribute. Nevertheless, a similar linked-list data structure (connecting rows which are used to update the same later row) can be used. Here each row of $\mathbf{U}$ is labelled by its first 'active' entry (Gustavson 1972). An entry is called **active** at stage k if it lies in columns k to n of $\mathbf{U}$.

```
      1 2 3 4 5 6 7 8 9 10
 1  u    u u    u        u
 2    u      u    u u
 3    u u    u        u u
 4        u    u u    u u
 5          u    u u u
 6            a a
 7            a    a
 8              a
 9                a
10                  a
```

Figure 10.11.1. Symmetric Gaussian elimination just before calculating row 6 of $\mathbf{U}$.

This concept and its implementation are most easily seen by looking at an example. In Figure 10.11.1, access is needed to rows 1, 3, and 4 of $\mathbf{U}$ when calculating row 6. The arithmetic is normally performed by loading row k of $\mathbf{A}$ ($k=6$ in the example) into a full vector (as in Section 9.3) which is then successively modified by multiples of the later parts of the rows with entries in column k (rows 1, 3, and 4 in the example). We leave the coding of this as an exercise for the reader (Exercise 10.9). It is usual to link all rows with first active entries in the same column, so that rows 1, 3, and 4 would be so linked.

Before continuing to process the next row $(k+1)$, all that is necessary is to update the linked lists since column k is no longer active. This update is easily effected when scanning the row (1, 3, or 4 in our example) since the next entry in the row will be identified and the row can be added to the appropriate list. In our example row 1 will be added to the list for column 10, row 3 to the list for column 9, and row 4 to the list for column 7. Additionally, the newly generated row 6 of U must be unloaded from the full vector and added to the list for column 7. Notice that the order in which the rows are held in the linked lists is unimportant.

If only the pivotal sequence is given, it is possible to generate the structure for U while performing the numerical factorization, but it is more efficient to perform the symbolic factorization a priori and use a static data structure, particularly if further factorizations are needed.

The numerical SOLVE phase is exactly as for the general case described in Section 9.4, except that only U (or only L) is stored and is used for both the forward substitution and the back-substitution.

10.12 Accommodating numerical pivoting within static data structures

As we saw in Sections 10.10 and 10.11, significant benefits and simplifications stem from the use of a static data structure during numerical factorization. Unfortunately, it is hard to achieve this and at the same time allow for numerical pivoting. Section 9.8 illustrates some of the penalties when the matrix is particularly sensitive to pivoting for numerical stability. However, some recent work allows numerical pivoting by setting up a data structure that includes entries in all positions that could conceivably fill-in during a factorization with numerical pivoting.

George and Ng (1985) have shown that for any row permutation P and factorization

$$PA = LU, \qquad (10.12.1)$$

the patterns of L^T and U are contained in the pattern of the Choleski factor $\bar{U}$, where

$$A^T A = \bar{U}^T \bar{U}. \qquad (10.12.2)$$

This suggests choosing a column permutation Q for A by an ANALYSE on the structure of $A^T A$. P can then be chosen to maintain stability in the knowledge that the pattern of $\bar{U}$ will certainly suffice to hold L^T and U. Thus a static data structure can be used for L and U, partial pivoting can be used without any further increase in storage, and the fast analysis available for the structure of a symmetric matrix can be used. The principal defect of the method is that $\bar{U}$ may be very dense compared with L and U of other methods (see Exercise 10.10) and indeed it is always likely to be worse since its structure must accommodate all row orderings **P**. George and Ng (1985) suggest some ways of trying to overcome this deficiency, involving update schemes where not all of the matrix is used in (10.12.2).

More recently, George and Ng (1987) suggest preordering the columns of **A** and at each stage of symbolic FACTORIZE they set the patterns of all the rows with entries in the pivot column equal to their union. The resulting data structure clearly includes the pattern of the L\U factorization that results from any choice of row interchanges. They show that the resulting data structure is usually smaller and is never larger than that of the George and Ng (1985) approach. However, it is still possible that it may seriously overestimate the storage needed for **L** and **U**.

10.13   Nested dissection, refined quotient tree, and one-way dissection

Nested dissection can be regarded as just another ordering of the sparse matrix. Hence once the ordering is determined, it may be passed directly to the multifrontal code of Section 10.8 or the symbolic FACTORIZE of Section 10.10. We discussed in Section 8.7 how the nested dissection ordering can be chosen.

Similarly, the refined quotient tree ordering or the one-way dissection ordering can be passed to the multifrontal or symbolic FACTORIZE code, but their success depends on the use of implicit factorization, as discussed in Sections 8.5 and 8.6. For a fuller discussion, see the book of George and Liu (1981). Note that one-way dissection is designed to exploit band or variable-band solutions of the diagonal blocks.

10.14   Effects of computer architecture

Throughout this and the previous chapter we have made several remarks about the influence of computer architectures. This section aims to draw these remarks together, summarize them, and add a few more comments. In particular we are concerned with architectures that include vector or parallel processing.

A characteristic of all vector and parallel processors is their ability to perform calculations on full matrices (matrix by matrix multiplication, matrix by vector multiplication, and Gaussian elimination) efficiently. With a little care it is possible to achieve both vectorization and parallelism when solving banded systems. Another possibility, when several separate systems are to be solved simultaneously (for example when solving partial differential equations in two or three space dimensions), is to vectorize across the systems (for example, Jordan 1979, Calahan 1982). This was mentioned for banded systems in Section 10.4. Clearly parallelism, too, can be exploited here. In the general sparse case, however, indirect addressing and the lack of long vectors or regular data patterns can cause problems. The use of special kernels for this loop can provide some worthwhile savings (Dodson 1981). With the introduction of the hardware gather-scatter feature on computers like the CRAY X-MP and Fujitsu FACOM VP/400, greater savings can be obtained, although the start-up times are high and the asymptotic rate is two to four times slower than for direct addressing.

Another way of exploiting vector architectures is to design codes so that the use of indirect addressing is avoided. We have already discussed some strategies which do this. Clearly the switch to full code (Section 9.5) will be particularly beneficial. Overall gains in speed of about a factor of four for unsymmetric ANALYSE and two for FACTORIZE and SOLVE can be obtained on the CRAY-1 by switching at very low densities (Duff 1984a). This causes little or no increase in overall storage. The frontal method of Sections 10.5 and 10.6 and the multifrontal method of Sections 10.7 and 10.8 also use direct addressing in their innermost loops and so do well on vector machines. For instance, inner loop speeds of 135 Megaflops and overall speeds of 109 Megaflops have been achieved on the CRAY-1 with experimental versions of the Harwell frontal code MA32. The multifrontal codes, however, have much more data handling outside the inner loop, which loop accounts for only about half of the execution time on a scalar machine. Thus the gains from vectorization are significant but not dramatic in this case.

If one examines the tree of Figure 10.7.3, possibilities of implementing multifrontal methods on multiprocessor machines suggest themselves (see Duff 1986, for example). At all levels of the tree beneath the root, there are two or more independent computations which could be performed by different processors. Of course, the amount of parallelism from this source decreases significantly as we approach the root. Fortunately, the frontal matrices become larger at the same time, so providing an alternative source for parallelism.

Unfortunately, we do not have good comparisons on common problem sets between the various implementations on different architectures. Because of the rapidly changing technologies, these comparisons are in a state of flux now anyway. For example, Lewis and Simon (private communication 1986) report that the hardware gather-scatter feature on the CRAY X-MP improved the speed of their minimum degree factorization of a structures problem of order 15 439 by the factor 3.4. This is enough to make it faster by a worthwhile margin than their implementation of the Reverse Cuthill-McKee algorithm, whereas the opposite was the case for the CRAY-1.

Exercises

10.1 Show how Wang's algorithm can be applied to a tridiagonal matrix whose order is not a product kp of two integers k and p of sizes that suit the parallelism available.

10.2 Indicate the amount of work and storage used by the frontal method when the front is of uniform width d.

10.3 Consider the triangulated region

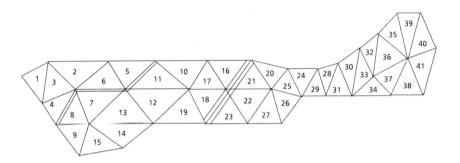

where there are single variables associated with each vertex, edge, and interior. If the elements are assembled in the order shown and no numerical pivoting is required, what is the order of the frontal matrix after assembly of element 6 and element 19? What is the maximum size of the frontal matrix and when is it achieved?

10.4 Perform a depth-first search (postorder) of the tree

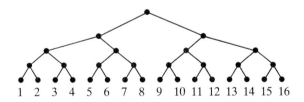

labelling the nodes in the assembly order thereby determined. If a stack is used to hold the intermediate results, what are its contents just before each time the top two elements are assembled? Show that the postordering of any tree will enable stack storage to be used for the generated elements.

10.5 Given the matrix whose pattern is

$$\begin{bmatrix} \times & & \times\times\times & & & & \\ & \times\times\times\times & & & & \\ \times\times\times & & \times & & & \\ \times\times & & \times\times & & & \\ \times\times\times\times\times\times & & \times\times & \\ & \times\times\times\times & & & \\ & & \times\times\times\times & \\ & & \times\times\times\times\times & \\ & & \times & \times\times\times \end{bmatrix},$$

show how the implementation of the minimum degree algorithm in Section 10.9 makes use of element absorption and recognition of identical rows. Assume that the matrix is in pivotal order.

10.6 Use a matrix argument to prove the validity of the observation, given in Section 10.10, that only previous columns of **L** whose *first* off-diagonal entry is in row k are required when calculating fill-in for column k of **L\U**.

10.7 If the number of off-diagonal entries in row i of **U** is n_i, $i = 1, 2, ..., n$, compute a formula for the number of floating-point operations involved in the numerical factorization.

10.8 Show that with Sherman's compressed index storage scheme, the integer storage needed for the triangular factors of the $n \times n$ matrix arising from the five-point discretization of the Laplacian operator on a square grid is $O(n)$ if pivots are chosen in the natural order.

10.9 Write Fortran code to convert a column of **A** to a column of **L** using the strategy outlined in Section 10.11. Include code to readjust the linked lists of columns whose first entries lie in the same row.

10.10 Give an example where the $\bar{\mathbf{U}}$ of (10.12.2) is significantly denser than the **L\U** factors.

11 Partitioning, matrix modification, and tearing

> We consider the solution of really huge systems, using the techniques of the previous chapters on subsystems or modified problems. We concentrate on the techniques used in the partitioning or modification of the huge system. We explore the relationship of these ideas to those of tearing (dividing the problem into parts).

11.1 Introduction

Some systems of equations are just too large to handle as one sparse matrix problem using the approach of the previous chapters. In this chapter, we show how a problem may be subdivided so that sparse matrix methods may be applied to the subproblems. We will call such problems **huge**. In some cases, this strategy results in a more efficient solution even if the problem is small enough to be treated as a single sparse matrix problem.

Quite different in motivation but related in solution technique is the problem of handling a perturbed system of equations. Here, having already solved $\mathbf{Ax} = \mathbf{b}$, we seek to solve $(\mathbf{A} + \mathbf{\Delta A})\mathbf{x} = \mathbf{b}$ where $\mathbf{\Delta A}$ is a matrix of low rank. This problem arises in contingency analysis, for example, where a sequence of slightly different problems is solved to evaluate alternatives. Of course, each problem could be treated as an independent sparse matrix problem, but in this chapter we show that much more efficient solution approaches are possible.

This suggests the related idea of solving $\mathbf{Ax} = \mathbf{b}$ by finding an artificial perturbation $\mathbf{\Delta A}$ chosen so that $(\mathbf{A} + \mathbf{\Delta A})\mathbf{x} = \mathbf{b}$ is easy to solve, then applying perturbation techniques to get the solution of the original problem from that of the perturbed one (Sections 11.7, 11.9, and 11.10). In some cases this is indeed the most efficient way to solve the original problem. Usually the artificial problem corresponds to removing (tearing out) part of the physical problem and hence the technique is called **tearing**. One of the application areas in which this technique has been widely used is that of power distribution. Here the term 'diakoptics' (Kron 1963) is normally used instead of tearing. We discuss tearing in Sections 11.11 to 11.16.

Two techniques are applicable to these situations, namely the use of matrix partitioning (Sections 11.2 and 11.3) and the use of the matrix modification formula (Sections 11.4 to 11.6). We will show that while they are closely related mathematically (Section 11.8), each offers advantages and disadvantages in practice.

Tearing can also be applied directly to nonlinear problems. We consider this case briefly in Section 11.16 and summarize our comments on this area in Section 11.17.

11.2 Exploiting the partitioned form

In Section 3.12 we explained that partitioning $\mathbf{A}$ into the 2×2 block form

$$\mathbf{A} = \begin{pmatrix} \mathbf{A}_{11} & \mathbf{A}_{12} \\ \mathbf{A}_{21} & \mathbf{A}_{22} \end{pmatrix} \qquad (11.2.1)$$

allowed the alternative factorizations

$$\begin{pmatrix} \mathbf{L}_{11} & \\ \mathbf{L}_{21} & \mathbf{L}_{22} \end{pmatrix} \begin{pmatrix} \mathbf{U}_{11} & \mathbf{U}_{12} \\ & \mathbf{U}_{22} \end{pmatrix}, \qquad (11.2.2)$$

$$\begin{pmatrix} \mathbf{A}_{11} & \\ \mathbf{A}_{21} & \bar{\mathbf{A}}_{22} \end{pmatrix} \begin{pmatrix} \mathbf{I} & \bar{\mathbf{U}}_{12} \\ & \mathbf{I} \end{pmatrix}, \qquad (11.2.3)$$

and

$$\begin{pmatrix} \mathbf{I} & \\ \ddot{\mathbf{L}}_{21} & \mathbf{I} \end{pmatrix} \begin{pmatrix} \mathbf{A}_{11} & \mathbf{A}_{12} \\ & \ddot{\mathbf{A}}_{22} \end{pmatrix}. \qquad (11.2.4)$$

It is easy to see that these partitioned forms can be useful for huge problems, since each step of the block forward substitution or block back-substitution can exploit the sparse matrix techniques of the previous chapters and each involves only a part of the whole problem. The gains associated with implicit factorization were discussed in Section 8.5. Furthermore the computation of $\mathbf{U}_{12}$ in (11.2.2) can be accomplished by solving the equation

$$\mathbf{L}_{11}\mathbf{U}_{12} = \mathbf{A}_{12} \qquad (11.2.5)$$

a few columns at a time. The columns of $\bar{\mathbf{U}}_{12}$ and rows of $\mathbf{L}_{21}$ or $\ddot{\mathbf{L}}_{21}$ can be generated in a similar fashion. Which of these three forms will be the most satisfactory for a sparse matrix will depend on the underlying problem. We illustrate this with two simple examples.

For the matrix partitioned in Figure 11.2.1, using the factorization (11.2.2) results in both $\mathbf{L}_{21}$ and $\mathbf{U}_{12}$ being dense. If (11.2.3) is used, $\mathbf{A}_{21}$ will remain sparse while the use of (11.2.4) will leave $\mathbf{A}_{12}$ sparse. This will cause either (11.2.3) or (11.2.4) to require both less storage and less computation when using the factors in forward substitution and back-substitution, see Exercise 11.1.

For the matrix partitioned in Figure 11.2.2, the opposite is true. The

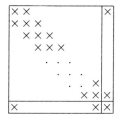

Figure 11.2.1. A partitioned sparse matrix.

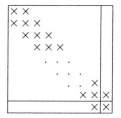

Figure 11.2.2. A second partitioned matrix.

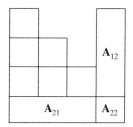

Figure 11.2.3. Partitioned $\mathbf{A}$ with $\mathbf{A}_{11}$ reducible.

standard factorization leaves both $\mathbf{L}_{21}$ and $\mathbf{U}_{12}$ sparse, while either of the other forms produces a dense off-diagonal block.

One interesting observation about these partitioned factorizations is that we may save computation on a symmetric matrix problem by ignoring some of the symmetry, as we found for Figure 11.2.1. George (1974) includes examples of these factorizations, including some where ignoring symmetry is advantageous. He also presents a particular class of matrices for which the best factorization can be predicted, but this class is so limited that we do not discuss it here. Rather, we leave the choice to special investigation for each case.

One strength of the partitioned approach arises when it is easy to solve sets of equations associated with the matrix $\mathbf{A}_{11}$. For example, partitions are often chosen so that $\mathbf{A}_{11}$ is block triangular, as illustrated in Figure 11.2.3. In this case, subproblems with matrix $\mathbf{A}_{11}$ can be solved by block forward substitutions (see Chapter 6) which need involve only one block at a time. Such a form is called bordered block triangular. We discussed one method of obtaining this form in Sections 8.8 and 8.9.

For the diagonal blocks we may use sparse **LU** factorization but there are other possibilities. For instance we may partition the blocks, that is construct a nested bordered block triangular form, or the blocks may correspond to problems with sufficient regularity for a fast-direct method to be applied (see Buzbee and Dorr 1974 and Swarztrauber 1977) for example). As a simple example of the latter, consider the discretization of a partial differential equation over an L-shaped region as shown in Figure 11.2.4. If we put into the border all the variables associated with nodes on the thin line, then the diagonal blocks will correspond to a discretization on a regular grid in a rectangle and fast direct solution is applicable.

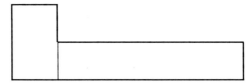

Figure 11.2.4. An L-shaped region.

Sometimes an iterative method may be a good way to solve the equations associated with A_{11}. This can happen if there are a small number of irregularities in a problem which otherwise is easy to solve; in this case the irregularities are placed in the border consisting of A_{21}, A_{22}, and A_{12}. Another case is an optimization problem with a small number of constraints; here, when using the Lagrangian formulation, the overall matrix is indefinite but with the constraints in the border we have a positive-definite A_{11} for which a good iterative method may be available.

11.3 Using partitioning to handle low-rank perturbations

Now consider the case where we have already solved $Ax = b$ and wish to solve a perturbed problem

$$(A + \Delta A)x = b \qquad (11.3.1)$$

where ΔA has low rank.

Suppose, first of all, that the perturbation can be arranged to have the form

$$\Delta A = \begin{pmatrix} 0 & 0 \\ 0 & \Delta A_{22} \end{pmatrix}, \qquad (11.3.2)$$

that is, the given problem can be permuted so that all the changed entries are confined to the lower right-hand corner of the matrix. If $\mathbf{A}$ is partitioned as in (11.3.2), the partitioned approach may work very well for this perturbed problem. All we have to do is to factorize the Schur complement, which is now $\mathbf{A}_{22} - \mathbf{A}_{21}\mathbf{A}_{11}^{-1}\mathbf{A}_{12} + \Delta\mathbf{A}_{22}$. If the changes cannot be constrained to the lower corner, but can be confined to one or both borders, for example

$$\Delta\mathbf{A} = \begin{pmatrix} \mathbf{0} & \Delta\mathbf{A}_{12} \\ \mathbf{0} & \Delta\mathbf{A}_{22} \end{pmatrix}, \tag{11.3.3}$$

this can also be exploited to simplify the cost of solving $(\mathbf{A} + \Delta\mathbf{A})\mathbf{x} = \mathbf{b}$, see Exercise 11.4. Note, however, that frequently a change in an off-diagonal entry is associated with changes in the diagonal entries in its row and column, and if this is the case for all changes, we have form (11.3.2) rather than (11.3.3).

There are some hidden costs with this way of solving the perturbed problem. First, the perturbation $\Delta\mathbf{A}$ constrains the choice of partition and this may well conflict with other objectives such as out-of-core treatment of large problems or improving efficiency in the solution of $\mathbf{A}\mathbf{x} = \mathbf{b}$. Second, we may want to solve a sequence of perturbed problems

$$(\mathbf{A} + \Delta\mathbf{A}^{(i)})\mathbf{x} = \mathbf{b}, \quad i = 1, 2,\dots . \tag{11.3.4}$$

If each $\Delta\mathbf{A}^{(i)}$ corresponded to changes in a different submatrix of $\mathbf{A}$, the partition would have to be such that $\mathbf{A}_{22}$ (or $\mathbf{A}_{12}$, $\mathbf{A}_{21}$, and $\mathbf{A}_{22}$) contained *all* of these changes. In the next section we show that the matrix modification formula (11.4.1) can be used to solve the perturbed problem *without* constraining the way in which we solve the original problem $\mathbf{A}\mathbf{x} = \mathbf{b}$.

11.4 The matrix modification formula

The matrix modification formula was defined by Sherman and Morrison (1949) for rank-one changes and was generalized by Woodbury (1950). Expressed in terms of matrix inverses it has the form

$$(\mathbf{A} + \mathbf{V}\mathbf{S}\mathbf{W}^T)^{-1} = \mathbf{A}^{-1} - \mathbf{A}^{-1}\mathbf{V}(\mathbf{S}^{-1} + \mathbf{W}^T\mathbf{A}^{-1}\mathbf{V})^{-1}\mathbf{W}^T\mathbf{A}^{-1}. \tag{11.4.1}$$

In this formula, $\mathbf{V}\mathbf{S}\mathbf{W}^T$ is a rank-k matrix, where $\mathbf{V}$ and $\mathbf{W}$ are $n \times k$ matrices and $\mathbf{S}$ is $k \times k$, with k usually much less than n. This formula may be readily verified and its verification is left to Exercise 11.5. To help the reader appreciate the shapes of the matrices we show equation (11.4.1) diagrammatically in Figure 11.4.1.

Working with explicit inverses is not sensible in our case since the inverse of a sparse matrix is usually full (see Section 12.6). However, we may multiply the right-hand side vector, $\mathbf{b}$, by (11.4.1) to yield the relation

$$(\mathbf{A} + \mathbf{V}\mathbf{S}\mathbf{W}^T)^{-1}\mathbf{b} = \mathbf{A}^{-1}\mathbf{b} - \mathbf{A}^{-1}\mathbf{V}(\mathbf{S}^{-1} + \mathbf{W}^T\mathbf{A}^{-1}\mathbf{V})^{-1}\mathbf{W}^T\mathbf{A}^{-1}\mathbf{b}. \tag{11.4.2}$$

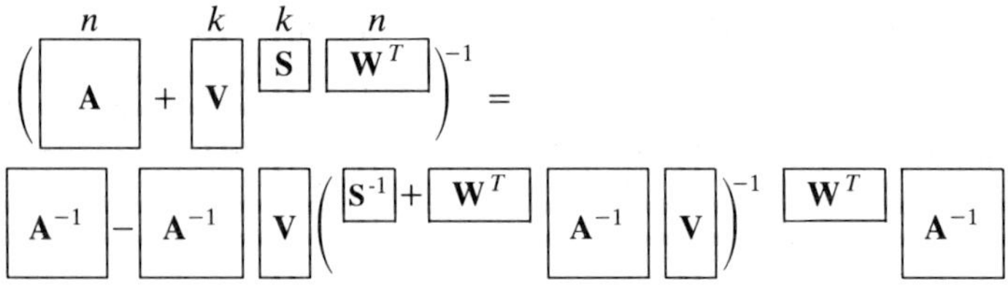

Figure 11.4.1. The matrix modification formula.

We see that the solution to the modified problem $(\mathbf{A} + \mathbf{VSW}^T)^{-1}\mathbf{b}$ is given as the solution to the original problem $\mathbf{A}^{-1}\mathbf{b}$ and a correction term. We discuss the representation of a rank-k modification $\mathbf{\Delta A}$ as $\mathbf{VSW}^T$ in the next section and examine several ways of computing the correction term in Section 11.6.

11.5 Low-rank modifications

A rank-one modification to $\mathbf{A}$ can be one entry, a_{ij} say, in which case it is easily represented in $\mathbf{VSW}^T$ form by

$$\mathbf{S} = (a_{ij}) \qquad 1 \times 1 \qquad\qquad (11.5.1a)$$

$$\mathbf{V} = \mathbf{e}_i \qquad n \times 1 \qquad\qquad (11.5.1b)$$

$$\mathbf{W} = \mathbf{e}_j \qquad n \times 1 \qquad\qquad (11.5.1c)$$

where $\mathbf{e}_k$ is column k of the identity matrix. A rank-one matrix could also be an entire column $\mathbf{v}$ which modifies the j-th column of $\mathbf{A}$, so that $\mathbf{V} = \mathbf{v}$, $\mathbf{W} = \mathbf{e}_j$, $\mathbf{S} = (1)$ and

$$\mathbf{\Delta A} = \mathbf{ve}_j^T. \qquad\qquad (11.5.2)$$

More generally, if $\mathbf{\Delta A}$ corresponds to modifying columns $i_1, i_2,\ldots, i_k$ of $\mathbf{A}$ by $\mathbf{v}_1, \mathbf{v}_2,\ldots, \mathbf{v}_k$, then $\mathbf{\Delta A}$ may be represented as

$$\mathbf{\Delta A} = \begin{pmatrix} \mathbf{v}_1 & \cdots & \mathbf{v}_k \end{pmatrix}(\mathbf{I}_k)\begin{pmatrix} \mathbf{e}_{i_1}^T \\ \cdot \\ \cdot \\ \cdot \\ \mathbf{e}_{i_k}^T \end{pmatrix} \qquad\qquad (11.5.3)$$

where $\mathbf{I}_k$ is the $k \times k$ identity matrix. Modifications of rows can be represented similarly.

Further examples can be found in Exercises 11.6 and 11.7. We note that a matrix can be very dense and still have low rank. For example, the matrix all of whose entries are 1 has rank one (see Exercise 11.8).

11.6 Use of the modification formula to solve equations

All of the implementations of equation (11.4.2) make use of the fact that we have already solved equations with $\mathbf{A}$, that is we have computed $\mathbf{x} = \mathbf{A}^{-1}\mathbf{b}$. The correction to $\mathbf{x}$, $\mathbf{w}$ say, may be computed using the following five steps :

$$\text{Solve}\quad \mathbf{AX} = \mathbf{V}. \tag{11.6.1a}$$

$$\text{Form}\quad \mathbf{Z} = \mathbf{W}^T\mathbf{X}. \tag{11.6.1b}$$

$$\text{Form}\quad \mathbf{d} = \mathbf{W}^T\mathbf{x}. \tag{11.6.1c}$$

$$\text{Solve}\quad (\mathbf{S}^{-1} + \mathbf{Z})\mathbf{y} = \mathbf{d}. \tag{11.6.1d}$$

$$\text{Form}\quad \mathbf{w} = \mathbf{Xy}. \tag{11.6.1e}$$

Since most authors follow these steps, we refer to (11.6.1) as the **natural implementation** of (11.4.2). It is illustrated diagrammatically in Figure 11.6.1.

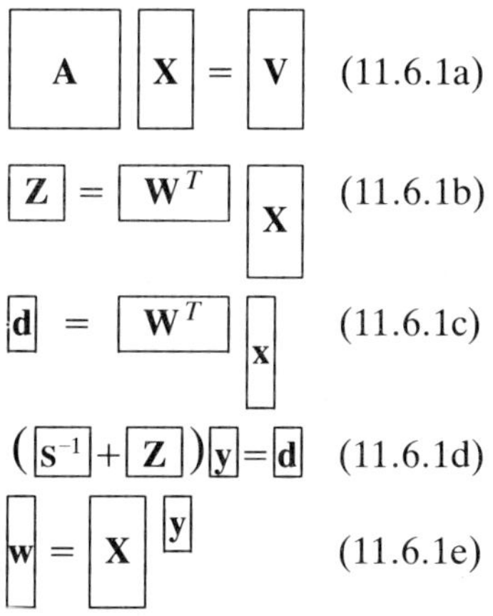

Figure 11.6.1. The natural implementation of the matrix modification formula for solving an equation.

In the usual case with $k \ll n$, steps (11.6.1a) and (11.6.1b) require the most computation. Step (11.6.1a) corresponds to solving k sets of equations whose coefficient matrix is $\mathbf{A}$, and step (11.6.1b) requires the product of a sparse $k \times n$ matrix and an $n \times k$ matrix that is probably full. Note that further sets of equations with matrix $\mathbf{A} + \mathbf{VSW}^T$ can be solved using steps (11.6.1c) to (11.6.1e) only.

We have now removed the constraint that the partition must be in the lower right-hand corner (see equation 11.3.2). Indeed, we have also allowed a general rank-k perturbation rather than one confined to a submatrix. On the other hand, the solution of k sets of equations in step (11.6.1a) represents considerably more work than simply refactorizing a $k \times k$ submatrix. We seek alternative implementations which will reduce this work.

Two obvious alternatives to steps (11.6.1) are based on different ways of computing Z. In (11.6.1) we computed Z as $W^T(A^{-1}V)$, but we can compute instead $(W^T A^{-1})V$ which gives the steps:

$$\text{Solve} \quad A^T Y = W. \quad [Y^T = W^T A^{-1}] \quad (11.6.2a)$$

$$\text{Form} \quad Z = Y^T V. \quad (11.6.2b)$$

$$\text{Form} \quad d = W^T x. \quad (11.6.2c)$$

$$\text{Solve} \quad (S^{-1} + Z)y = d. \quad (11.6.2d)$$

$$\text{Form} \quad c = V y. \quad (11.6.2e)$$

$$\text{Solve} \quad A w = c. \quad (11.6.2f)$$

In addition to replacing (11.6.1a) by (11.6.2a), we have had to replace (11.6.1e) by (11.6.2e) and (11.6.2f). The matrix V is probably sparser than X but, on the other hand, we must add the forward substitution and back-substitution steps on the single vector c in (11.6.2f).

If the factorization $A = LU$ is available, another way of computing $W^T A^{-1} V$ is to compute $(W^T U^{-1})(L^{-1}V)$. This suggests the steps:

$$\text{Solve} \quad LX^{(1)} = V. \quad [X^{(1)} = L^{-1}V]$$

$$\text{and} \quad U^T X^{(2)} = W. \quad [X^{(2)\,T} = W^T U^{-1}] \quad (11.6.3a)$$

$$\text{Form} \quad Z = X^{(2)\,T} X^{(1)}. \quad (11.6.3b)$$

$$\text{Form} \quad d = W^T x. \quad (11.6.3c)$$

$$\text{Solve} \quad (S^{-1} + Z)y = d. \quad (11.6.3d)$$

$$\text{Form} \quad c = X^{(1)}y. \quad (11.6.3e)$$

$$\text{Solve} \quad U w = c. \quad (11.6.3f)$$

In this case, the k forward substitution and k back-substitution steps in (11.6.1a) and (11.6.2a) are replaced by $2k$ forward substitutions in (11.6.3a). This may be an advantage since right-hand side sparsity can be exploited during forward substitution but is probably absent during back-substitution (see Sections 7.9 and 9.4). $X^{(1)}$ in (11.6.3e) is probably less sparse than V in (11.6.2e) but probably more sparse than X in (11.6.1e). We add the single back-substitution step (11.6.3f).

These three variations are similar to the variations in the partitioned solution discussed in Sections 3.12 and 11.2. In fact, we will show in Section 11.8 that there is a rather simple relationship between the matrix modification formula and the partitioned solution. Which variation of the modification formula is best depends on the problem, as for the partitioned solution.

We now illustrate the advantage in step (11.6.3a) of avoiding back-substitutions. Consider again the special case

$$\Delta \mathbf{A} = \begin{pmatrix} 0 & 0 \\ 0 & \Delta \mathbf{A}_{22} \end{pmatrix},$$
(11.6.4)

where we can make the choices

$$\mathbf{V} = \begin{pmatrix} 0 \\ \mathbf{I} \end{pmatrix} = \mathbf{W}, \quad \mathbf{S} = \Delta \mathbf{A}_{22}.$$
(11.6.5)

Step (11.6.3a) involves only forward substitutions and so reduces to a $k \times k$ problem. These steps then become very similar to those in the partitioned solution for this special case. More work would be involved in the implementation of either algorithm (11.6.1) or algorithm (11.6.2).

Unfortunately, to achieve this computational efficiency we have still had to constrain the reordering of $\mathbf{A}$ so that $\Delta \mathbf{A}$ only affects the lower corner $k \times k$ submatrix. In the next section we show another implementation of (11.4.2). This provides similar computational efficiency in the case where $\Delta \mathbf{A}$ corresponds to a $k \times k$ submatrix but it is not restricted to a modification of only the lower corner.

11.7 Perturbations that are full-rank submatrices

In this section we consider the common case where $\Delta \mathbf{A}$ has rank k and has nonzeros in only k rows and columns. In this case it can be expressed in the form

$$\Delta \mathbf{A} = \mathbf{V} \mathbf{S} \mathbf{W}^T,$$
(11.7.1)

where $\mathbf{V}$ and $\mathbf{W}$ are $n \times k$ matrices containing columns of the identity matrix. Specifically, $\mathbf{V}$ has a 1 in each row corresponding to a nonzero row of $\Delta \mathbf{A}$, and $\mathbf{W}$ has a 1 in each row corresponding to a nonzero column of $\Delta \mathbf{A}$. $\mathbf{S}$ contains the nonzero entries of $\Delta \mathbf{A}$.

In this case, $\mathbf{Z} = \mathbf{W}^T \mathbf{A}^{-1} \mathbf{V}$ is just *the $k \times k$ submatrix of $\mathbf{A}^{-1}$ corresponding to the positions changed by $\Delta \mathbf{A}^T$*. In Section 12.7 we show that entries of $\mathbf{A}^{-1}$ within the sparsity pattern of the factors of $\mathbf{A}^T$ may be computed very economically. Where several perturbations $\Delta \mathbf{A}$ are anticipated, for example in a contingency analysis, it is probably best to compute all the required entries in $\mathbf{A}^{-1}$ first. We therefore assume that $\mathbf{Z}$ is available so that steps (11.6.2a) and (11.6.2b) may be omitted. Notice, too,

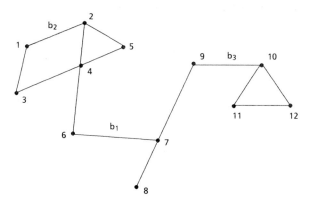

Figure 11.7.1. Example network.

that steps (11.6.2c) and (11.6.2e) involve no computation at all. $\mathbf{W}^T\mathbf{x}$ simply selects k entries from $\mathbf{x}$ (a gather) while $\mathbf{V}\mathbf{y}$ places the components of $\mathbf{y}$ into an n vector of zeros in the appropriate k positions (a scatter). Given the appropriate entries of $\mathbf{A}^{-1}$, we have only to solve the $k \times k$ system (11.6.2d) and perform a single forward substitution and back-substitution (11.6.2f). Further, no ordering constraint has been placed on $\mathbf{A}$ to anticipate this modification.

We illustrate these ideas with the network shown in Figure 11.7.1. This could correspond to an electric power system network where we want to study the impact of modifying b_1 (changing matrix entries a_{66}, a_{77}, a_{67}, and a_{76}), b_2 (changing matrix entries a_{11}, a_{22}, a_{12}, and a_{21}), or b_3 (changing matrix entries a_{99}, $a_{10,10}$, $a_{9,10}$, and $a_{10,9}$) — a contingency analysis. In this case, the matrix has a symmetric pattern whose graph is shown in Figure 11.7.1. Perturbing b_1 corresponds to changes in rows and columns 6 and 7, so a suitable choice is given by $\mathbf{V} = \mathbf{W} = (\mathbf{e}_6\ \mathbf{e}_7)$ and $\mathbf{Z}$ is the submatrix of rows and columns 6 and 7 of $\mathbf{A}^{-1}$. This is also the case in the admittance formulation of the electrical network problem.

Similarly, perturbing b_2 requires the submatrix of rows 1 and 2 of $\mathbf{A}^{-1}$ and perturbing b_3 requires the submatrix of rows 9 and 10. In view of symmetry we therefore need inverse entries only in the (1,1), (1,2), (2,2), (6,6), (6,7), (7,7), (9,9), (9,10), and (10,10) positions, provided we never need to consider modifying more than one of b_1, b_2, and b_3 at once.

11.8 The equivalence of modification and partitioning

The solution of any partitioned system

$$\begin{pmatrix} \mathbf{A}_{11} & \mathbf{A}_{12} \\ \mathbf{A}_{21} & \mathbf{A}_{22} \end{pmatrix} \begin{pmatrix} \mathbf{x}_1 \\ \mathbf{x}_2 \end{pmatrix} = \begin{pmatrix} \mathbf{b}_1 \\ \mathbf{b}_2 \end{pmatrix} \tag{11.8.1}$$

can be obtained by solving the system that results from changing the matrix
to the block form

$$\bar{\mathbf{A}} = \begin{pmatrix} \mathbf{A}_{11} & \\ & \mathbf{I} \end{pmatrix} \tag{11.8.2}$$

and then applying the modification matrix

$$\mathbf{VSW}^T \tag{11.8.3}$$

with

$$\mathbf{V} = \begin{pmatrix} \mathbf{A}_{12} & \mathbf{0} \\ \mathbf{0} & \mathbf{I} \end{pmatrix}, \mathbf{S} = \begin{pmatrix} \mathbf{I} & \mathbf{0} \\ \mathbf{0} & \mathbf{I} \end{pmatrix}, \mathbf{W}^T = \begin{pmatrix} \mathbf{0} & \mathbf{I} \\ \mathbf{A}_{21} & \mathbf{A}_{22}-\mathbf{I} \end{pmatrix}. \tag{11.8.4}$$

If the matrix modification formula (11.4.2) is used to solve the system with
modified coefficient matrix

$$\bar{\mathbf{A}} + \mathbf{VSW}^T, \tag{11.8.5}$$

then the formulae reduce to those for the solution of the partitioned system
(11.8.1). The amount of algebra involved in reducing the formulae (see
Exercise 11.9) clearly indicates that this route is unlikely to be the best
approach for solving (11.8.1); that is, when the perturbation can be
confined to border partitions, the partitioned form is the most reasonable
to use. When the ordering constraint caused by partitioning the problem is
too severe, or when a sequence of changes in different parts of the matrix is
desired, the use of the modification formula is more appropriate.

A general modification cannot be viewed as a matrix partition.
However, it is possible to establish an equivalence by considering an
augmented problem (Bunch and Rose 1974). The perturbed problem

$$(\mathbf{A} + \mathbf{VSW}^T)\mathbf{x} = \mathbf{b} \tag{11.8.6}$$

is equivalent to the equation

$$\begin{pmatrix} \mathbf{A} & \mathbf{V} \\ \mathbf{W}^T & -\mathbf{S}^{-1} \end{pmatrix} \begin{pmatrix} \mathbf{x} \\ \mathbf{y} \end{pmatrix} = \begin{pmatrix} \mathbf{b} \\ \mathbf{0} \end{pmatrix}. \tag{11.8.7}$$

This can be readily verified by adding $\mathbf{VS}$ times the second block row to the
first block row to give the relation (11.8.6).

Now suppose we use one of the alternative methods discussed in Section

11.2 for solving this partitioned problem. In each case there results one of the Section 11.6 implementations of the modification formula. For instance (11.2.3) corresponds exactly to (11.6.1), (11.2.4) corresponds exactly to (11.6.2), and (11.2.2) corresponds to (11.6.3) with some minor rearrangements of terms. This incidentally provides an alternative proof of the validity of the modification formula. For the solution of perturbed problems, we have thus shown that careful implementations of both the modification formula (11.4.2) and the partitioned form (11.8.7) lead to equivalent computation.

It is no accident that we have focused on modifying the solution. Only in the special case, highlighted in Section 11.3, where the changes could be effectively confined to the later columns or the lower corner do we compute the **LU** factors of $\mathbf{A} + \mathbf{\Delta A}$. Though research has been done in this area, see Law and Fenves (1981) for example, it appears that in general modifying the **LU** factors of $\mathbf{A}$ is more costly than simply modifying the solution to $\mathbf{Ax} = \mathbf{b}$. An alternative factorization is used for linear and nonlinear programming (see Reid 1982, for example), but its description is beyond the scope of this book.

11.9 Modifications that change the matrix order

For simplicity, we have assumed so far that the perturbation matrix has the same dimension as $\mathbf{A}$, but this need not be the case. For example, we might wish to replace branch b_1 in Figure 11.7.1 by an entire subnetwork as shown in Figure 11.9.1. Similar methods may be applied in this case.

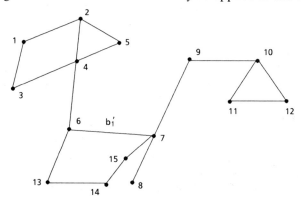

Figure 11.9.1. A perturbation of the Figure 11.7.1 network that changes the dimension.

The idea is most clearly seen from the partitioned form induced by the perturbation. Referring to Figure 11.7.1, changing the value of b_1

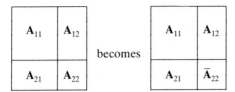

Figure 11.9.2. Partitioned view of a perturbation that does not change the dimension.

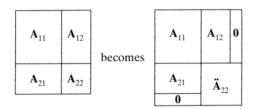

Figure 11.9.3. Partitioned view of a perturbation that changes the dimension.

corresponds to changing $\mathbf{A}_{22}$ as shown in Figure 11.9.2. Now if b_1 is changed to the subnetwork shown in Figure 11.9.1, this corresponds to the partitioned form change shown in Figure 11.9.3. Since $\mathbf{A}_{11}$, $\mathbf{A}_{12}$, and $\mathbf{A}_{21}$ all remain the same, all of the advantage of the previous strategies may be achieved in spite of the fact that $\ddot{\mathbf{A}}_{22}$ and $\mathbf{A}_{22}$ have different dimensions.

Similarly, the matrix modification formula can be applied to a problem where the perturbation is of different dimension. In this case, the original matrix can be artificially extended by a simple matrix (for example, the identity matrix) corresponding to the newly introduced nodes and then the perturbation can be treated as before.

11.10 Artificial perturbations

We have shown that, given the solution to one problem, we can easily find the solution of a perturbed problem when the perturbation has low rank.

The methods for solving perturbed problems suggest another idea. Perhaps an artificial perturbation of low rank can be introduced which would make the perturbed problem 'easy' to solve. Then it might be possible that computational savings could be achieved, compared with solving the given problem, from a two-step process: solve the 'easy' problem and then use the perturbation techniques to solve the original problem. In this case, the success of the idea will depend on being able to find the perturbation which will cause the two-step process to be less expensive than solving the original problem.

That such a process could be effective is illustrated by the example

$$
\mathbf{A} = \begin{pmatrix}
2 & 1 & 1 & 1 & 1 & 1 \\
1 & 2 & 1 & 1 & 1 & 1 \\
1 & 1 & 2 & 1 & 1 & 1 \\
1 & 1 & 1 & 2 & 1 & 1 \\
1 & 1 & 1 & 1 & 2 & 1 \\
1 & 1 & 1 & 1 & 1 & 2
\end{pmatrix} . \tag{11.10.1}
$$

Since the matrix is dense and symmetric, straightforward direct solution would need $\frac{1}{3}n^3 + O(n^2)$ floating-point operations. But by choosing

$$
\mathbf{\Delta A} = \begin{pmatrix} 1 \\ 1 \\ 1 \\ 1 \\ 1 \\ 1 \end{pmatrix} (1) \, (1\ 1\ 1\ 1\ 1\ 1) , \tag{11.10.2}
$$

we have

$$
\mathbf{A} - \mathbf{\Delta A} = \begin{pmatrix}
1 & & & & \\
& 1 & & & \\
& & 1 & & \\
& & & 1 & \\
& & & & 1
\end{pmatrix} . \tag{11.10.3}
$$

Thus the solution to $\mathbf{Ax} = \mathbf{b}$ using modification can be computed in $2n$ operations.

We note in passing that this use of the modification formula is equivalent (see equation (11.8.7)) to solving

$$
\begin{pmatrix}
1 & & & & & & 1 \\
& 1 & & & & & 1 \\
& & 1 & & & & 1 \\
& & & 1 & & & 1 \\
& & & & 1 & & 1 \\
& & & & & 1 & 1 \\
1 & 1 & 1 & 1 & 1 & 1 & -1
\end{pmatrix}
\begin{pmatrix} \mathbf{x} \\ \mathbf{y} \end{pmatrix}
=
\begin{pmatrix} \mathbf{b} \\ \mathbf{0} \end{pmatrix} \tag{11.10.4}
$$

We could certainly not have obtained this by partitioning the original problem.

11.11 Branch tearing

If removing a small number of branches from the graph of the matrix yields a graph that is disconnected into parts, then it is natural to consider the problem resulting after their removal. It will be easier to solve because each part may be treated independently. Such a solution is of no interest to us except in that by using our techniques for treating low-rank perturbations we can solve the problem that really interests us. An example is illustrated in Figures 11.11.1 and 11.11.2. Note that the artificial perturbations consist of the removal of pairs of off-diagonal entries. The technique is called **branch tearing**.

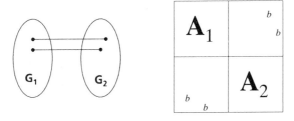

Figure 11.11.1. Two interconnected subsystems.

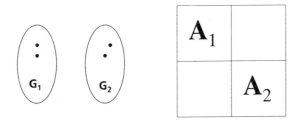

Figure 11.11.2. The torn problem.

For illustration we have used undirected graphs corresponding to symmetrically patterned matrices but the concepts apply equally well to directed graphs.

If we consider branch tearing from a physical system viewpoint, as Kron (1968) and Happ (1980) do, rather than a matrix viewpoint, some differences result. For them, tearing branches corresponds to analysing the model which would result from tearing the branches from the physical system rather than just removing the off-diagonal entries from the matrix. The difference this makes is that their model may change the diagonal matrix entries where these branches connect, as in the case of the admittance matrix mentioned in Section 11.7. In this case branch tearing would be as shown in Figure 11.11.3.

As a historical note, the motivation for Kron's tearing came from using inverses explicitly in solving equations. The inverse of the original matrix is probably dense while the inverse of the reducible torn matrix preserves the off-diagonal zero blocks.

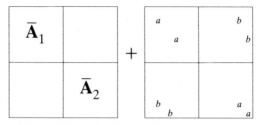

Figure 11.11.3. Physical branch tearing.

11.12 Node tearing

Another way to disconnect a graph into parts is to remove some nodes together with all the branches that end on them. This is known as **node tearing**.

To illustrate node tearing, consider Figure 11.12.1, which is the same as Figure 11.11.1 except that we have shown some of the internal structure of the graph G_2. We have shown other nodes within G_2 which connect to those nodes that are connected to G_1. Node tearing for this example is illustrated in Figures 11.12.1 and 11.12.2. In the second figure, we have separated out the two tear nodes into a new subgraph G_3 and the remainder of G_2, which we label as $\overline{G}_2$. For the matrix form, we reorder the rows and columns of A_2 so that the two rows and columns corresponding to the tear are last and then partition it to give the form shown in Figure 11.12.2.

Node tearing can also be viewed as analysing the nodal equivalent to a torn physical system. As in the branch tearing case, this interpretation would lead to changes in both A_1 and $\overline{A}_2$.

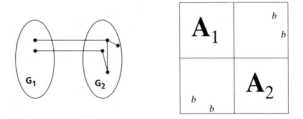

Figure 11.12.1. Another view of Figure 11.11.1.

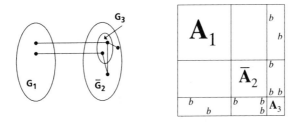

Figure 11.12.2. An alternative view of Figure 11.12.1.

11.13 Implementation of node and branch tearing

There are several ways, using either the partitioned approach or the matrix modification formula, to implement a tearing solution. We could, for example, apply the partitioned solution ideas directly to the matrix in Figures 11.11.1 and 11.12.1. In this case, the advantage which comes from tearing is simply the partitioned form which it induces.

Using the matrix modification formula, we could use the rank-four change consisting of the four nonzeros in the off-diagonal blocks of Figure 11.11.1. However, a simpler and more effective modification is the rank-two change consisting of the two nonzeros in one of the off-diagonal blocks. The modified problem is block triangular instead of block diagonal, but this makes it no harder to solve the associated sets of equations and we can save work overall because of the reduced rank of the change.

11.14 Manual choice of partitions and tear sets

So far we have taken the partitions or tear sets to be given and have discussed their use. We now consider how they might be chosen. Physical insight is often available and should be used when it is. Many large physical models are constructed by interconnecting submodels, and this submodel structure is just what is potentially useful in partitioning the problem. Figure D.10 in Appendix D contains an example of a 1624-order matrix corresponding to an interconnected power system. A potential sub-structure is obvious and, in fact, is available from the data which was entered by subsystems. We do not know, however, whether the use of this substructure would be of benefit in solving equations.

In the case of node tearing, an objective is to find a small number of nodes which when removed, along with the branches to and from them, results in a decomposable problem. This suggests the aim of finding a small

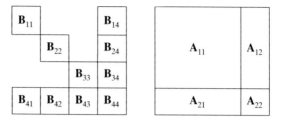

Figure 11.14.1. Bordered block diagonal form partition.

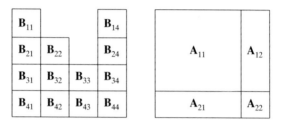

Figure 11.14.2. Bordered block triangular form partition.

number of nodes whose removal would result in a decomposition of the remaining graph into pieces. The partition resulting from such a strategy would have one of the forms given in Figures 11.14.1 and 11.14.2, if three pieces resulted.

The work in solving the partitioned equations will be modest if A_{11} is 'very reducible', that is there are many blocks B_{ii} of similar size rather than one large block, and A_{22} is as small as possible. This is because the factorization work for A_{11} is just that involved in the factorization of the diagonal blocks.

Other objectives have also been considered, such as the bordered triangular form shown in Figure 11.14.3. The advantage of this is that no factorization work is required for A_{11}. The disadvantage is that for many practical problems the order of A_{22} is very large and the savings in the A_{11} partition do not compensate for the loss in the border.

Of course the block triangular form and the block diagonal form are desirable. These reducible forms, illustrated in Figure 11.14.4, were considered in Chapter 5.

Finally, we emphasize again that these partitioning techniques are not necessarily numerically stable. The discussion of Sections 4.4 and 8.9 is important.

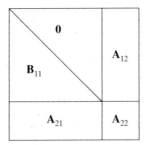

Figure 11.14.3. Partition corresponding to bordered triangular form.

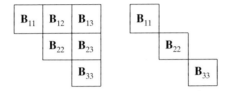

Figure 11.14.4. The block triangular form and the block diagonal form.

11.15 Automatic choice of partitions and tear sets

Given the objectives of the previous section, we now seek an algorithm which can achieve the desired partitions based on information contained in the graph. While there is a great deal of literature on this subject, many of the proposed algorithms are heuristic with very few test results. Further, there have been few comparisons of one method with another. This is still a very active research area and we simply cite some references.

Two categories of algorithms have potential for dealing with this problem:

(a) sparse matrix reorderings that can be interpreted as partitionings, and

(b) algorithms that address this objective directly.

In Chapter 8 we considered several ordering heuristics that produce partitioned forms: nested dissection, one-way dissection, the refined quotient tree, and the Hellerman-Rarick algorithm. The various implicit strategies for dealing with partitioned forms discussed in Section 3.12 can thus be applied to these orderings. George and Liu (1978a) have shown that the implicit partitioned factorization applied to the one-way dissection ordering can be particularly effective.

In generalizing the one-way and nested-dissection orderings to irregular graphs, George and Liu (1978b) suggested using the level set concept to find separators in the 'middle' of the graph (see Sections 8.6 and 8.7). While this works well on fairly regular graphs arising from partial differential equations or structural analysis, it is much less effective on very irregular graphs from networks. On these, we have found it worthwhile to move the 'separator' set slightly away from the middle if a smaller set can be found. This is in line with the objective of making $\mathbf{A}_{22}$ as small as possible while achieving a maximum breakup of the rest of the graph.

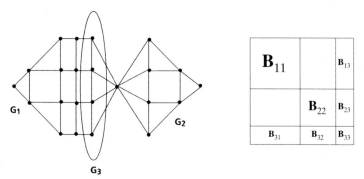

Figure 11.15.1. Partition induced by the middle separator. $\mathbf{B}_{11}$, $\mathbf{B}_{22}$, and $\mathbf{B}_{33}$ have sizes 11×11, 8×8, and 4×4, respectively.

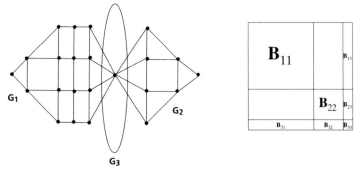

Figure 11.15.2. Partition induced by a small separator near the middle. $\mathbf{B}_{11}$, $\mathbf{B}_{22}$, and $\mathbf{B}_{33}$ have sizes 15×15, 7×7, and 1×1, respectively.

This idea is illustrated in Figure 11.15.1 and Figure 11.15.2. The first shows the effect of selecting a 'middle' separator while the second shows the benefit of finding a small near-the-middle separator.

Research work on the use of graphs has been done by Steward (1965,

1969), Tarjan (1976), and Sangiovanni-Vincentelli (1976). Their objective was to develop algorithms for finding small separators to achieve reducible remaining graphs. Lin and Mah (1977) similarly propose a heuristic to select a row and column from the matrix and then, effectively, apply an algorithm to reduce the remaining submatrix to see if decomposition is possible. Cheung and Kuh (1974) proposed an algorithm for minimizing A_{22} in finding the bordered triangular form, see Figure 11.14.3. Barnes (1981) looked at the graph partitioning problem, motivated by the layout problem for circuit design. Because a good solution could save manufacturing costs, not just computing costs, he could afford a much more elaborate ordering. These references just scratch the surface, but in spite of all of them we are not familiar with any that are presently in standard use with a sparse matrix package.

11.16 Nonlinear tearing

Usually nonlinear systems of equations are solved with some variant of Newton's method. Thus, there is an underlying linear problem to be solved and the exploitation of sparsity there accomplishes the exploitation of sparsity in the nonlinear analysis. Curtis, Powell, and Reid (1974) show that we can also exploit sparsity in computing the Jacobian, and this is discussed in Section 12.3.

In chemical engineering (see Westerberg and Berna 1979), electric power system analysis (see Erisman 1981), and other applications, there is a different approach to solving nonlinear equations through tearing. We illustrate that approach to show that it is different from simply exploiting the sparsity in the Jacobian.

Suppose the nonlinear system is of the form

$$x = f(x), \qquad (11.16.1)$$

where f is a nonlinear vector-valued function of the vector variable x. If each f_i is a function of only some of the variables x_k, then the relationship is sparse. Consider the example

$$x_1 = f_1(x_2), \qquad (11.16.2a)$$
$$x_2 = f_2(x_3), \qquad (11.16.2b)$$
$$x_3 = f_3(x_4), \qquad (11.16.2c)$$
$$x_4 = f_4(x_5), \qquad (11.16.2d)$$
$$x_5 = f_5(x_1). \qquad (11.16.2e)$$

Rather than considering the Jacobian matrix of the whole system, we might try to find one torn variable (or, in general, a small number of variables) the determination of whose value would permit the values of the other variables to be found directly from equations (11.16.2). The equations (11.16.2) may be represented in tableau form as shown in Figure 11.16.1, where the explicit variable in each equation is shown in bold. The leading 4×4 submatrix is triangular, so we can treat x_5 as the torn variable. Given a value for x_5, we can determine in turn x_4, x_3, x_2, and x_1. This permits us to treat f_5 as a function of x_5 and iterate to solve $x_5 = f_5$ by direct iteration, $x_5^{(n+1)} = f_5(x_5^{(n)})$, or by some Newton-like method.

Figure 11.16.1. The tableau for the equations (11.16.2).

In general, a permutation will be needed to get a suitable form (called precedence ordered). We need to form a bordered triangular system and choose appropriate variables for the nonlinear iteration. We may also aim for a bordered block triangular form, giving rise to small simultaneous problems. Further discussion is outside the scope of this book, and the interested reader is referred to Westerberg and Berna (1979). Decomposition algorithms used in linear programming (see, for example, Dantzig 1963) are similar. We have introduced the topic mainly to emphasize that nonlinear tearing may be very distinct from linear tearing, both in terms of solving the nonlinear equation and in terms of the sparsity strategy.

When solving nonlinear systems it is usually much more efficient to follow Eriksson (1976) in performing the tearing before rather than after linearization. The exploitation of sparsity by algorithms like that of Curtis *et al.* can then be restricted to smaller subproblems. In the block triangular case, which Eriksson discusses, the Curtis-Powell-Reid algorithm is performed only within the blocks on the diagonal.

11.17 Conclusions

The use of artificial perturbations can lead to very efficient overall solution. Unfortunately we do not know of any automatic means of creating such an artificial perturbation, but the example in Section 11.10 illustrated the potential gains.

Given that an artificial modification has been constructed for a given problem, either from physical considerations or an automatic algorithm, the solution of the perturbed problem can be achieved by a partitioned approach or using the modification formula.

Comparisons have been made in the past between direct factorization and the 'natural' implementation of the modification formula associated with tearing. Rose and Bunch (1972) did this with the matrix of Figure 11.2.1 and concluded that, for this problem, tearing is more efficient than the direct solution. In fact, the example simply shows that the natural implementation of the modification formula (11.6.1) leads to less arithmetic than the natural factorization (11.2.2). The same saving associated with tearing would be achieved using one of the alternative partitioned solutions (11.2.3) or (11.2.4). Alternatively, the same additional arithmetic associated with the natural factorization would be required with the other implementations, (11.6.2) and (11.6.3), of the modification formula.

Figure 11.17.1. Independent submodels.

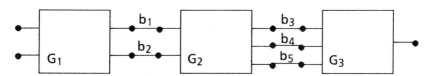

Figure 11.17.2. The interconnected model.

As a concluding comment, we should put the discussion of node and branch tearing in context with the way large problems are often solved. We commented that node tearing is generally more efficient than branch tearing (Section 11.13) because there is less arithmetic associated with the corresponding partitioned matrix problem. This is true if the only objective is to solve the overall problem.

Often, however, a large model is built from the interconnection of submodels. In the course of building such a model, the submodels are analysed independently. For example, we may have already analysed the submodels G_1, G_2, and G_3 illustrated in Figure 11.17.1. Now if the interconnected model is as illustrated in Figure 11.17.2, then branch tearing is most efficient since the 'torn' problem, Figure 11.17.1, has already been analysed. In this case, we may interpret the problem in Figure 11.17.1 as a perturbation of the one in Figure 11.17.2 where the solution of both problems is desired. With this interpretation the question is not whether node tearing or branch tearing should be used but how to solve two problems which differ by a low-rank matrix.

Exercises

11.1 Find the sparsity patterns of the matrix factorizations (11.2.2), (11.2.3), and (11.2.4) for the tridiagonal circulant matrix of Figure 11.2.1. In the general case of order n with leading submatrix tridiagonal and of order $(n-1)$, count the number of floating-point operations needed to form these factorizations and to use them to solve for a problem with a full right-hand side vector $\mathbf{b}$.

11.2 For the matrix of Figure 11.2.2 and the partitioned factorization (11.2.3), how much storage does the implicit representation of $\mathbf{A}_{11}^{-1}\mathbf{A}_{12}$ save and what is the effect on operation count for subsequent solution?

11.3 Consider matrices $\mathbf{A} = \begin{pmatrix} \mathbf{A}_{11}\mathbf{A}_{12} \\ \mathbf{A}_{21}\mathbf{A}_{22} \end{pmatrix}$ with $\mathbf{A}_{11}$ tridiagonal and $\mathbf{A}_{12}$ and $\mathbf{A}_{21}$ very sparse. How large must the order of $\mathbf{A}_{22}$ be if implicit representation of $\mathbf{A}_{11}^{-1}\mathbf{A}_{12}$ in the partitioned factorization (11.2.3) saves work for subsequent solutions?

11.4 Illustrate how matrix partitioning can be used to advantage when solving a perturbed problem with changes confined to a border as in (11.3.3).

11.5 Verify the matrix modification formula (11.4.1).

11.6 For the symmetric matrix $\mathbf{A} = \begin{pmatrix} \mathbf{A}_{11}\mathbf{A}_{12} \\ \mathbf{A}_{12}^{T}\mathbf{A}_{22} \end{pmatrix}$ with $\mathbf{A}_{22}$ of order 1, determine a rank-two matrix $\mathbf{V}$ and a 2×2 matrix $\mathbf{S}$ such that

$$\mathbf{A} = \begin{bmatrix} & & & 0 \\ & \mathbf{A}_{11} & & \cdot \\ & & & \cdot \\ & & & 0 \\ 0 & \cdot & \cdot & \cdot & \cdot & 01 \end{bmatrix} + \mathbf{VSV}^{T}$$

11.7 Generalize the result of Exercise 11.6 for $\mathbf{A}_{22}$ of order k.

11.8 Determine a rank-one tearing of the matrix $\begin{bmatrix} 3 & 1 & -1 & -1 \\ 1 & 4 & -1 & -1 \\ -1 & -1 & 5 & 1 \\ -1 & -1 & 1 & 6 \end{bmatrix}$ which puts it in a simpler form for solving equations.

11.9 Show that the solution to the perturbed problem with coefficient matrix (11.8.5) using the matrix modification formula (11.4.2) is the same as solving the equivalent partitioned system (11.8.1).

11.10 Find a rank-one tearing which puts the following matrix in a simpler form for solving equations

$$
\begin{bmatrix}
\times\ \times\ \times & & & & & & & & & \times\ \times\ \times\ \times \\
\times\ \times\ \times & & & & & & & & & \times \\
\times\ \times\ \times\ \times & & & & & & & & & \times \\
& \times\ \times\ \times\ \times & & & & & & & & \times \\
& & \times\ \times\ \times & & & & & & & \times \\
& & \times\ \times\ \times\ \times & & & & & & & \times \\
& & & & \times\ \times\ \times\ \times & & & & & \times \\
& & & & & \times\ \times\ \times & & & & \times \\
& & & & & \times\ \times\ \times & & & & \times \\
\times & & & & & & & & \times\ \times\ \times\ \times \\
\times & & & & & & & & \times\ \times\ \times\ \times \\
\times & & & & & & & & \times\ \times\ \times\ \times \\
\times\ \times\ \times\ \times\ \times\ \times\ \times\ \times\ \times\ \times\ \times\ \times\ \times\ \times
\end{bmatrix}
$$

Note that a standard partitioned solution fills in three columns.

12 Other sparsity-oriented issues

> We discuss sparsity outside the immediate context of solving
> linear equations. Consideration is given to issues in sparse
> nonlinear problems, the approximation of a sparse matrix by a
> positive-definite matrix, obtaining entries of the inverse,
> backward error analysis for sparse problems, the assembly of
> large finite-element matrices, and the use of hypermatrices.

12.1 Introduction

Much of this book is concerned with preserving sparsity in the factorization
of a sparse matrix $\mathbf{A}$. In this chapter, we discuss some other issues:
preserving sparsity in forming $\mathbf{A}$, the approximation of $\mathbf{A}$ by a
positive-definite matrix, computing entries of the inverse of $\mathbf{A}$ in positions
determined by sparsity, the interpretation of backward error analysis in a
sparse context, and the treatment of very large finite-element matrices.

In the next three sections, we deal with some computations involved in
the solution of nonlinear equations and in mathematical optimization. We
include an efficient way to compute a sparse Jacobian and a way to update
a sparse Hessian approximation while preserving the sparsity. In Section
12.5, we discuss the problem of approximating a sparse symmetric matrix
by a positive-definite one with the same sparsity pattern. Though the
inverse of a sparse matrix is generally dense (Section 12.6), there is an
economical way of computing entries of the inverse in positions within the
pattern of $(\mathbf{L} \backslash \mathbf{U})^T$. This is discussed in Section 12.7.

In Section 12.8, we discuss the problem of backward error analysis for
sparse problems. In particular, it would be desirable to be able to find a
matrix $\mathbf{H}$ such that

$$\mathbf{A} + \mathbf{H} = \mathbf{LU}, \tag{12.1.1}$$

where $\mathbf{L}$ and $\mathbf{U}$ are the computed factors of $\mathbf{A}$ and where $\mathbf{H}$ has the same
sparsity pattern as $\mathbf{A}$. The motivation for this is the backward error analysis
interpretation in the dense matrix case: the computed factors are the exact
factors of a perturbed matrix. We want $\mathbf{H}$ with the same sparsity as $\mathbf{A}$
because it is often true that the zeros are not subject to perturbation. We
discuss the recent work of Arioli, Demmel, and Duff (1988).

One of the major application areas for sparse matrix techniques is
structural analysis or, more generally, analyses using the finite-element
method. We consider the assembly of large finite-element matrices in
Section 12.9 and the use of hypermatrix storage schemes in Section 12.10.

12.2 Sparsity in nonlinear computations

Sparsity plays a critical role in the computational methods of large-scale nonlinear optimization. A review of computational methods for nonlinear optimization is outside the scope of this book. We refer the reader to the books of Fletcher (1980, 1981); Gill, Murray, and Wright (1981); Dennis and Schnabel (1983); and (for the sparse case) Coleman (1984). It is clear that the sparse matrix solution techniques developed earlier in this volume are applicable to solving the sparse linear equations that arise in nonlinear optimization. Therefore, in this and the next two sections, we confine our attention to the generation of the sparse matrices involved.

In the solution of a nonlinear system of equations

$$\mathbf{F}(\mathbf{x}) = \mathbf{0}, \tag{12.2.1}$$

where $\mathbf{F}$ is a vector-valued function of the vector variable $\mathbf{x}$, an estimate of the Jacobian matrix

$$\mathbf{J} = \nabla\mathbf{F}(\mathbf{x}) = \left[\frac{\partial F_i(\mathbf{x})}{\partial x_j} \right] \tag{12.2.2}$$

is generally used. If $\mathbf{x}$ and $\mathbf{F}(\mathbf{x})$ are n-component vectors and n is large, $\mathbf{J}$ is usually sparse. This sparsity is useful not only in solving the equations efficiently, but also in reducing the number of function evaluations necessary to estimate $\mathbf{J}$ through finite differences. This can be accomplished by an algorithm due to Curtis, Powell, and Reid (1974), which is discussed in Section 12.3. For the case where the matrix is known to be symmetric, Powell and Toint (1979) have extended the algorithm and we describe this work, too, in Section 12.3.

In the solution of a nonlinear optimization problem we seek a point $\mathbf{x}^*$ satisfying

$$f(\mathbf{x}^*) = \min_{\mathbf{x} \in N(\mathbf{x}^*)} f(\mathbf{x}), \tag{12.2.3}$$

where $N(\mathbf{x}^*)$ is some neighbourhood of $\mathbf{x}^*$ and f is a twice-differentiable function. The point $\mathbf{x}^*$ is a local minimum of f. Powell's (1970) algorithm for computing $\mathbf{x}^*$ makes use of an approximation $\mathbf{B}_k$ to the Hessian

$$\nabla^2 f(\mathbf{x}_k) = \left[\frac{\partial f(\mathbf{x}_k)}{\partial x_i \partial x_j} \right], \tag{12.2.4}$$

and, in the sparse case, the algorithm of Powell and Toint (1979) may be used to compute it by differencing the gradients.

Rather than approximating $\nabla^2 f(\mathbf{x}_k)$ afresh at every iteration, Powell updates the approximation at most iterations. The updated approximation $\mathbf{B}_{k+1}$ is required to be symmetric and satisfy the equation

$$\mathbf{B}_{k+1}(\mathbf{x}_{k+1} - \mathbf{x}_k) = \mathbf{g}(\mathbf{x}_{k+1}) - \mathbf{g}(\mathbf{x}_k), \tag{12.2.5}$$

where $g(x)$ is the gradient of $f(x)$ at x. Subject to these conditions, B_{k+1} is chosen to minimize the quantity

$$\| B_k - B_{k+1} \|_F, \tag{12.2.6}$$

where $\| \ \|_F$ is the Frobenius norm (possibly weighted). The condition (12.2.5) is known as the quasi-Newton or secant equation. Unfortunately this update destroys the sparsity of the approximate Hessian.

Toint (1977) proposes adding the constraint

$$[B_{k+1}]_{ij} = 0 \quad \text{if} \quad [\nabla^2 f(x)]_{ij} \equiv 0. \tag{12.2.7}$$

The algorithm for preserving the sparsity in the update B_{k+1} through the constraint (12.2.7) is reviewed in Section 12.4.

12.3 Estimating a sparse Jacobian matrix

In this section we describe the **CPR algorithm** (Curtis, Powell, and Reid 1974) for approximating the $n \times n$ Jacobian (12.2.2) of the system (12.2.1) by finite differences and the extension of this algorithm by Powell and Toint (1979) to the symmetric case. The simplest way to approximate J is to use the finite difference

$$\frac{\partial F_i}{\partial x_j} \simeq \frac{F_i(x + h_j e_j) - F_i(x)}{h_j} \tag{12.3.1}$$

where e_j is column j of the identity matrix and h_j is a suitable step-length. A straightforward implementation of (12.3.1) requires computing $F(x + h_j e_j)$ for each j; that is n evaluations of F at displacements from x. Because of sparsity, each F_i depends on only a few x_j, and we use this fact to reduce the number of function evaluations.

We begin by considering the case where the relations

$$\frac{\partial F_i(x)}{\partial x_j} = 0, \quad |i - j| \geq 2 \tag{12.3.2}$$

hold, so that the matrix J is tridiagonal. In this case, we may compute the approximations to columns 1, 4, 7,... simultaneously by using the displacement $h_1 e_1 + h_4 e_4 + \dots$. Similarly the columns 2, 5, 8,... and 3, 6, 9,... can be found together, so that only three function evaluations at displacements from x are needed.

What this example shows for the more general sparse case is that we need to find a set of columns of J which are such that no two have entries in the same row. A set of such columns of J can be computed simultaneously with a single function evaluation. Finding an optimal set is an NP-complete problem (McCormick 1983), but Curtis *et al.* (1974) propose a heuristic for grouping columns of J for any given sparsity pattern.

The first group is formed by starting with the first column and combining with it the next column that has no entries in common with it. The pattern of this next column is merged with the first, and other columns are examined in turn seeking additional columns with no entries in common with the merged pattern. The first group is terminated when the merged pattern is full or all columns have been examined. The second and successive groups are formed in the same way from remaining columns. Such an algorithm is called a **greedy** algorithm.

The experience of Curtis *et al.* with this heuristic indicates that it can greatly reduce the number of function evaluations, though it need not be optimal. An a priori permutation of columns (to alter the order in which they are examined) or an application of the algorithm to $\mathbf{J}^T$ rather than $\mathbf{J}$ sometimes changes the number of groups slightly, but the significance of the change is minor. Coleman and Moré (1983) have gone to great lengths to improve the CPR algorithm, but their further reduction in the number of function evaluations is usually marginal. Software for their algorithms is given by Coleman, Garbow, and Moré (1984).

These schemes partition the columns of the Jacobian matrix and thus have great potential for exploiting parallel architectures, since each finite difference and each set of columns of $\mathbf{J}$ can be evaluated independently on separate processors.

In general, the CPR algorithm is very effective, but one problem where the above algorithms are totally ineffective is shown in Figure 12.3.1 (see Exercise 12.1).

Figure 12.3.1. Nonzero pattern where algorithm is ineffective.

Now consider the case where the matrix $\mathbf{J}$ is known to be symmetric. Once estimates for the coefficients in column j are available, they may be used for row j too. Not only does this ensure that the estimated $\mathbf{J}$ is symmetric, but also having some rows already known may allow more columns to be estimated at once. For example, if the matrix of Figure 12.3.1 is symmetric, we may first estimate the last column (and row) by changing x_n alone and then estimate the rest of the diagonal by changing $x_1, x_2,..., x_{n-1}$ all together. This is the basis for the 'direct' method of Powell and Toint (1979). In general they group the columns as in the CPR algorithm except that they allow columns in a group to have entries in the same row if that row corresponds to a column that has already been treated. To enhance this effect they order the columns by decreasing numbers of entries when choosing which to place first in a group.

Powell and Toint's second suggestion is to substitute an estimate for J_{ij} whenever J_{ji} is needed to calculate another coefficient. For example, this allows a symmetric tridiagonal matrix to be estimated with two instead of three groups of columns. The two groups consist of the odd-numbered and the even-numbered columns, respectively. The first group provides estimates for J_{11}, $J_{21} + J_{23}$, J_{33}, $J_{43} + J_{45}$,... and the second for J_{12}, J_{22}, $J_{32} + J_{34}$, J_{44}, $J_{54} + J_{56}$,... . The diagonal entries are thus all available directly, as is J_{12}. Using the relation $J_{21} = J_{12}$ allows us to calculate J_{23} from $J_{21} + J_{23}$, and we then calculate J_{34} from $J_{32} + J_{34}$, and so on. In general, Powell and Toint recommend applying the CPR algorithm to the lower triangular part of a permutation of the structure of $\mathbf{J}$. Their algorithm again allows a Hessian matrix of the structure of Figure 12.3.1 to be estimated with 3 function evaluations.

The substitution method of the last paragraph is usually more efficient than the direct method of the last-but-one paragraph. For example, it needs only $r+1$ groups for a symmetric band matrix of bandwidth $2r+1$, whereas the direct method needs $2r+1$ groups. Admittedly it is less accurate, though this is unlikely to be worrying unless the quantities $|J_{ij}h_j|$ vary widely. Both methods have been improved slightly by Coleman and Moré (1984), with corresponding software given by Coleman, Garbow, and Moré (1985).

12.4 Updating a sparse Hessian matrix

The sparsity constraint (12.2.7) on the updating of the Hessian subject to minimum-norm change (12.2.6) while maintaining symmetry and satisfying the quasi-Newton equation (12.2.5) represents a nontrivial modification to the standard update. Simply updating $\mathbf{B}_k$ to $\mathbf{B}_{k+1}$ using a standard update and then imposing the sparsity pattern of the Hessian on $\mathbf{B}_{k+1}$ (sometimes called the 'gangster operator' because it fills $\mathbf{B}_{k+1}$ with holes) is not adequate because $\mathbf{B}_{k+1}$ would no longer satisfy (12.2.5). It is interesting to note that the work associated with updating $\mathbf{B}_{k+1}$ using the algorithm of this section is often increased rather than reduced by sparsity. The saving comes from being able to operate with a sparse $\mathbf{B}_{k+1}$, not in computing it.

Toint's (1977) algorithm for solving this problem is now described. Let $\mathbf{X}$ be the matrix with entries

$$x_{ij} = \begin{cases} (\mathbf{x}_{k+1} - \mathbf{x}_k)_j & \text{if } b_{ij} \text{ is an entry} \\ 0 & \text{otherwise.} \end{cases} \qquad (12.4.1)$$

Let $\mathbf{Q}$ be the matrix with entries

$$
q_{ij} = \begin{cases} x_{ij}x_{ji} & i \neq j \\ x_{ii}^2 + \displaystyle\sum_{k=1}^{n} x_{ki}^2 & i = j \end{cases} . \tag{12.4.2}
$$

If no column of $\mathbf{X}$ is identically zero, then $\mathbf{Q}$ is symmetric and positive definite.

Using $\mathbf{Q}$, we solve the system of equations

$$
\mathbf{Q}\lambda = \mathbf{g}(\mathbf{x}_{k+1}) - \mathbf{g}(\mathbf{x}_k) - \mathbf{B}_k(\mathbf{x}_{k+1} - \mathbf{x}_k) \tag{12.4.3}
$$

for λ. Then it can be shown that $\mathbf{B}_{k+1}$, defined by

$$
\mathbf{B}_{k+1} = \mathbf{B}_k + \mathbf{E}, \tag{12.4.4}
$$

where

$$
[\mathbf{E}]_{ij} = \begin{cases} \lambda_i x_j + \lambda_j x_i & \text{if } b_{ij} \text{ is an entry} \\ 0 & \text{otherwise}, \end{cases} \tag{12.4.5}
$$

satisfies all of the constraints, but unfortunately $\mathbf{B}_k$ is not positive definite and methods based on this approach (for example, Toint 1981) have been found often to be less satisfactory than fresh calculation of the approximation (see Thapa 1983). Sorensen (1981) has found cases where imposing the additional constraint of positive definiteness leads to arbitrarily large entries in $\mathbf{E}$ and a poor approximate Hessian.

The critical part of the computation lies in building and factorizing $\mathbf{Q}$. But since $\mathbf{Q}$ is a sparse positive-definite matrix of the same structure as $\mathbf{B}_k$, the ANALYSE used for $\mathbf{B}_k$ can be used to solve this system.

The proof that $\mathbf{B}_{k+1}$ satisfies all of the constraints is given by Toint (1977).

12.5 Approximating a sparse matrix by a positive-definite one

In this section we consider another sparse matrix approximation problem: for a given sparse symmetric matrix find the closest positive-definite matrix with the same sparsity pattern. There are at least three cases where a solution to this problem is important.

The first case arises in the application which we were discussing in the previous section, namely sparse optimization. Here it is normally desirable to keep $\mathbf{B}_k$ positive definite and one way of doing this would be to replace the matrix generated by equation (12.4.4) by a nearby positive-definite matrix of the same sparsity pattern. We are unaware of work using this approach, although we believe the challenge to be a legitimate one.

The second case is encountered in building a large circuit analysis model from measured data. The positive definiteness of the matrix is necessary to preserve the physical passiveness of the model: a negative eigenvalue would imply positive exponentials in the solution of the differential equations. Yet, since the measurements are subject to uncorrelated error, it frequently happens that the resulting matrix is indefinite. The problem is to find a nearby matrix with the same sparsity that gives a good representation of the measured data. As an aside, note that if there is a negative eigenvalue that is not comparable with the data errors for the largest entries of **A**, this probably indicates a blunder (a wrong entry, not a rounded one) and this process is useful for identifying them.

The problem can arise in a quite different way when 'drop tolerances' are used (see Section 9.9). One difficulty is that the tolerance is often based on the magnitude of uncertainty in the data. Thus a matrix can change from positive definite and reasonably well-conditioned to an indefinite or ill-conditioned one.

Ignoring the sparsity constraint, the closest positive-semidefinite matrix (in the Frobenius norm) to a given symmetric one is known to be the one with the same eigenvectors but with the negative eigenvalues changed to zero. Similarly the closest positive-definite matrix with minimum eigenvalue greater than or equal to σ is obtained in the same way except all eigenvalues less than σ are replaced by σ. Unfortunately this approximation usually destroys sparsity.

Computing eigensystems is impractical if we were working with very large matrices. But where the matrix is a sum

$$\mathbf{A} = \sum \mathbf{A}^{[l]} \qquad (12.5.1)$$

of matrices with entries only in dense submatrices, this approach is practical for each $\mathbf{A}^{[l]}$ and does not destroy the sparsity of the overall matrix.

An alternative is to select a value δ such that $\lambda_{min} + \delta = \sigma$, where λ_{min} is the minimum eigenvalue. Then the matrix $\mathbf{A} + \delta\mathbf{I}$ would have the same off-diagonal sparsity as the original matrix, would have its minimum eigenvalue σ, and would be close to **A** if **A** is 'almost' positive definite.

12.6 The inverse of a sparse matrix

In this section we consider structural properties of an irreducible matrix **A**. The **structural inverse** of a matrix is defined to be the pattern which is the union of all patterns for $\mathbf{A}^{-1}$ generated by choosing different values for entries of **A**. A structural **L**, **U**, and solution vector **x** are defined similarly. In this section, all references to **L**, **U**, $\mathbf{A}^{-1}$, and **x** are to their structures in this sense.

We show that, if **A** is irreducible, **L** has at least one entry beneath the diagonal in each column (except the last), **U** has at least one entry to the right of the diagonal in each row (except the last), **x** is full, and $\mathbf{A}^{-1}$ is full.

We first show that if nodes i and j in the digraph (Section 1.2) of the matrix are joined by a path (i, k_1), $(k_1, k_2), \ldots, (k_s, j)$, where $k_1, k_2, \ldots, k_s$ are all less than i and j (called a **legal path** by Richard Karp), then (i, j) will be an entry in L\U. To show this, suppose that $k_m = \min_i k_i$. Just before elimination step k_m, there are entries in positions (k_{m-1}, k_m) and (k_m, k_{m+1}). Therefore, following it, there is an entry in position (k_{m-1}, k_{m+1}) so the digraph of the reduced submatrix contains the (legal) path (i, k_1), (k_1, k_2), ..., $(k_{m-1}, k_{m+1}), \ldots, (k_s, j)$. Continuing the argument similarly, we eventually find that (i, j) is an entry in L\U.

Next we show that there is at least one entry to the right of the diagonal in every row of U (except the last) and at least one entry beneath the diagonal in every column of L (except the last). We prove this by showing that if there is a row k of U, $k < n$, with no entries to the right of the diagonal, then A is reducible. The proof for L is similar. Suppose there is a path in the digraph of A from node k to a node after k and let l be the first such node on the path. The path from k to l is a legal path, so (k, l) must be an entry in L\U, contrary to our assumption. Therefore there can be no path in the digraph of A from k to a later node. Let S be the set consisting of node k and all nodes to which there is a path from k. The complement of S is not empty because it contains all the nodes after k. If we reorder the matrix so that the nodes of S precede those of its complement, we find a 2×2 block lower triangular form, that is A is reducible. This proves the statement in the first sentence of this paragraph.

Notice that some permutations of a reducible matrix may give factors L and U possessing the property of the last paragraph. For example, interchanging the columns of the matrix

$$\begin{pmatrix} \times & \times \\ 0 & \times \end{pmatrix} \tag{12.6.1}$$

leads to full matrices L and U. On the other hand, any permutation of an irreducible matrix is irreducible, so the property of the last paragraph holds for all permutations of irreducible matrices.

We next show that if L is a lower triangular factor of an irreducible matrix and $\mathbf{b} \neq \mathbf{0}$, the solution of the equation

$$\mathbf{Ly} = \mathbf{b} \tag{12.6.2}$$

has an entry in its last position. To prove this, suppose b_k is the last nonzero of $\mathbf{b}$. Since the corresponding component of the solution is found from the equation

$$y_k = b_k - \sum_{s=1}^{k-1} l_{ks} y_s, \tag{12.6.3}$$

it is an entry in $\mathbf{y}$. If $k = n$, this is our desired result. Otherwise, since L is a factor of an irreducible matrix, it has an entry l_{jk}, $j>k$; by replacing k by j in equation (12.6.3), we conclude that the solution has an entry in position j. If $j=n$, this is our desired result. If not we continue the argument, finding successive entries in the solution until position n is reached.

If $\mathbf{U}$ is an upper triangular factor of an irreducible matrix and y_n is an entry, the solution of the equation

$$\mathbf{U}\mathbf{x} = \mathbf{y} \tag{12.6.3}$$

is dense. To prove this, we remark that the equation for x_k is

$$u_{kk}x_k = y_k - \sum_{j=k+1}^{n} u_{kj}x_j. \tag{12.6.4}$$

Clearly x_n is an entry in $\mathbf{x}$. For $k<n$, suppose $x_{k+1},\ldots,x_n$ are all entries; since $\mathbf{U}$ is a factor of an irreducible matrix, it has an entry u_{kj}, $j>k$, so we deduce from equation (12.6.4) that the solution has an entry in position k, too. Hence, by induction, $\mathbf{x}$ is dense.

Together, the last two paragraphs tell us that for any set of equations

$$\mathbf{A}\mathbf{x} = \mathbf{b} \tag{12.6.5}$$

with an irreducible matrix $\mathbf{A}$, the solution vector $\mathbf{x}$ is dense if it is computed with Gaussian elimination and any cancellations are ignored. This is also true for $\mathbf{A}^{-1}$ if it is computed in the usual way by solving the equation

$$\mathbf{A}\mathbf{X} = \mathbf{I} \tag{12.6.6}$$

by Gaussian elimination.

It might be thought that there could be some sparsity in the inverse that is not exposed by this treatment, but this is not the case. Duff, Erisman, Gear, and Reid (1988) show that for every position (i,j) there is a matrix with the given irreducible structure that has a nonzero in position (i,j) of its inverse.

12.7 Computing entries of the inverse of a sparse matrix

The standard approach to computing entries of the inverse of a matrix $\mathbf{A}$ involves solving the equation

$$\mathbf{A}\mathbf{X} = \mathbf{I}, \tag{12.7.1}$$

using the $\mathbf{L}\mathbf{U}$ factorization of $\mathbf{A}$. Since this approach computes the entries of $\mathbf{A}^{-1}$ a column at a time, the entire lower triangle of $\mathbf{A}^{-1}$ would have to be computed in order to get all of the diagonal entries, for example. We review here an algorithm for computing just the entries in the sparsity pattern of $(\mathbf{L}\backslash\mathbf{U})^T$. This algorithm is discussed further by Erisman and Tinney (1975).

There are many applications where entries of the inverse are useful. In the least-squares data-fitting problem, the diagonal entries of the inverse of the normal matrix $\mathbf{A}^T\mathbf{A}$ have particular significance since they yield estimates of the variances of the fitted parameters. Entries of the inverse can also be used to compute sensitivity information. Thus we are motivated to compute a portion of $\mathbf{A}^{-1}$, not to use it as an operator, but for the entries themselves.

Let $\mathbf{Z} = \mathbf{A}^{-1}$ and suppose we have computed a sparse factorization of $\mathbf{A}$,

$$\mathbf{A} = \mathbf{LDU}, \tag{12.7.2}$$

where $\mathbf{L}$ is unit lower triangular, $\mathbf{D}$ is diagonal and $\mathbf{U}$ is unit upper triangular. The formulae

$$\mathbf{Z} = \mathbf{D}^{-1}\mathbf{L}^{-1} + (\mathbf{I} - \mathbf{U})\mathbf{Z} \tag{12.7.3a}$$

and

$$\mathbf{Z} = \mathbf{U}^{-1}\mathbf{D}^{-1} + \mathbf{Z}(\mathbf{I} - \mathbf{L}) \tag{12.7.3b}$$

of Takahashi, Fagan, and Chin (1973), may be readily verified. Since $(\mathbf{I} - \mathbf{U})$ is strictly upper triangular and $(\mathbf{I} - \mathbf{L})$ is strictly lower triangular, we may conclude that the following relations are true:

$$z_{ij} = [(\mathbf{I} - \mathbf{U})\mathbf{Z}]_{ij}, \quad i < j, \tag{12.7.4a}$$

$$z_{ij} = [\mathbf{Z}(\mathbf{I} - \mathbf{L})]_{ij}, \quad i > j, \tag{12.7.4b}$$

$$z_{ii} = d_{ii}^{-1} + [(\mathbf{I} - \mathbf{U})\mathbf{Z}]_{ii}, \tag{12.7.4c}$$

and

$$z_{ii} = d_{ii}^{-1} + [\mathbf{Z}(\mathbf{I} - \mathbf{L})]_{ii}. \tag{12.7.4d}$$

Using the sparsity of $\mathbf{U}$ and $\mathbf{L}$, these formulae provide a means of computing particular entries of $\mathbf{Z}$ from previously computed ones.

In particular, all entries of $\mathbf{Z}$ in the sparsity pattern of $(\mathbf{L}\backslash\mathbf{U})^T$ can be computed without calculating any entries outside this pattern. To see this, let z_{ij} be one such entry. Suppose $i < j$. Then from (12.7.4a)

$$z_{ij} = -\sum_{k=i+1}^{n} u_{ik} z_{kj}. \tag{12.7.5}$$

The entries z_{kj} which are needed to evaluate this formula are those corresponding to entries u_{ik}. But if l_{ji} and u_{ik} are entries, so is $(\mathbf{L}\backslash\mathbf{U})_{jk}$, which implies that z_{kj} is an entry that has been computed (see also Exercise 12.4). This suggests that entries of the inverse may be computed starting with z_{nn} and working up through the sparsity pattern of $(\mathbf{L}\backslash\mathbf{U})^T$. It shows, for example, that the diagonal entries of the inverse of a symmetric tridiagonal matrix may be computed with just the entries $z_{ii}, i = 1, 2,\dots n$, and $z_{i,i+1}, i = 1, 2,\dots, n-1$.

In general, a computational sequence can be developed by calculating entries of $\mathbf{Z}$ lying in the pattern of $(\mathbf{L}\backslash\mathbf{U})^T$ in reverse Crout order, starting with z_{nn}. Thus, when calculating entry z_{ij}, all entries z_{st} $(s>i, t>j)$ have already been calculated and equation (12.7.5) and its counterpart for $i>j$ (Exercise 12.4) allow z_{ij} to be computed immediately.

While this computational sequence will always work, the following example shows that it may not be the most efficient. Let $\mathbf{A}$ have the pattern

$$
\begin{bmatrix}
\times & 0 & 0 & 0 & \times \\
0 & \times & \times & \times & 0 \\
0 & \times & \times & \times & 0 \\
0 & 0 & 0 & \times & \times \\
\times & 0 & \times & \times & \times
\end{bmatrix}. \tag{12.7.6}
$$

In this case, $\mathbf{L}\backslash\mathbf{U}$ has the same pattern. The sequence

$$
\begin{align}
z_{55} &= d_{55}^{-1}, & \text{using (12.7.4c)} & \tag{12.7.7a} \\
z_{51} &= z_{55} l_{51}, & \text{using (12.7.4b) and (12.7.7a)} & \tag{12.7.7b} \\
z_{11} &= d_{11}^{-1} - u_{15} z_{51}, & \text{using (12.7.4c) and (12.7.7b).} & \tag{12.7.7c}
\end{align}
$$

can be used to compute z_{11} without previously calculating all other z_{ij} in the pattern of $(\mathbf{L}\backslash\mathbf{U})^T$.

The minimum number of computations required to compute any entry of $\mathbf{A}^{-1}$ in the sparsity pattern of $(\mathbf{L}\backslash\mathbf{U})^T$ is unknown. Another unknown question is the best way of computing other entries of $\mathbf{A}^{-1}$, though the formulae (12.7.4) still apply.

12.8 Sparsity constrained backward error analysis

In Chapter 4, we discussed the backward error analysis of Gaussian elimination. The computation is considered stable if the computed factors $\mathbf{L}$ and $\mathbf{U}$ of $\mathbf{A}$ satisfy

$$
\mathbf{A} + \mathbf{H} = \mathbf{LU} \tag{12.8.1}
$$

and $\|\mathbf{H}\|$ is small compared with $\|\mathbf{A}\|$. For dense matrices, a bound for $\|\mathbf{H}\|$ was given in inequality (4.3.14). In Section 5.4, we showed that for sparse $n \times n$ matrices the corresponding bound depends on p rather than n, where p is the maximum number of entries in any column of $\mathbf{U}$.

A justification for backward error analysis is that $\mathbf{H}$ can be regarded as a perturbation of the original problem for which $\mathbf{L}$ and $\mathbf{U}$ are the exact factors. In the sparse matrix problem, this interpretation may not be valid. The reason is that small perturbations to zeros in $\mathbf{A}$ may not make sense physically. For example, if $\mathbf{A}$ is the matrix in a model of a water-distribution network, a zero may mean no direct connection between two points in the network; 'small' connections in this context make no sense.

Since $\mathbf{H}$ is defined by equation (12.8.1), its nonzeros are not restricted to the sparsity pattern of $\mathbf{A}$. We might consider determining a matrix $\mathbf{F}$ with the sparsity pattern of $\mathbf{A}$, $\|\mathbf{F}\|$ small, and satisfying the equation

$$(A + F)\tilde{x} = b, \tag{12.8.2}$$

where $\tilde{x}$ is the computed solution. Gear (1975) provided a counter example to this objective and we present it here.

Gear's example is the problem

$$A = \begin{bmatrix} 1 & 1 & -1 & -1 \\ 0 & \delta & 0 & 0 \\ 0 & 0 & \delta & 0 \\ 1 & 0 & 0 & 1 \end{bmatrix}, \quad b = \begin{bmatrix} 0 \\ 1 \\ 1 \\ 2 \end{bmatrix}, \tag{12.8.3}$$

with solution $(1 \ \ 1/\delta \ \ 1/\delta \ \ 1)^T$. Consider the candidate solution

$$\tilde{x} = \begin{bmatrix} (\delta-\sigma)/\delta \\ 1/\delta \\ 1/\delta \\ (\delta-\sigma)/\delta \end{bmatrix}, \tag{12.8.4}$$

which satisfies the equation

$$\begin{bmatrix} 1 & 1 & -1 & -1 \\ 0 & \delta & 0 & 0 \\ 0 & 0 & \delta & 0 \\ 1 & 2\sigma & 0 & 1 \end{bmatrix} \tilde{x} = \begin{bmatrix} 0 \\ 1 \\ 1 \\ 2 \end{bmatrix}. \tag{12.8.5}$$

For small σ, this represents a small perturbation of the original problem. It can be shown that the minimum perturbation within the sparsity pattern of A requires adding $\sigma/(\delta-\sigma)$ to both the (4,1) and (4,4) entries of A. If δ is also small, then $\sigma/(\delta-\sigma)$ can be large, so that it is not possible to satisfy (12.8.2) with $\|F\|$ small. Note, however, that if δ is small then A is badly scaled.

The problem is tackled by Arioli, Demmel, and Duff (1988) using the notion of componentwise relative error (Oettli and Prager 1964) and allowing b to be perturbed as well as A. They show that the computed solution $\tilde{x}$ is the exact solution of the perturbed problem

$$(A + F)\tilde{x} = b + \delta b, \tag{12.8.6}$$

with perturbations satisfying the inequalities

$$|f_{ij}| \leq \omega |a_{ij}|, \quad i, j, = 1, 2, \dots, n \tag{12.8.7}$$

and

$$|\delta b_i| \leq \omega f_i, \quad i = 1, 2, \dots, n, \tag{12.8.8}$$

where

$$\omega = \max_i \frac{|A\tilde{x} - b|_i}{(|A| \, |\tilde{x}|)_i + f_i}. \tag{12.8.8}$$

Normally f_i is equal to $|b_i|$, but when the denominator of expression (12.8.8) is near zero, they choose f_i equal to $\sum_j |a_{ij}| \, \|\tilde{x}\|_\infty$. Thus ω represents a relative error for the entries of A and many of the entries of b, but an absolute error for the other entries of b (different for each component). It may be necessary to perform one or more iterations of iterative refinement to obtain an approximate solution $\tilde{x}$ for which ω, calculated by expression 12.8.8, is at the level of roundoff. The experiments of Arioli *et al.* (1988) suggest that more than one iteration is rarely needed. For Gear's example, working with $\delta = 10^{-8}$ and $\sigma = 10^{-15}$ on a machine having precision $16^{-13} \approx 2 \times 10^{-16}$, they found one step of iterative refinement gave an improved $\tilde{x}$ for which $\omega \leq 10^{-16}$. In fact, this is a case where the condition number $\|A^{-1}\| \, \|A\|$ (see Section 4.9) is large only because of poor scaling. A better condition number in such a case is that of Skeel (1979), $\| \, |A^{-1}| \, |A| \, \|$. Here the classical condition number is $1 + \delta^{-1}$ whereas the Skeel condition number is 4. Unfortunately, discussion of Skeel's work is beyond the scope of this book.

12.9 Assembling large finite-element matrices to band form

In Sections 10.2 and 10.3, we remarked that it is convenient to process large positive-definite band and variable-band matrices from auxiliary storage provided we have enough main storage for the active part. If the greatest semibandwidth is m and we have main storage for $\tfrac{1}{2}m(m+1)$ variables and a reasonably large input-output buffer, then it is straightforward to organize the elimination without writing entries of the factorized form out to auxiliary storage more than once. This is possible because the k-th stage of the elimination involves at most entries in the square submatrix of rows and columns k to $k+m-1$. Entries in earlier rows and columns have already been written to auxiliary storage and entries in later rows and columns are not yet required. Similar remarks apply to the refined quotient tree algorithm of George (1977b) (see Section 8.5), if m is now the order of the largest block.

For finite-element problems

$$A = \sum_l A^{[l]} \tag{12.9.1}$$

(see Section 10.5), the assembly of the individual stiffness matrices $A^{[l]}$ into the overall band or variable-band storage pattern is reasonably straightforward. If auxiliary storage is not in use then we may hold the rows in sequence, with storage allocated for all positions between the first entry and the diagonal and pointers to the positions of the diagonal entries. The stiffness matrices $A^{[l]}$ may be added easily into this structure in any order. If auxiliary storage is in use, it is usual to read a file containing the stored matrices $A^{[l]}$ in compact form several times. On each pass we accumulate as many of the rows of the overall matrix as will fit in store at once. Figure 12.9.1 illustrates this with an example needing three passes, each of which is used to accumulate one of the partitions shown.

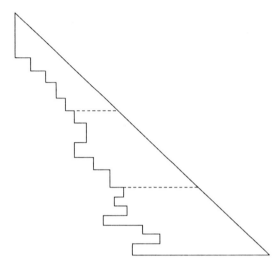

Figure 12.9.1. Lower triangular part of a variable-band matrix requiring to be read three times for the assembly.

If the number of times the file has to be read is high, this can be alleviated by accumulating into a more compact representation of the matrix using the techniques for handling fill-in discussed in Chapter 2. Alternatively (see Exercise 12.6), provided a direct-access file is in use, matrices $A^{[l]}$ needed for the same block of rows may be linked together so that only those needed on each pass are actually read.

12.10 Hypermatrices

An approach that has been used with success by Argyris and his collaborators (see, for example, Schrem 1971) is to store matrices by blocks, using a full matrix of pointers to indicate where the blocks are actually stored. A simple example is shown in Figure 12.10.1. Any blocks that are zero are not stored and have a corresponding pointer 0, but the whole of any block that has a nonzero is held. In the example of Figure 12.10.1, the zeros marked explicitly are held explicitly and the pointers indicate that A_{11} is held from location 1, A_{33} is held from location 7, etc.

The technique can be nested so that the first level pointers indicate where matrices of second level pointers are held, and so on. Only the last level holds real numbers. Three levels are probably as many as are ever likely to be useful.

```
                              × × 0          × 0 0 ×
                                × ×          0 0 × ×
                                  ×          × × 0 0
          1  0 17               × × ×
                                 × 0
          29  0
                                  ×
           7
                                        × × × 0
                                        × × ×
                                          × ×
                                            ×
```

Figure 12.10.1. Upper triangular blocked matrix and an associated matrix of pointers.

The hypermatrix approach permits really huge problems to be handled with modest demands on main storage, but care is needed to avoid excessive input-output operations. For this reason the block sizes are usually chosen to suit hardware characteristics.

Exercises

12.1 Why is the CPR algorithm ineffective on the matrix in Figure 12.3.1?

12.2 Show that Q of (12.4.2) is symmetric and positive definite if no column of X defined by (12.4.1) is identically zero.

12.3 Verify the formulae (12.7.3).

12.4 Using the notation of Section 12.7, show that any entry z_{ij} of Z with $i > j$ and u_{ji} an entry can be computed without calculating an inverse entry outside the sparsity pattern of $(L \backslash U)^T$.

12.5 Show that the minimum perturbation, F, to the matrix A of (12.8.3) where F lies within the sparsity pattern of A and is such that (12.8.4) is a solution to $(A+F)x = b$, is given by

$$F = \begin{pmatrix} 0 & 0 & 0 & 0 \\ 0 & 0 & 0 & 0 \\ 0 & 0 & 0 & 0 \\ \alpha & 0 & 0 & \alpha \end{pmatrix}$$

where $\alpha = \sigma/(\delta - \sigma)$.

12.6 Show that it is possible to arrange to read only those matrices $A^{[k]}$ actually needed for assembling a particular set of rows in the overall matrix $A = \Sigma A^{[k]}$.

Appendix A: Matrix and vector norms

A norm provides a means of measuring the size of a vector or matrix. Formally, a norm $\|\mathbf{x}\|$ of the vector $\mathbf{x}$ is a non-negative real value having the properties

$$\|\mathbf{x}\| = 0 \text{ if and only if } \mathbf{x} = \mathbf{0}, \tag{A.1a}$$

$$\|\alpha\mathbf{x}\| = |\alpha| \, \|\mathbf{x}\| \text{ for any scalar } \alpha, \text{ and} \tag{A.1b}$$

$$\|\mathbf{x} + \mathbf{y}\| \leqslant \|\mathbf{x}\| + \|\mathbf{y}\| \text{ for any two vectors } \mathbf{x} \text{ and } \mathbf{y}. \tag{A.1c}$$

Norms of interest in this book are $\|\cdot\|_p$, for $p = 1, 2$ and ∞, defined for real or complex vectors of length n by the equations

$$\|\mathbf{x}\|_1 = \sum_{i=1}^{n} |x_i| \tag{A.2a}$$

$$\|\mathbf{x}\|_2 = \sqrt{\sum_{i=1}^{n} |x_i|^2} \tag{A.2b}$$

$$\|\mathbf{x}\|_\infty = \max_i |x_i| \tag{A.2c}$$

It is straightforward to deduce from these definitions that the inequalities

$$0 \leqslant n^{-\frac{1}{2}}\|\mathbf{x}\|_2 \leqslant \|\mathbf{x}\|_\infty \leqslant \|\mathbf{x}\|_2 \leqslant \|\mathbf{x}\|_1 \leqslant n^{\frac{1}{2}}\|\mathbf{x}\|_2 \tag{A.3}$$

are true for any vector $\mathbf{x}$. These inequalities allow us to choose p on the basis of convenience in most instances.

Since matrices are used as operators, a natural measure for the size of a matrix is the maximum ratio of the size of $\mathbf{Ax}$ to the size of $\mathbf{x}$. This leads to the definition

$$\|\mathbf{A}\| = \max_{\|\mathbf{x}\|=1} \|\mathbf{Ax}\|. \tag{A.4}$$

For the infinity norm and any vector $\mathbf{x}$ such that $\|\mathbf{x}\|_\infty = 1$, we find the relations

$$\|\mathbf{Ax}\|_\infty = \max_i \left| \sum_j a_{ij}x_j \right| \leqslant \max_i \sum_j |a_{ij}|. \tag{A.5}$$

If the latter maximum is attained for $i=k$ and $\mathbf{v}$ is the vector with components $v_j = \text{sign}(a_{kj})$, $j=1, 2,\ldots, n$, then the following relations hold

$$\|\mathbf{v}\|_\infty = 1, \quad (\mathbf{Av})_k = \sum_j |a_{kj}|. \tag{A.6}$$

In view of the definition (A.4), it follows that $\mathbf{A}$ has norm

$$\|\mathbf{A}\|_\infty = \max_i \sum_j |a_{ij}| \tag{A.7}$$

and that $\mathbf{v}$ is a vector attaining the maximum (A.4). In view of relation (A.7) the infinity norm is also known as the row norm.

For the one norm and any vector $\mathbf{x}$ such that $\|\mathbf{x}\|_1 = 1$, we find the relations

$$
\begin{aligned}
\|\mathbf{Ax}\|_1 &= \sum_i \left| \sum_j a_{ij} x_j \right| \\
&\leq \sum_j \sum_i |a_{ij}| \, |x_j| \\
&\leq \sum_j \left(\max_j \sum_i |a_{ij}| \right) |x_j| \\
&= \max_j \sum_i |a_{ij}| \, .
\end{aligned}
\tag{A.8}
$$

If the latter maximum is attained for $j=k$ and $\mathbf{v}$ is column k of $\mathbf{I}$, the following equalities hold:

$$
\|\mathbf{v}\|_1 = 1 \quad \text{and} \quad \|\mathbf{Av}\|_1 = \sum_i |a_{ik}|.
\tag{A.9}
$$

It follows that $\mathbf{A}$ has norm

$$
\|\mathbf{A}\|_1 = \max_j \sum_i |a_{ij}|
\tag{A.10}
$$

and that $\mathbf{v}$ is the vector attaining the maximum (A.4). In view of relation (A.10), the one norm is also known as the column norm.

For the two norm, maximizing $\|\mathbf{Ax}\|_2 / \|\mathbf{x}\|_2$ corresponds to maximizing its square, whose value is $\mathbf{x}^T \mathbf{A}^T \mathbf{Ax} / \mathbf{x}^T \mathbf{x}$ and, by the Rayleigh-Ritz principle, this has value equal to the largest eigenvalue of $\mathbf{A}^T \mathbf{A}$. The maximum is attained for $\mathbf{x}$ equal to the corresponding eigenvector. If the eigenvalues of $\mathbf{A}^T \mathbf{A}$ are σ_i^2, $i=1, 2,..., n$, the norm is therefore

$$
\|\mathbf{A}\|_2 = \max_i \sigma_i.
\tag{A.11}
$$

The numbers σ_i are known as the singular values of $\mathbf{A}$.

In connection with conditioning, the minimization of $\|\mathbf{Ax}\| / \|\mathbf{x}\|$ is of interest. Writing $\mathbf{Ax}$ as $\mathbf{v}$, we seek to minimize $\|\mathbf{v}\| / \|\mathbf{A}^{-1}\mathbf{v}\|$. This minimum occurs for the vector $\mathbf{v}$ that maximizes $\|\mathbf{A}^{-1}\mathbf{v}\| / \|\mathbf{v}\|$. It follows that

$$
\min \frac{\|\mathbf{Ax}\|}{\|\mathbf{x}\|} = \frac{1}{\|\mathbf{A}^{-1}\|}
\tag{A.12}
$$

and that the minimum is attained for the vector

$$
\mathbf{x} = \mathbf{A}^{-1}\mathbf{v}
\tag{A.13}
$$

where $\mathbf{v}$ is the vector that attains $\max \|\mathbf{A}^{-1}\mathbf{v}\| / \|\mathbf{v}\|$.

In the case of the one norm, if column k attains $\max_j \sum_i |(\mathbf{A}^{-1})_{ij}|$, then $\mathbf{v}$ is column k of $\mathbf{I}$ and $\mathbf{x}$ is therefore the column of $\mathbf{A}^{-1}$ whose norm is greatest.

In the case of the infinity norm, if row k attains $\max_i \sum_j |(\mathbf{A}^{-1})_{ij}|$ then $\mathbf{v}$

has components $v_j = \text{sign}\big((A^{-1})_{kj}\big)$ and x is therefore the linear combination of columns of A^{-1} with multipliers ± 1 that maximizes $\|x\|_\infty$.

A more direct approach to minimizing $\|Ax\|/\|x\|$ is available for the two norm since the Rayleigh-Ritz principle immediately yields the result

$$\min \frac{\|Ax\|_2}{\|x\|_2} = \min_i \sigma_i \qquad (A.14)$$

by the same argument as that used to show the truth of (A.11). The minimum is attained for the eigenvector of $A^T A$ corresponding to its least eigenvalue.

Corresponding to the properties (A.1) satisfied for vector norms, the matrix norms (A.4) all satisfy the equations

$$\|A\| = 0 \text{ if and only if } A = 0, \qquad (A.15a)$$

$$\|\alpha A\| = |\alpha| \, \|A\| \text{ for any scalar } \alpha, \text{ and} \qquad (A.15b)$$

$$\|A + B\| \leq \|A\| + \|B\| \text{ for any two } m\times n \text{ matrices } A \text{ and } B. \qquad (A.15c)$$

Further, they satisfy the inequalities

$$\|Ax\| \leq \|A\| \, \|x\|, \qquad (A.16)$$

for any $m\times n$ matrix A and any vector x of length n, and

$$\|AB\| \leq \|A\| \, \|B\|, \qquad (A.17)$$

for any $m\times n$ matrix A and any $n\times p$ matrix B. Corresponding to inequalities (A.3) the inequalities

$$0 \leq n^{-1}\|A\|_2 \leq n^{-\frac{1}{2}}\|A\|_\infty \leq \|A\|_2 \leq n^{\frac{1}{2}}\|A\|_1 \leq n\|A\|_2 \qquad (A.18)$$

hold. They are a simple deduction from (A.3) and (A.4).

Because of the difficulty of evaluating the two norm (see (A.11)), the alternative

$$\|A\|_F = \sqrt{\sum_{i=1}^m \sum_{j=1}^n |a_{ij}|^2}, \qquad (A.19)$$

called the Frobenius norm, is often used. Note that, trivially, $\|x\|_F = \|x\|_2$ for vectors. It can be shown that $\|\cdot\|_F$ satisfies all the properties (A.15) – (A.17), but it is not so useful because the bounds (A.16) and (A.17) are not so tight.

For proofs of these results and a more detailed treatment, we refer the reader to Chapter 4 of Stewart (1973). Note that scaling is important when norms are used. For example, for the matrices

$$A = \begin{pmatrix} 0.001 & 2.42 \\ 1.00 & 1.58 \end{pmatrix}, \quad B = \begin{pmatrix} 1.00 & 2420 \\ 1.00 & 1.58 \end{pmatrix}, \text{ and } H = \begin{pmatrix} 0.00 & 0.00 \\ 0.00 & 1.58 \end{pmatrix}, \qquad (A.20)$$

$\|\mathbf{H}\|$ is small compared with $\|\mathbf{B}\|$, but is not small compared with $\|\mathbf{A}\|$. The scaling that produces $\mathbf{B}$ from $\mathbf{A}$ leaves $\mathbf{H}$ unchanged, so care must be exercised in the use of norms.

Appendix B: The LINPACK condition number estimate

Cline, Moler, Stewart, and Wilkinson (1979) suggested estimating $\|\mathbf{A}^{-1}\|_1$ by calculating the vector $\mathbf{x}$ satisfying the equation

$$\mathbf{A}^T \mathbf{x} = \mathbf{b} \qquad (\text{B.1})$$

for a specially constructed vector $\mathbf{b}$, then solving the equation

$$\mathbf{A}\mathbf{y} = \mathbf{x} \qquad (\text{B.2})$$

and using $\|\mathbf{y}\|_1 / \|\mathbf{x}\|_1$ as an estimate for $\|\mathbf{A}^{-1}\|_1$. A good choice for $\mathbf{b}$ is one that leads to $\|\mathbf{x}\| / \|\mathbf{b}\|$ being large. We show that this is the case for the two norm and, in view of relations (A.18), it will be true for the one norm, too.

If $\mathbf{A}$ has the singular value decomposition

$$\mathbf{A} = \mathbf{V}\,\mathbf{diag}(\sigma_i)\mathbf{W}^T, \qquad (\text{B.3})$$

where $\mathbf{V}$ and $\mathbf{W}$ are orthogonal matrices with columns $\mathbf{v}_i$ and $\mathbf{w}_i$, $i = 1, 2,\ldots, n$ and $\mathbf{b}$ has the expansion $\mathbf{b} = \sum_{i=1}^{n} \alpha_i \mathbf{w}_i$, then $\mathbf{x}$ is the vector

$$\mathbf{x} = \mathbf{A}^{-T}\mathbf{b} = \mathbf{V}\,\mathbf{diag}(\sigma_i^{-1})\mathbf{W}^T \sum_{i=1}^{n} \alpha_i \mathbf{w}_i = \sum_{i=1}^{n} \sigma_i^{-1} \alpha_i \mathbf{v}_i \qquad (\text{B.4})$$

and $\mathbf{y}$ is the vector

$$\mathbf{y} = \mathbf{A}^{-1}\mathbf{x} = \mathbf{W}\,\mathbf{diag}(\sigma_i^{-1})\mathbf{V}^T \sum_{i=1}^{n} \sigma_i^{-1} \alpha_i \mathbf{v}_i = \sum_{i=1}^{n} \sigma_i^{-2} \alpha_i \mathbf{w}_i. \qquad (\text{B.5})$$

It follows that the ratio

$$\frac{\|\mathbf{y}\|_2}{\|\mathbf{x}\|_2} = \frac{\left\| \sum_{i=1}^{n} \sigma_i^{-2} \alpha_i \mathbf{w}_i \right\|_2}{\left\| \sum_{i=1}^{n} \sigma_i^{-1} \alpha_i \mathbf{v}_i \right\|_2} = \sqrt{\frac{\sum_{i=1}^{n} \sigma_i^{-4} \alpha_i^2}{\sum_{i=1}^{n} \sigma_i^{-2} \alpha_i^2}} \qquad (\text{B.6})$$

is likely to be a good estimate for $\|\mathbf{A}^{-1}\|_2 = \sigma_1^{-1}$ unless $|\alpha_1|$ is small compared with $|\alpha_i|$, $i>1$. Therefore a good choice for $\mathbf{b}$ is one that leads to $\|\mathbf{x}\| / \|\mathbf{b}\|$ being large.

Cline et al. (1979) choose $\mathbf{b}$ to have components ± 1 with signs chosen to make $\|\mathbf{z}\|$ large where $\mathbf{z}$ is the intermediate vector found when solving (B.1) by forward substitution and back-substitution. That is, $\mathbf{z}$ satisfies the equation

$$\mathbf{U}^T \mathbf{z} = \mathbf{b}. \qquad (\text{B.7})$$

At a typical stage two components

$$z_k^+ = (1 - \sum_{i=1}^{k-1} u_{ik} z_i) / u_{kk} \qquad \text{(B.8)}$$

and

$$z_k^- = (-1 - \sum_{i=1}^{k-1} u_{ik} z_i) / u_{kk} \qquad \text{(B.9)}$$

are calculated. Cline *et al.* (1979) originally chose the sign to maximize $|z_k|$, but obtained improved results by maximizing the expression

$$|u_{kk} z_k| + \sum_{j=k+1}^{n} |\sum_{i=1}^{k-1} u_{ij} z_i + u_{kj} z_k|. \qquad \text{(B.10)}$$

Adding the second term in (B.10) provides a measure of look ahead, aiming also for large $|z_i|$, $i > k$. It improves the robustness of the algorithm at the expense of approximately doubling the work in solving (B.7).

Appendix C: Fortran conventions

In all the Fortran examples, we use implicit typing and continuation lines always have * in column 6. CONTINUE statements are always used for DO loop ends and GO TO loop starts and they are not used otherwise. The following is a list of the variable and array names that we use.

ALPHA	Multiple of one vector added to another
I	Vector component or matrix row index
IAPTR (N)	Positions of the first entries of the columns of **A**
IHEAD	Linked list header
IHEAD (N)	Linked list headers
ILPTR (N)	Positions of the first entries of the columns (or rows) of **L**
INC (.)	Array holding interpretative code
IND (NZ)	Indices of components of a vector
INDX (NZX)	Indices of components of vector **x**
INDY (NZY)	Indices of components of vector **y**
INDZ (NZZ)	Indices of components of vector **z**
IPOS (N)	Positions of components after a permutation
IRN (NZ)	Row indices for a matrix
IROWST (N)	Positions of the first entries of the rows of a matrix
ISWAP (N)	Vector of interchanges
ITEMP (N)	Temporary array
IUPTR (N)	Positions of the first entries of the columns of **U**
IW (N)	Work array
J	Vector component or matrix column index
JCN (NZ)	Column indices for a matrix
JCNA (NZ)	Column indices for **A**
JCOLST (N)	Positions of the first entries of the columns of a matrix
JDUMMY	DO loop index that is not used within the loop
JJ	Temporary, usually a column index
K	Row index or loop index over NZ
K1	Lower limit for K
K2	Upper limit for K
KC	Temporary
KDUMMY	DO loop index that is not used within the loop
KK	Loop index over NZ
KX	Loop over NZX
KY	Loop over NZY
L	Temporary
LAST	Previous entry in linked list
LENCOL (N)	Numbers of entries in columns
LENR	Number of entries in a row
LENROW (N)	Numbers of entries in rows
LINK (NZ)	Next entry in linked list
LINKBK (N)	Previous entry in linked list (backward link)
LINKCL (NZ)	Next entry in linked list by columns

LINKFD (N)	Next entry in linked list (forward link)
LINKRW (NZ)	Next entry in linked list by rows
N	Matrix or vector order
NCOL	Number of columns in block pivot
NEXT	Next entry in linked list
NPIV	Number of pivots in block pivot
NZ	Number of entries
NZ1	Number of entries
NZX	Number of entries in x
NZY	Number of entries in y
NZZ	Number of entries in z
PROD	Scalar for accumulating inner product
TEMP	Temporary scalar
VAL (NZ)	Values held in packed form
VALX (NZX)	Values of entries of x held in packed form
VALY (NZY)	Values of entries of y held in packed form
VALZ (NZZ)	Values of entries of z held in packed form
W (N)	Work vector
X (N)	Vector x
XJ	X (J)
XNEXT	Next entry in linked list for x
Y (N)	Vector y

Appendix D: Pictures of sparse matrices

Throughout the book, we have illustrated most of the important points by using matrices from a wide range of application areas. We present the structure of several of these matrices in this appendix. We have intentionally chosen only one or two matrices from each of the major application areas in order to stress the quite different characteristics of sparse systems in each area. We present these patterns in Figures D.1 to D.11. In each case, the application area is given in the caption. Further details on each matrix, including its source, are given in Table 1.6.1 or Table 1.6.2. A discussion of the effect of structure on sparse matrix algorithms is given by Duff (1981b) and Erisman (1981). Additionally, Duff (1981b) shows actual geometric structures which give rise to given matrices. Many of our examples are fairly small by the current limits on sparse solution. This is largely because we feel that the various structures are more clearly displayed using small examples. It is also true that, in some application areas, typical sparse matrices are not very large although many solutions may be required.

In Figures D.12 to D.15, we show the pattern of the L\U form obtained after using different minimum degree codes on a five-diagonal matrix arising from the five-point discretization of the Laplacian operator on a 20×20 grid. We do this to illustrate the effect of tie-breaking on the minimum degree ordering (Section 7.7) and to show the effect of postordering the assembly tree in the multifrontal approach (Section 10.9).

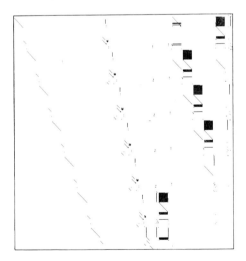

Figure D.1. Linear programming basis. Order of matrix is 363 and number of nonzeros is 2454.

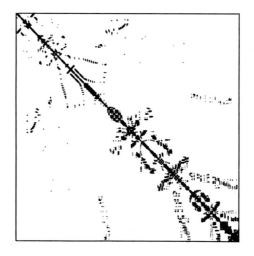

Figure D.2. Symmetric pattern of stiffness matrix from dynamic analysis. Order of matrix is 2003 and number of nonzeros is 83883.

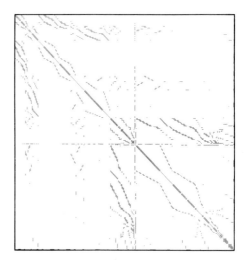

Figure D.3. Symmetric pattern of structures matrix from aerospace. Order of matrix is 256 and number of nonzeros is 2916.

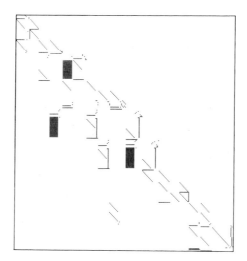

Figure D.4. Unsymmetric pattern from hydrocarbon separation problem in chemical engineering. Order of matrix is 225 and number of nonzeros is 1308.

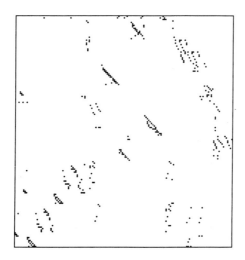

Figure D.5. Unsymmetric pattern from simple chemical plant model. Order of matrix is 156 and number of nonzeros is 371.

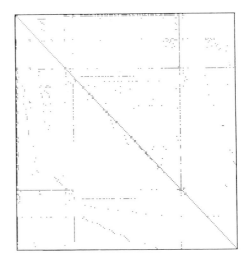

Figure D.6. Unsymmetric pattern from solution of ordinary differential equations in photochemical smog. Order of matrix is 183 and number of nonzeros is 1069.

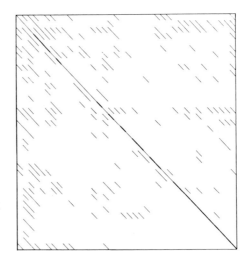

Figure D.7. Unsymmetric pattern from solution of ordinary differential equations in ozone depletion studies. Order of matrix is 760 and number of nonzeros is 5976.

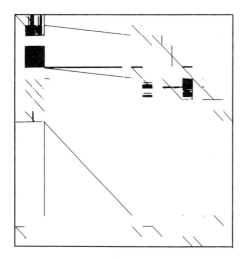

Figure D.8. Unsymmetric pattern from economic modell-ing. Order of matrix is 2529 and number of nonzeros is 90158.

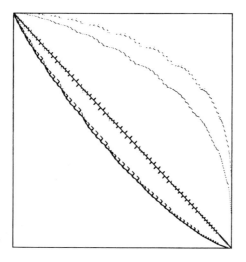

Figure D.9. Unsymmetric pattern from simulation of computing systems. Order of matrix is 1107 and number of nonzeros is 5664.

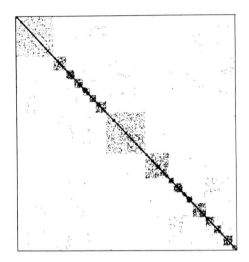

Figure D.10. Symmetric pattern from representation of Western USA power system. Order of matrix is 1624 and number of nonzeros is 6050.

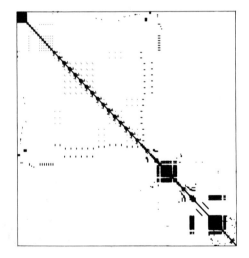

Figure D.11. Symmetric pattern from electric circuit modelling. Order of matrix is 1176 and number of nonzeros is 18552.

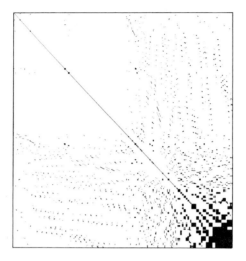

Figure D.12. Pattern of $L\backslash L^T$ factors with minimum degree ordering from SPARSPAK. Five-point discretization of the Laplacian operator on a 20×20 grid.

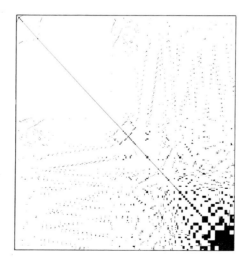

Figure D.13. Pattern of $L\backslash L^T$ factors with minimum degree ordering from YSMP. Five-point discretization of the Laplacian operator on a 20×20 grid.

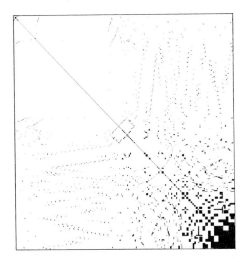

Figure D.14. Pattern of $\mathbf{L}\backslash\mathbf{L}^T$ factors with minimum degree ordering from MA27 (prior to postorder of tree). Five-point discretization of the Laplacian operator on a 20×20 grid.

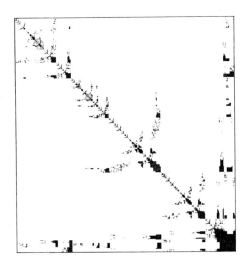

Figure D.15. Pattern of $\mathbf{L}\backslash\mathbf{L}^T$ factors from MA27 after postordering of assembly tree. Five-point discretization of the Laplacian operator on a 20×20 grid.

Solutions to selected exercises

1.1 The performance for a vector of length n is given by

$$r = \frac{n}{(s+n)c}.$$

Clearly as $n \to \infty$, $r \to 1/c$, yielding the required maximum rate. For half this rate the equation

$$\frac{n}{(s+n)c} = \frac{1}{2c}$$

must be true. The solution to this equation is $n=s$.

1.4 No; any choice for the first elimination introduces a new entry.

2.1
```
          NZZ = 0
          KY = 1
          DO 70 KX = 1,NZX
              IF (KY.GT.NZY) GO TO 30
              K = KY
              DO 20 KY = K,NZY
                  NZZ = NZZ + 1
                  IF (INDY(KY)-INDX(KX)) 10,50,40
    10            VALZ(NZZ) = ALPHA*VALY(KY)
                  INDZ(NZZ) = INDY(KY)
    20        CONTINUE
              KY = NZY + 1
    30        NZZ = NZZ + 1
    40        VALZ(NZZ) = VALX(KX)
              GO TO 60
    50        VALZ(NZZ) = VALX(KX) + ALPHA*VALY(KY)
              KY = KY + 1
    60        INDZ(NZZ) = INDX(KX)
    70    CONTINUE
          IF (KY.GT.NZY) GO TO 90
    C Deal with y components beyond last x component
          K = KY
          DO 80 KY = K,NZY
              NZZ = NZZ + 1
              VALZ(NZZ) = ALPHA*VALY(KY)
              INDZ(NZZ) = INDY(KY)
    80    CONTINUE
    90    . . . . . . . .
```

2.2
```
          DO 10 K = 1,NZY
              I = INDY(K)
              IW(K) = ITEMP(I)
              ITEMP(I) = -K
    10    CONTINUE
          DO 20 K = 1,NZX
              I = INDX(K)
              IF (ITEMP(I).GE.0) GO TO 20
              KY = -ITEMP(I)
              VALX(K) = VALX(K) + ALPHA*VALY(KY)
              ITEMP(I) = IW(KY)
    20    CONTINUE
```

```
        DO 30 K = 1,NZY
          I = INDY(K)
          IF(ITEMP(I).GE.0)GO TO 30
          NZX = NZX + 1
          KY = -ITEMP(I)
          VALX(NZX) = ALPHA*VALY(KY)
          INDX(NZX) = I
          ITEMP(I) = IW(KY)
     30 CONTINUE
```

2.6
```
        LAST = LINKBK(I)
        NEXT = LINKFD(I)
        IF(LAST.EQ.0)IHEAD = NEXT
        IF(LAST.NE.0)LINKFD(LAST) = NEXT
        IF(NEXT.NE.0)LINKBK(NEXT) = LAST
```

2.8
```
        DO 10 J = 1,N
          JCOLST(J) = 0
          IROWST(J) = 0
     10 CONTINUE
        DO 20 K = 1,NZ
          J = JCN(K)
          JCN(K) = JCOLST(J)
          JCOLST(J) = K
     20 CONTINUE
        DO 40 J = N,1,-1
          K = JCOLST(J)
          DO 30 KDUMMY = 1,N
            IF(K.EQ.0)GO TO 40
            I = IRN(K)
            LINK(K) = IROWST(I)
            IROWST(I) = K
            NEXT = JCN(K)
            JCN(K) = J
            K = NEXT
     30    CONTINUE
     40 CONTINUE
```

2.10
```
     C Count the entries in each column
        DO 10 J = 1,N
          LENCOL(J) = 0
     10 CONTINUE
        DO 30 I = 1,N
          K1 = IROWST(I)
          K2 = LENROW(I) + K1-1
          DO 20 K = K1,K2
            J = JCN(K)
            LENCOL(J) = LENCOL(J) + 1
     20    CONTINUE
     30 CONTINUE
     C Set JCOLST(J) to point just beyond the last entry of
     C    column J, J = 1,2,...,N
        JCOLST(1) = LENCOL(1) + 1
        DO 40 J = 2,N
          JCOLST(J) = JCOLST(J-1) + LENCOL(J)
     40 CONTINUE
```

```
C Set up column collection by scanning rows
C      in reverse order
      DO 60 I = N,1,-1
         K1 = IROWST(I)
         K2 = LENROW(I) + K1-1
         DO 50 K = K1,K2
            J = JCN(K)
            KC = JCOLST(J)-1
            JCOLST(J) = KC
            IRN(KC) = I
50       CONTINUE
60    CONTINUE
```

2.11
```
      DO 10 J = 1,N
         JCOLST(J) = -J
10    CONTINUE
      DO 30 I = N,1,-1
         K = IROWST(I)
         DO 20 KDUMMY = 1,N
            IF (K.LE.0) GO TO 30
            J = JCN(K)
            JCN(K) = JCOLST(J)
            JCOLST(J) = K
            IF (LINK(K).LE.0)LINK(K) = -I
            K = LINK(K)
20       CONTINUE
30    CONTINUE
```

2.13 The sign of IPOS(I) can be used to flag whether X(I) contains the original or permuted value.
```
      DO 30 I = 1,N
C Skip if X(I) contains the permuted value
      IF(IPOS(I).LT.0) GO TO 20
C Perform the cycle of permutations that starts at I
         J = I
         XJ = X(J)
         DO 10 JDUMMY = 1,N
            NEXT = IPOS(J)
            IPOS(J) = -NEXT
            XNEXT = X(NEXT)
            X(NEXT) = XJ
            IF(NEXT.EQ.I)GO TO 20
            XJ = XNEXT
            J = NEXT
10       CONTINUE
C Restore the sign of IPOS(I)
20       IPOS(I) = -IPOS(I)
30    CONTINUE
```

3.1 Since each $\mathbf{P}^{(k)}$ represents a single interchange, the relations

$$\mathbf{P}^{(k)}\mathbf{P}^{(k)} = \mathbf{I}, \quad k=1,2,\ldots,n-1 \tag{3S.1.1}$$

are true. Therefore if we substitute expression (3X.1.3) for $\bar{\mathbf{L}}^{(k)}$, $k=1,2,\ldots,n-1$, in equation (3X.1.2) and use relation (3S.1.1) repeatedly, we find equation (3X.1.1), as required. To interpret equation (3X.1.2), we merely need to remark that $\bar{\mathbf{L}}^{(k)}$ is equal to $\mathbf{L}^{(k)}$ except for the subdiagonal entries of column k which have been permuted by the accumulated effects of successive premultiplications by $\mathbf{P}^{(k+1)}$, $\mathbf{P}^{(k+2)},\ldots,\mathbf{P}^{(n-1)}$.

3.3 $\mathbf{L} = \begin{pmatrix} 1 & 0 & 0 \\ 4 & -3 & 0 \\ 7 & -6 & 1 \end{pmatrix}$ $\mathbf{U} = \begin{pmatrix} 1 & 2 & 3 \\ 0 & 1 & 2 \\ 0 & 0 & 1 \end{pmatrix}$

3.5 $\begin{bmatrix} 1 & 0 & 0 & 0 \\ -2 & 1 & 0 & 0 \\ -3 & 0 & 1 & 0 \\ -4 & 0 & 0 & 1 \end{bmatrix}$

3.7 $2n^3$.

3.9 The forward substitution phase for column $n-j+1$ requires

$$2(j-1) + 2(j-2) + \ldots + 2 = j(j-1)$$

operations. Therefore the whole forward substitution phase needs

$$\sum_{j=1}^{n} j(j-1) = \tfrac{1}{3}(n+1)n(n-1) = \tfrac{1}{3}n^3 + O(n^2)$$

operations in all. Both phases therefore need

$$\tfrac{1}{3}n^3 + n^3 + O(n^2) = \tfrac{4}{3}n^3 + O(n^2)$$

operations.

3.12 Suppose on the contrary that the inequality $d_{jj} \leq 0$ is true for some j. Let $\mathbf{x}$ be the vector such that $\mathbf{L}^T\mathbf{x} = \mathbf{e}_j$, where $\mathbf{e}_j$ is zero except for component j which is one. In this case the relations

$$\mathbf{x}^T\mathbf{A}\,\mathbf{x} = \mathbf{x}^T\mathbf{L}\mathbf{D}\mathbf{L}^T\mathbf{x} = \mathbf{e}_j^T\mathbf{D}\mathbf{e}_j = d_{jj} \leq 0$$

are true, which contradicts the positive-definiteness of $\mathbf{A}$.

3.14 From (3.12.6) we find the relation

$$\bar{\mathbf{A}}_{22} = \mathbf{A}_{22} - \mathbf{A}_{21}(\mathbf{A}_{11}^{-1}\mathbf{A}_{12})$$

and from (3.12.7) we find

$$\ddot{\mathbf{A}}_{22} = \mathbf{A}_{22} - (\mathbf{A}_{21}\mathbf{A}_{11}^{-1})\mathbf{A}_{12}.$$

$\dot{\mathbf{A}}_{22}$ is also identical (apart from roundoff) since from (3.12.3), we find

$$\dot{\mathbf{A}}_{22} = \mathbf{A}_{22} - (\mathbf{A}_{21}\mathbf{U}_{11}^{-1})(\mathbf{L}_{11}^{-1}\mathbf{A}_{12})$$

$$= \mathbf{A}_{22} - \mathbf{A}_{21}(\mathbf{U}_{11}^{-1}\mathbf{L}_{11}^{-1})\mathbf{A}_{12}$$

$$= \mathbf{A}_{22} - \mathbf{A}_{21}\mathbf{A}_{11}^{-1}\mathbf{A}_{12} \ .$$

3.15 If we write $\mathbf{A} = \bar{\mathbf{L}}\bar{\mathbf{U}}$ and solve $\mathbf{A}\mathbf{x} = \mathbf{b}$ by the solution of

$$\bar{\mathbf{L}}\mathbf{y} = \mathbf{b} \tag{3S.15.1a}$$

and then

$$\bar{\mathbf{U}}\mathbf{x} = \mathbf{y} \tag{3S.15.1b}$$

then with the partitioned form (3.12.2) for $\mathbf{A}, \bar{\mathbf{L}}, \bar{\mathbf{U}}$ we get

$$\mathbf{L}_{11}\mathbf{y}_1 = \mathbf{b}_1$$
$$\mathbf{L}_{22}\mathbf{y}_2 = \mathbf{b}_2 - \mathbf{L}_{21}\mathbf{y}_1$$
$$\mathbf{U}_{22}\mathbf{x}_2 = \mathbf{y}_2$$
$$\mathbf{U}_{11}\mathbf{x}_1 = \mathbf{y}_1 - \mathbf{U}_{12}\mathbf{x}_2$$

which is simply effected through the solution of four triangular systems and the substitution operations shown above. With the partitioned form (3.12.5), equations (3S.15.1) become

$$\mathbf{y}_1 = \mathbf{b}_1$$
$$\mathbf{y}_2 = \mathbf{b}_2 - \ddot{\mathbf{L}}_{21}\mathbf{y}_1$$
$$\ddot{\mathbf{A}}_{22}\mathbf{x}_2 = \mathbf{y}_2$$
$$\mathbf{A}_{11}\mathbf{x}_1 = \mathbf{y}_1 - \mathbf{A}_{12}\mathbf{x}_2$$

which, apart from simple substitution, involves the solution (by any means) of systems with coefficient matrices $\mathbf{A}_{11}$ and $\ddot{\mathbf{A}}_{22}$.

4.1 Choosing the (1,1) pivot involves the multipliers 491 and -149 and produces the 2×2 system

$$688 \ x_2 - 464 \ x_3 = 737$$
$$-212 \ x_2 + 141 \ x_3 = -224$$

and choosing the (2,2) pivot involves the multiplier 0.308 and produces the 1×1 system

$$-2 \ x_3 = 3.0.$$

This gives the computed solution

$$x_1 = 0.0,$$
$$x_2 = 0.0596,$$
$$x_3 = -1.5.$$

The instability was manifest in the large growth which took place when -0.0101 was used as a pivot.

4.2 With partial pivoting in use, we have the relations

$$|a_{ij}^{(k+1)}| = |a_{ij}^{(k)} + l_{ik}a_{kj}^{(k)}|$$
$$\leq |a_{ij}^{(k)}| + |a_{kj}^{(k)}|$$

Hence the inequality

$$\max_{i,j} |a_{ij}^{(k+1)}| \leq 2 \max_{i,j} |a_{ij}^{(k)}|$$

holds, from which we find the inequalities

$$\max_{i,j} |a_{ij}^{(k+1)}| \leq 2^k \max_{i,j} |a_{ij}^{(1)}| = 2^k \max_{i,j} |a_{ij}|.$$

The result is immediate.

4.3 (i) Using the given formulae for **H** and **r** gives the relations

$$(A+H)\bar{x} = A\bar{x} + r\bar{x}^T \frac{\bar{x}}{\|\bar{x}\|_2^2}$$

$$= A\bar{x} + r$$

$$= A\bar{x} + b - A\bar{x}$$

$$= b.$$

(ii) For the second part we find the relations

$$\|H\|_2 \leqslant \frac{\|r\|_2 \|\bar{x}\|_2}{\|\bar{x}\|_2^2} = \frac{\|r\|_2}{\|\bar{x}\|_2} = \alpha \|A\|_2.$$

(iii) If $\|r\|_2$ is small compared with $\|A\|_2 \|\bar{x}\|_2$, then α is small so $\|H\|_2$ is small compared with $\|A\|_2$, that is we have solved a nearby problem. Conversely, assume that the computed solution satisfies the equation

$$(A+H)\bar{x} = b$$

with $\|H\|_2 = \alpha \|A\|_2$, $\alpha \ll 1$. We have

$$\|r\|_2 = \|b - A\bar{x}\|_2 = \|H\bar{x}\|_2 \leqslant \|H\|_2 \|\bar{x}\|_2 = \alpha \|A\|_2 \|\bar{x}\|_2,$$

so $\|r\|_2$ is small compared with $\|A\|_2 \|\bar{x}\|_2$.

(iv) Finally, if $\|r\|_2$ is small compared with $\|A\bar{x}\|_2$, it is certainly small compared with $\|A\|_2 \|\bar{x}\|_2$ since $\|A\bar{x}\|_2 \leqslant \|A\|_2 \|\bar{x}\|_2$. And if $\|r\|_2$ is small compared with $\|b\|_2$, say $\|r\|_2 = \beta \|b\|_2$, then the inequality

$$\beta^{-1}\|r\|_2 = \|b\|_2 = \|r + A\bar{x}\|_2 \leqslant \|r\|_2 + \|A\bar{x}\|_2$$

is true, from which the inequality

$$\|r\|_2 \leqslant \frac{1}{\beta^{-1}-1} \|A\bar{x}\|_2 \leqslant \frac{\beta}{1-\beta} \|A\|_2 \|\bar{x}\|_2$$

follows, provided $\beta < 1$. Again $\|r\|_2$ is small compared with $\|A\|_2 \|\bar{x}\|_2$.

4.4 The factors are

$$\begin{pmatrix} 1 & \\ 81.74 & 1 \end{pmatrix} \begin{pmatrix} 1.134 & 2.183 \\ & -83.22 \end{pmatrix}$$

so ρ has the value 95.18. The Erisman–Reid bound is $82.74 \times 83.22 \simeq 6886$, which is much too pessimistic.

4.5 Since **A** is symmetric, it has an orthonormal set of eigenvectors v_i, $i=1,2,\ldots,n$. Let the corresponding eigenvalues be λ_i, $i=1,2,\ldots,n$. Any vector **x** may be expressed in the form

$$x = \sum_i \alpha_i v_i$$

from which we deduce the relations

$$\mathbf{x}^T \mathbf{A} \mathbf{x} = \left(\sum_i \alpha_i \mathbf{v}_i^T \right) \mathbf{A} \left(\sum_j \alpha_j \mathbf{v}_j \right)$$

$$= \left(\sum_i \alpha_i \mathbf{v}_i^T \right) \left(\sum_j \alpha_j \mathbf{A} \mathbf{v}_j \right)$$

$$= \left(\sum_i \alpha_i \mathbf{v}_i^T \right) \left(\sum_j \alpha_j \lambda_j \mathbf{v}_j \right)$$

$$= \sum_i \alpha_i^2 \lambda_i.$$

If $\lambda_i > 0$, $i=1,2,...,n$ then $\mathbf{x}^T \mathbf{A} \mathbf{x} > 0$ unless $\alpha_i = 0$, $i=1,2,...,n$, that is unless $\mathbf{x} = \mathbf{0}$. Conversely if $\lambda_k < 0$, we may choose $\mathbf{x} = \mathbf{v}_k$ and find $\mathbf{x}^T \mathbf{A} \mathbf{x} = \lambda_k < 0$.

4.6 If $\mathbf{A}$ is diagonally dominant by rows, the following relations are true.

$$|a_{ii}^{(2)}| - \sum_{\substack{j=2 \\ j \neq i}}^n |a_{ij}^{(2)}| = \left| a_{ii} - \frac{a_{i1} a_{1i}}{a_{11}} \right| - \sum_{\substack{j=2 \\ j \neq i}}^n \left| a_{ij} - \frac{a_{i1} a_{1j}}{a_{11}} \right|$$

$$\geq |a_{ii}| - \sum_{\substack{j=2 \\ j \neq i}}^n |a_{ij}| - \sum_{j=2}^n \frac{|a_{i1}| |a_{1j}|}{|a_{11}|}$$

$$\geq |a_{ii}| - \sum_{\substack{j=2 \\ j \neq i}}^n |a_{ij}| - |a_{i1}|$$

$$= |a_{ii}| - \sum_{\substack{j=1 \\ j \neq i}}^n |a_{ij}|$$

$$\geq 0$$

Thus $\mathbf{A}^{(2)}$ is diagonally dominant. Further, the relations

$$\sum_{j=2}^n |a_{ij}^{(2)}| = \sum_{j=2}^n \left| a_{ij} - \frac{a_{i1} a_{1j}}{a_{11}} \right|$$

$$\leq \sum_{j=2}^n |a_{ij}| + \sum_{j=2}^n \frac{|a_{i1}| |a_{1j}|}{|a_{11}|}$$

$$\leq \sum_{j=1}^n |a_{ij}|$$

all hold, which is the next result required. It follows by induction that all $\mathbf{A}^{(k)}$ are diagonally dominant and that the inequalities $\sum_{j=k}^n |a_{ij}^{(k)}| \leq \sum_{j=1}^n |a_{ij}|$ are all true. The inequalities

$$|a_{ij}^{(k)}| \leq \sum_{j=k}^n |a_{ij}^{(k)}| \leq \sum_{j=1}^n |a_{ij}| \leq 2|a_{ii}|$$

all hold, which is the required result.

For diagonal dominance by columns we remark that the numbers $a_{ij}^{(k)}$ $(i \geq k, j \geq k)$ generated when Gaussian elimination is applied to $\mathbf{A}^T$ are exactly the same as the numbers generated when Gaussian elimination is applied to $\mathbf{A}$ (it is just that each number is stored in the transposed position). Therefore the result is also true for matrices that are diagonally dominant by columns.

4.8 There is a nonzero vector **x** such that the equation

$$(A - a_k^{(k)} e_k^T) x = 0.$$

This may be rearranged to the equation

$$x = A^{-1} a_k^{(k)} e_k^T x.$$

Taking norms, we find the inequality

$$\|x\|_p \leq \|A^{-1}\|_p \|a_k^{(k)}\|_p \|x\|_p,$$

from which the result follows.

4.16 For the first problem we find the relations

$$r = \begin{pmatrix} 3.57 \\ 1.47 \end{pmatrix} - \begin{pmatrix} 0.001 & 2.42 \\ 1.00 & 1.58 \end{pmatrix} \begin{pmatrix} -0.863 \\ 1.48 \end{pmatrix} = \begin{pmatrix} -0.0107 \\ -0.0054 \end{pmatrix}$$

and

$$\|r\|_\infty = 0.0107.$$

For the second problem we find the relations

$$r = \begin{pmatrix} 0.00215 \\ 3.59 \end{pmatrix} - \begin{pmatrix} 0.001 & 2.42 \\ 1.00 & 1.58 \end{pmatrix} \begin{pmatrix} 3.59 \\ -0.001 \end{pmatrix} = \begin{pmatrix} 0.00098 \\ 0.00158 \end{pmatrix}$$

and

$$\|r\|_\infty = 0.00158.$$

Since in both cases $\|r\|_\infty$ is small compared with $\|b\|_\infty$, both can be considered to be solutions to nearby problems.

4.17 For the first problem we find the relations

$$r = \begin{pmatrix} 3570 \\ 1.47 \end{pmatrix} - \begin{pmatrix} 1.00 & 2.42 \\ 1.00 & 0.00158 \end{pmatrix} \begin{pmatrix} -10.0 \\ 1480 \end{pmatrix} = \begin{pmatrix} -1.6 \\ 9.13 \end{pmatrix}$$

and

$$\|r\|_\infty = 9.13.$$

For the second problem we find the relations

$$r = \begin{pmatrix} 2.15 \\ 3.59 \end{pmatrix} - \begin{pmatrix} 1.00 & 2.42 \\ 1.00 & 0.00158 \end{pmatrix} \begin{pmatrix} 3.59 \\ -0.595 \end{pmatrix} = \begin{pmatrix} -0.0001 \\ 0.00094 \end{pmatrix}$$

and

$$\|r\|_\infty = 0.00095.$$

Since in both cases $\|r\|_\infty$ is small compared with $\|b\|_\infty$, both can be considered to be solutions to nearby problems.

4.18 The rescaled solutions are

$$\begin{pmatrix} -0.863 \\ 1480 \end{pmatrix} \text{ and } \begin{pmatrix} 3.59 \\ -1.00 \end{pmatrix}.$$

For the first problem we find the relations

$$\mathbf{r} = \begin{pmatrix} 3570 \\ 1.47 \end{pmatrix} - \begin{pmatrix} 1.00 & 2.42 \\ 1.00 & 0.00158 \end{pmatrix} \begin{pmatrix} -0.863 \\ 1480 \end{pmatrix} = \begin{pmatrix} -10.7 \\ -0.0054 \end{pmatrix}$$

and

$$\|\mathbf{r}\|_\infty = 10.7.$$

For the second problem we find the relations

$$\mathbf{r} = \begin{pmatrix} 2.15 \\ 3.59 \end{pmatrix} - \begin{pmatrix} 1.00 & 2.42 \\ 1.00 & 0.00158 \end{pmatrix} \begin{pmatrix} 3.59 \\ -1.00 \end{pmatrix} = \begin{pmatrix} 0.98 \\ 0.00158 \end{pmatrix}$$

and

$$\|\mathbf{r}\|_\infty = 0.98.$$

The first $\|\mathbf{r}\|_\infty$ is small compared with the corresponding $\|\mathbf{b}\|_\infty$, but the second $\|\mathbf{r}\|_\infty$ is of comparable size with $\|\mathbf{A}\|_\infty \|\bar{\mathbf{x}}\|_\infty$ and $\|\mathbf{b}\|_\infty$. Hence the first can be considered to be a solution to a nearby problem while the second cannot.

4.19 The rescaled solutions are

$$\begin{pmatrix} 10.0 \\ 1.48 \end{pmatrix} \text{ and } \begin{pmatrix} 3.59 \\ -0.000595 \end{pmatrix}.$$

For the first problem we find the relations

$$\mathbf{r} = \begin{pmatrix} 3.57 \\ 1.47 \end{pmatrix} - \begin{pmatrix} 0.001 & 2.42 \\ 1.00 & 1.58 \end{pmatrix} \begin{pmatrix} -10.0 \\ 1.48 \end{pmatrix} = \begin{pmatrix} -0.0016 \\ 9.13 \end{pmatrix}$$

and

$$\|\mathbf{r}\|_\infty = 9.13.$$

For the second problem we find the relations

$$\mathbf{r} = \begin{pmatrix} 0.00215 \\ 3.59 \end{pmatrix} - \begin{pmatrix} 0.001 & 2.42 \\ 1.00 & 1.58 \end{pmatrix} \begin{pmatrix} 3.59 \\ -0.000595 \end{pmatrix} = \begin{pmatrix} -0.0000001 \\ 0.00094 \end{pmatrix}$$

and

$$\|\mathbf{r}\|_\infty = 0.00094.$$

The first $\|\mathbf{r}\|_\infty$ is large compared with $\|\mathbf{A}\|_\infty \|\bar{\mathbf{x}}\|_\infty$ and $\|\mathbf{b}\|_\infty$, but the second is small compared with $\|\mathbf{b}\|_\infty$. Hence the first cannot be considered to be a solution to a nearby problem while the second can.

4.20 This is an immediate consequence of the fact that the eigenvalues of $\mathbf{A}^T\mathbf{A}$ are the squares of the singular values of $\mathbf{A}$.

5.1 The form (3.2.5) is suitable if access to $\mathbf{U}$ is by rows. With access by columns, we use the steps, for $k = n, n-1, \ldots, 1$:

$$x_k = c_k / u_{kk}$$

and

$$c_i := c_i - u_{ik} x_k, \quad i = 1, 2, \ldots, k-1.$$

For forward substitution, the form (3.2.6) is suitable if $\mathbf{L}$ is stored by rows. With $\mathbf{L}$ stored by columns, we use the steps, for $k = 1, 2, \ldots, n$:

$$c_k = b_k / l_{kk}$$

and

$$b_i := b_i - l_{ik} c_k, \quad i = k+1, k+2, \ldots, n.$$

5.5 We need the assumption that there is limited fill-in, say that the number of entries in the rows of **U** averages less than c (that is that the number of fill-ins does not exceed the number of entries eliminated). The number of operations at a single step of the elimination then averages less than $2c^2$, and the total number of operations is less than $2c^2 n$.

6.1 Reversing the order of the rows and columns has the desired effect. This corresponds to

$$\mathbf{P} = \mathbf{Q} = \begin{pmatrix} & & & 1 \\ & & 1 & \\ & \cdot^{\displaystyle\cdot^{\displaystyle\cdot}} & & \\ 1 & & & \end{pmatrix}.$$

6.4 Using the notation of Exercise 6.3, the matrix

$$\begin{pmatrix} \mathbf{F} & \mathbf{0} & \mathbf{F} \\ \mathbf{0} & \mathbf{I} & \mathbf{I} \\ \mathbf{0} & \mathbf{F} & \mathbf{0} \end{pmatrix}$$

has columns ordered but requires $O(n^3)$ operations.

6.5 If none of the columns k to n have an entry in rows k to n, the matrix has the block form

$$\begin{pmatrix} \mathbf{A} & \mathbf{B} \\ \mathbf{C} & \mathbf{0} \end{pmatrix}$$

where **A** has order $k-1$. Such a matrix has rank at most $2k-2$, so is singular unless the inequality

$$2k-2 \geq n$$

holds, which is the required result.

6.9

Stack															
							8_1	8	8	8	8	8	8	8
						7	7	7_1	7	7	7	7	7	7
					6	6	6	6	6_1	6	6	6	6	6
				5	5	5	5	5	5	5_1	5	5	5	5
			4	4	4	4	4	4	4	4	4_1	4	4	4
		3	3	3	3	3	3	3	3	3	3	3_1	3	3
	2	2	2	2	2	2	2	2	2	2	2	2	2_1	2
1	1	1	1	1	1	1	1	1	1	1	1	1	1	1

| Step | 1 | 2 | 3 | 4 | 5 | 6 | 7 | 8 | 9 | 10 | 11 | 12 | 13 | 14 | 15 |

The excessive relabelling is completely avoided.

6.10 The matrix $\begin{pmatrix} 0 & 1 \\ 1 & 1 \end{pmatrix}$ cannot be reduced by a symmetric permutation but reordering the columns gives the trivially reducible matrix $\begin{pmatrix} 1 & 0 \\ 1 & 1 \end{pmatrix}$.

7.1 A simple example is the following pattern:

```
× ×
× × ×
  × × ×
```

7.2 Any leaf node has minimum degree (unity). Eliminating the corresponding variable gives no fill-in and the graph of the remaining submatrix is again a tree; it is the original tree less the chosen leaf node and its connecting edge. The root may also have degree unity and in this case it may be eliminated without fill-in and the new graph will again be a tree. Interior nodes must have degree at least two and so are not chosen. Thus minimum degree yields a subtree at every stage and there is no fill-in.

7.4 Since A is structurally irreducible, its digraph contains a path from any node to any other. If such a path passes through node 1, say $i \to 1 \to j$, there are entries in the positions a_{i1} and a_{1j} and so after the first elimination there will be an entry $a_{ij}^{(2)}$ which means that the corresponding new path will have a direct link $i \to j$. Thus the active part of $A^{(2)}$ is irreducible.

Similarly the active path of each successive $A^{(k)}$ is irreducible.

If row k of U has no off-diagonal entries, the active part of $A^{(k)}$ must have its first row zero except for the first entry, that is $A^{(k)}$ must be reducible. This contradicts what we have just proved.

7.7

```
× × × × × × ×
  ×
    ×
      ×
        ×
          ×
            ×
```

8.2 By hypothesis, a_{21} and a_{12} are nonzero. Therefore entries l_{21} and u_{12} of the **LU** factors of A are nonzero. Assume inductively that $l_{i,i-1}$ and $u_{i-1,i}$ are nonzero for $i=2,3,\ldots,k-1$. If a_{kj} is the first nonzero in row k, elimination of this will produce a fill-in in position $(k,j+1)$ if this is zero since $u_{j,j+1}$ is nonzero. Elimination of the resulting entry will fill in position $(k,j+2)$ if it is not already nonzero because $u_{j+1,j+2}$ is nonzero. Continuing, we find that the whole of row k from its first nonzero fills in. A similar argument shows that column k fills in from its first nonzero. Thus our hypothesis is true also for k and the result is proved.

8.4 Consider a node i in level set S_k. It is in S_k and in its position within it because of being a neighbour of a node of S_{k-1}, say node j, and not of any node in an earlier level set or earlier in S_{k-1}. The first nonzero of row i is thus a_{ij}. A node after node i will be in its position because of being a neighbour of node j or a node after node j. Therefore its first nonzero will be in column j or a column after j. Thus the leading nonzeros of the rows are in columns that form a monotonic increasing sequence.

8.5 The last node numbered is always a leaf node, hence will cause no fill-in when eliminated first. Remove this node and edge. The second last numbered node is a leaf node of the resulting tree. We continue similarly to establish the result.

9.1
```
C Row counts are in array LENROW.
C
C Link together rows with the same number of entries.
C IHEAD(K) is the first row with K entries.
C LINK(I) is the row after row I, or zero if I is last.
      DO 5 I = 1,N
         IHEAD(I) = 0
    5 CONTINUE
      DO 10 I = 1,N
         LENR = LENROW(I)
         J = IHEAD(LENR)
         IHEAD(LENR) = I
         LINK(I) = J
   10 CONTINUE
C Loop through the rows in order of increasing row
C count, overwriting LINK(I) by the position of
C row I in the ordered sequence.
      K = 1
      DO 20 LENR = 1,N
         I = IHEAD(LENR)
         DO 15 KDUMMY = 1,N
            IF (I.EQ.0) GO TO 20
            NEXT = LINK(I)
            LINK(I) = K
            K = K + 1
            I = NEXT
   15    CONTINUE
   20 CONTINUE
```

9.4 The matrix with the following pattern

$$
\begin{array}{llllll}
\times & \times & \times & & & \\
\times & \times & \times & & & \\
\times & \times & \times & \times & \times & \times \\
& & \times & \times & & \\
& & \times & \times & \times & \times & \times \\
& & \times & \times & \times & \times \\
\end{array}
$$

has lowest Markowitz count 4, which is attained by the entry $(1,1)$. Entries in row 4 have row count 2 but Markowitz count 5.

9.5 The 5×5 matrix

$$
\begin{array}{lllll}
1 & & & & 9 \\
& 1 & & & 9 \\
& & 1 & & 9 \\
& & & 1 & 9 \\
9 & 9 & 9 & 9 & 1 \\
\end{array}
$$

has diagonal entries which are unacceptable on stability grounds if $u>\frac{1}{9}$, where u is the threshold in inequality (9.2.2). Rows 1 to 4 and columns 1 to 4 will all be searched before one of the off-diagonal entries is accepted as the first pivot.

9.7
```
C Place the entries of row k of A in array W
C and make a linked list of their column indices
      IHEAD = N+1
      DO 10 KK = IAPTR(K+1)-1, IAPTR(K), -1
         J = JCNA(KK)
         W(J) = VAL(KK)
         IW(J) = IHEAD
         IHEAD = J
   10 CONTINUE
C Scan linked list adding multiples of appropriate
C rows of U to row k of A. The list is ordered.
      J = IHEAD
      DO 60 JDUMMY = 1,N
         IF(J.GE.K)GO TO 70
C Compute multiplier
         ALPHA = -W(J)/VAL(IUPTR(J))
         W(J) = ALPHA
C Add in multiple of row j of U
         LAST = J
         DO 50 KK = IUPTR(J)+1, ILPTR(J+1)-1
            JJ = JCN(KK)
            W(JJ) = W(JJ)+ALPHA*VAL(KK)
C If it is a fill-in, add it to linked list
   20       CONTINUE
               IF(JJ-IW(LAST))40,50,30
   30          LAST = IW(LAST)
               GO TO 20
   40          IW(JJ) = IW(LAST)
               IW(LAST) = JJ
   50    CONTINUE
         J = IW(J)
   60 CONTINUE
C Now set row k of L\U into packed form
   70 J = IHEAD
      KK = ILPTR(K)
      DO 80 JDUMMY = 1,N
         IF(J.GT.N)GO TO 90
         IF(J.EQ.K)IUPTR(K) = KK
         JCN(KK) = J
         VAL(KK) = W(J)
         W(J) = 0.0
         J = IW(J)
         KK = KK+1
   80 CONTINUE
   90 ILPTR(K+1)=KK
```

9.10
```
      DO 20 J = 1,N-1
         XJ = -X(JCN(ILPTR(J)))
         IF(XJ.EQ.0.0)GO TO 20
C Run down the column of L (sparse SAXPY)
         DO 10 K = ILPTR(J)+1, ILPTR(J+1)-1
            X(JCN(K)) = X(JCN(K))+XJ*VAL(K)
   10    CONTINUE
   20 CONTINUE
```

9.12 Keeping to the operation codes of Figure 9.7.1, we first copy the right-hand side vector **x** into positions 8 to 10 of array A. The data is then

$$3\ 10\ 5\ 8\ 3\ 10\ 7\ 9\ \mathbf{1}\ 10\ 6\ 1\ 9\ 4\ 3\ 8\ 3\ 10\ 3\ 8\ 2\ 9\ \mathbf{1}\ 8\ \mathbf{1}\ 4$$

9.13 If $\mathbf{x}^*$ is the true solution to (9.9.5), then from (9.9.6a) we have

$$\mathbf{r}^{(k)} = \mathbf{A}\mathbf{e}^{(k)} \qquad\qquad (9S.13.1)$$

where $\mathbf{e}^{(k)}$ is the error vector $\mathbf{x}^* - \mathbf{x}^{(k)}$ at stage k.

From (9.9.6c) we have

$$\mathbf{e}^{(k+1)} = \mathbf{e}^{(k)} - \Delta\mathbf{x}^{(k)}$$

so that, using (9.9.6b) and (9S.13.1) we have

$$\bar{\mathbf{L}}\bar{\mathbf{U}}\mathbf{e}^{(k+1)} = \bar{\mathbf{L}}\bar{\mathbf{U}}\mathbf{e}^{(k)} - \mathbf{A}\mathbf{e}^{(k)}.$$

Hence,

$$\mathbf{e}^{(k+1)} = (\mathbf{I} - (\bar{\mathbf{L}}\bar{\mathbf{U}})^{-1}\mathbf{A})\mathbf{e}^{(k)}.$$

9.14 If $|a_{ij}^{(k)}|$ is added, the perturbation $\Delta\mathbf{A}$ is diagonally dominant and the inequality

$$\mathbf{x}^T\Delta\mathbf{A}\mathbf{x} \geqslant 0$$

holds for any vector $\mathbf{x}$. The smallest eigenvalue of the new matrix is

$$\min_{\|\mathbf{x}\|=1} \mathbf{x}^T(\mathbf{A} + \Delta\mathbf{A})\mathbf{x} \geqslant \min_{\|\mathbf{x}\|=1} \mathbf{x}^T\mathbf{A}\mathbf{x} + \min_{\|\mathbf{x}\|=1} \mathbf{x}^T\Delta\mathbf{A}\mathbf{x} \geqslant \min_{\|\mathbf{x}\|=1} \mathbf{x}^T\mathbf{A}\mathbf{x}.$$

Hence the new matrix is positive definite if the old one is.

10.1 Choose a suitable integer $kp > n$ and solve the problem

$$\begin{pmatrix} \mathbf{A} & \mathbf{0} \\ \mathbf{0} & \mathbf{I} \end{pmatrix} = \begin{pmatrix} \mathbf{b} \\ \mathbf{0} \end{pmatrix},$$

of order kp.

10.3 7, 5, 10/11.

10.5 When two rows have the same structure, we group the corresponding variables into a **supervariable**.

(i) Eliminate variable 1 of degree 4. The fill-ins in rows 3 and 4 make them identical. Generated element 1 has supervariables (4,3), (5).

(ii) Eliminate variable 2 of degree 4. Element 2 has supervariables (4,3), (5) and can absorb element 1.

(iii) Eliminate supervariable (4,3) of degree 3, after assembling element 2. Element 4 has supervariable (5) only.

(iv) Eliminate variable 7 of degree 4. The fill-ins in rows 6 and 9 make rows 6, 8, and 9 identical. We suppose that 9 was found identical to 8 and then (8,9) to 6. Generated element 7 has supervariable (6,8,9) only.

(v) Eliminate supervariable (6,8,9) of degree 4 after assembling element 7. The element generated contains variable 5 only and absorbs element 4.

(vi) Eliminate variable 5 after assembling element 6.

10.6 If column i of $\mathbf{L}$ has entries in rows j and k, with $i<j<k$, then column j will contain the sparsity pattern of column i from row k onwards. Therefore the pattern of column i is not needed to determine the pattern of column k. Thus the pattern of column k is completely determined from the patterns of the columns whose first entry is in row k.

10.10 The arrowhead matrix

$$
A = \begin{pmatrix}
\times & & & & & \times \\
& \times & & & & \times \\
& & \times & & & \times \\
& & & \times & & \times \\
& & & & \times & \times \\
\times & \times & \times & \times & \times & \times
\end{pmatrix}
$$

gives a full $\bar{U}$ but L\U has the same structure as A.

11.1 In (11.2.2) L_{11} is lower bidiagonal, U_{11} is upper bidiagonal and L_{21} and U_{12} are full. In (11.2.3) $A_{11}^{-1}A_{12}$ is full and in (11.2.4) $A_{21}A_{11}^{-1}$ is full. The numbers of operations needed to form the factorizations are $8n-11$, $7n-8$, $7n-8$, respectively. To use them requires $9n-12$, $10n-12$, $10n-12$ operations, respectively.

11.6
$$
V = \begin{pmatrix} A_{12} & 0 \\ 0 & 1 \end{pmatrix} \qquad S = \begin{pmatrix} 0 & 1 \\ 1 & a_{22}-1 \end{pmatrix}
$$

11.8 If the rank 1 matrix is ee^T where e is a column vector of dimension 4 all of whose entries are one, then the torn system

$$
A - ee^T,
$$

where A is the given matrix, is

$$
\begin{pmatrix}
4 & 2 & & \\
2 & 5 & & \\
& & 6 & 2 \\
& & 2 & 7
\end{pmatrix}
$$

which is in much simpler form for solving equations.

11.9 Using the definitions in (11.8.4) in (11.4.2) we obtain

$$
\begin{pmatrix} x_1 \\ x_2 \end{pmatrix} = \begin{pmatrix} A_{11}^{-1}b_1 \\ b_2 \end{pmatrix} - \begin{pmatrix} A_{11}^{-1}A_{12} & 0 \\ 0 & I \end{pmatrix} \begin{pmatrix} I & I \\ A_{21}A_{11}^{-1}A_{12} & A_{22} \end{pmatrix}^{-1} \begin{pmatrix} 0 & I \\ A_{21} & A_{22}-I \end{pmatrix} \begin{pmatrix} A_{11}^{-1}b_1 \\ b_2 \end{pmatrix}
$$

$$
= \begin{pmatrix} A_{11}^{-1}b_1 \\ b_2 \end{pmatrix} - \begin{pmatrix} A_{11}^{-1}A_{12} & 0 \\ 0 & I \end{pmatrix} \begin{pmatrix} C^{-1}A_{21}A_{11}^{-1}b_1 - C^{-1}b_2 \\ b_2 - C^{-1}A_{21}A_{11}^{-1}b_1 + C^{-1}b_2 \end{pmatrix},
$$

where $C = A_{21}A_{11}^{-1}A_{12} - A_{22}$, which is the same as solving the partitioned system (11.8.1).

12.1 Since the last row of the matrix is full, the CPR algorithm will always yield n groups, where n is the order of the matrix, irrespective of the initial column ordering. Thus n additional vector-function evaluations are required and the CPR algorithm is ineffective.

12.4 From (12.7.4b)

$$
z_{ij} = -\sum_{k=j+1}^{n} z_{ik}l_{kj}
$$

and the entries of Z which we require to evaluate z_{ij} are those z_{ik} in row i for which $k \geqslant j+1$ and l_{kj} is nonzero.

But we have assumed that u_{ji} is an entry, so that when l_{kj} is nonzero, so must $(L\backslash U)_{ki}$ and entry (i,k) lies within the sparsity pattern of $(L\backslash U)^T$.

Since this is true for all k such that l_{kj} is nonzero, all of the entries of **Z** $(z_{ik} \mid l_{kj} \neq 0)$ which we need to compute z_{ij} lie in the sparsity pattern of $(\mathbf{L \backslash U})^T$.

12.6

(1) Read all the index lists in order to find the position of the first entry in each row.

(2) Accumulate the lengths of the rows to find where each diagonal entry is stored in the overall matrix and to choose suitable break points for blocks of rows to be assembled at once.

(3) Set pointers $p(b)$ for each block row to zero.

(4) Read the file. For each matrix $\mathbf{A}^{[k]}$:

    (a) determine the blocks $b_1, b_2, \ldots, b_l$ that will need it;

    (b) store with $\mathbf{A}^{[k]}$ the pairs $(b_i, p(b_i))$, $i=1, 2, \ldots, l$; and

    (c) set $p(b_i) = k$, $i=1, 2, \ldots, l$.

(5) For each block b, set $k_1 = p(b)$ and for $i=1, 2, \ldots$ until $k_i = 0$:

    (a) read matrix $\mathbf{A}^{[k_i]}$ and associated list of pairs of integers;

    (b) use $\mathbf{A}^{[k_i]}$ for assembly operations; and

    (c) look at pairs for the one with first entry b, then set k_{i+1} to the second entry.

References

Aho, A. V., Hopcroft, J. E., and Ullman, J. D. (1974). *The design and analysis of computer algorithms.* Addison-Wesley, Massachusetts, Palo Alto, and London.

Anderson, P. M. (ed.) (1977). Exploring applications of parallel processing to power systems applications. Report EL-566-SR, Electric Power Research Institute, California.

Arioli, M., Demmel, J. W., and Duff, I. S. (1988). Solving sparse linear systems with sparse backward error. Report CSS 214, Computer Science and Systems Division, Harwell Laboratory. To appear in *SIAM J. Matrix Anal. and Applics.*

Barker, V. A. (ed.) (1977). Sparse matrix techniques. Lecture notes in mathematics, 572. Springer-Verlag, Berlin, Heidelberg, New York, and Tokyo.

Barlow, J. L. (1986). A note on monitoring the stability of triangular decomposition of sparse matrices. *SIAM J. Sci. Stat. Comput.* 7, 166-168.

Barnes, E. R. (1981). An algorithm for partitioning the nodes of a graph. Report RC 8690 (♯37977), IBM Thomas J. Watson Research Center, Yorktown Heights, New York.

Barwell, V. and George, A. (1976). A comparison of algorithms for solving symmetric indefinite systems of linear equations. *ACM Trans. Math. Softw.* 2, 242-251.

Bauer, F. L. (1963). Optimally scaled matrices. *Numerische Math.* 5, 73-87.

Bending, M. J. and Hutchison, H. P. (1973). The calculation of steady state incompressible flow in large networks of pipes. *Chem. Eng. Sci.* 28, 1857-1864.

Brayton, R. K., Gustavson, F. G., and Willoughby, R. A. (1970). Some results on sparse matrices. *Math. Comp.* 24, 937-954.

Bunch, J. R. (1974). Partial pivoting strategies for symmetric matrices. *SIAM J. Numer. Anal.* 11, 521-528.

Bunch, J. R. and Parlett, B. N. (1971). Direct methods for solving symmetric indefinite systems of linear equations. *SIAM J. Numer. Anal.* 8, 639-655.

Bunch, J. R. and Rose, D. J. (1974). Partitioning, tearing and modification of sparse linear systems. *J. Math. Anal. and Appl.* 48, 574-594.

Bunch, J. R. and Rose, D. J. (eds.) (1976). *Sparse matrix computations.* Academic Press, New York and London.

Bunch, J. R., Kaufman, L., and Parlett, B. N. (1976). Decomposition of a symmetric matrix. *Numerische Math.* 27, 95-110.

Buneman, O. (1969). A compact non-iterative Poisson solver. Report 294, Institute for Plasma Research, Stanford University, California.

Businger, P. A. (1971). Monitoring the numerical stability of Gaussian elimination. *Numerische Math.* **16**, 360-361.

Buzbee, B. L. and Dorr, F. W. (1974). The direct solution of the biharmonic equation on rectangular regions and the Poisson equation on irregular regions. *SIAM J. Numer. Anal.* **11**, 753-763.

Calahan, D. A. (1982). Vectorized direct solvers for 2-D grids. Paper SPE 10522. Proceedings of the 6th symposium on reservoir simulation, New Orleans, Feb. 1-2, 1982.

Chang, A. (1969). Application of sparse matrix methods in electric power system analysis. In Willoughby (1969), 113-121.

Cheung, L. K. and Kuh, E. S. (1974). The bordered triangular matrix and minimum essential sets of a digraph. *IEEE Trans. Circuits and Systems* CAS-21, 633-639.

Chu, E., George, A., Liu, J., and Ng, E. (1984). SPARSPAK : Waterloo sparse matrix package user's guide for SPARSPAK-A. Report CS-84-36, Department of Computer Science, University of Waterloo, Ontario, Canada.

Cline, A. K. and Rew, R. K. (1983). A set of counter examples to three condition number estimators. *SIAM J. Sci. Stat. Comput.* **4**, 602-611.

Cline, A. K., Conn, A. R., and Van Loan, C. F. (1982). Generating the LINPACK condition estimator. In Hennart (1982), 73-83.

Cline, A. K., Moler, C. B., Stewart, G. W., and Wilkinson, J. H. (1979). An estimate for the condition number of a matrix. *SIAM J. Numer. Anal.* **16**, 368-375.

Coleman, T. F. (1984). Large sparse numerical optimization. Springer-Verlag, Berlin, Heidelberg, New York, and Tokyo.

Coleman, T. F. and Moré, J. J. (1983). Estimation of sparse Jacobian matrices and graph coloring problems. *SIAM J. Numer. Anal.* **20**, 187-209.

Coleman, T. F. and Moré, J. J. (1984). Estimation of sparse Hessian matrices and graph coloring problems. *Math. Programming* **28**, 243-270.

Coleman, T. F., Edenbrandt, A., and Gilbert, J. R. (1986). Predicting fill for sparse orthogonal factorization. *J. ACM* **33**, 517-532.

Coleman, T. F., Garbow, B. S., and Moré, J. J. (1984). Software for estimating sparse Jacobian matrices. *ACM Trans. Math. Softw.* **10**, 329-347.

Coleman, T. F., Garbow, B. S., and Moré, J. J. (1985). Software for estimating sparse Hessian matrices. *ACM Trans. Math. Softw.* **11**, 363-378.

Collins, R. J. (1973). Bandwidth reduction by automatic renumbering. *Int. J. Numer. Meth. Engng.* **6**, 345-356.

Cowell, W. R. (1984). *Sources and development of mathematical software.* Prentice-Hall, Englewood Cliffs, New Jersey.

Curtis, A. R. and Reid, J. K. (1971a). The solution of large sparse unsymmetric systems of linear equations. *J. Inst. Maths. Applics.* **8**, 344-353.

Curtis, A. R. and Reid, J. K. (1971b). Fortran subroutines for the solution of sparse sets of linear equations. Report AERE R6844, HMSO, London.

Curtis, A. R. and Reid, J. K. (1972). On the automatic scaling of matrices for Gaussian elimination. *J. Inst. Maths. Applics.* **10**, 118-124.

Curtis, A. R., Powell, M. J. D., and Reid, J. K. (1974). On the estimation of sparse Jacobian matrices. *J. Inst. Maths. Applics.* **13**, 117-120.

Cuthill, E. and McKee, J. (1969). Reducing the bandwidth of sparse symmetric matrices. Proceedings 24th National Conference of the Association for Computing Machinery, Brandon Press, New Jersey, 157-172.

Dantzig, G. B. (1963). *Linear programming and extensions.* Princeton University Press, Princeton, New Jersey.

Dembart, B. and Erisman, A. M. (1973). Hybrid sparse matrix methods. *IEEE Trans. Circuit Theory* CT–**20**, 641-649.

Dembart, B. and Neves, K. W. (1977). Sparse triangular factorization on vector computers. In Anderson (1977), 57-101.

Dennis, J. E. Jr. and Schnabel, R. B. (1983). *Numerical methods for unconstrained optimization and nonlinear equations.* Prentice-Hall, Englewood Cliffs, New Jersey.

Dodds, R. H., Jr and Lopez, L. A. (1980). Substructuring in linear and non-linear analysis. *Int. J. Numer. Meth. Engng.* **15**, 583-597.

Dodson, D. (1981). Preliminary timing study for the CRAYPACK library. Report G4550-CM-39, ETA Division, Boeing Computer Services, Seattle, Washington.

Dongarra, J. J. and Duff, I. S. (1987). Advanced architecture computers. Report AERE R 12415, HMSO, London.

Dongarra, J. J. and Johnsson, S. L. (1987). Solving banded systems on a parallel processor. *Parallel Computing* **5**, 219-246.

Dongarra, J. J., Gustavson, F. G., and Karp, A. (1984). Implementing linear algebra algorithms for dense matrices on a vector pipeline machine. *SIAM Review* **26**, 91-112.

Duff, I. S. (1972). *Analysis of sparse systems.* D.Phil. Thesis. Oxford University, England.

Duff, I. S. (1974). On the number of nonzeros added when Gaussian elimination is performed on sparse random matrices. *Math. Comp.* **28**, 219-230.

Duff, I. S. (1977a). A survey of sparse matrix research. *Proc. IEEE* **65**, 500-535.

Duff, I. S. (1977b). On permutations to block triangular form. *J. Inst. Maths. Applics.* **19**, 339-342.

Duff, I. S. (1977c). MA28 – a set of Fortran subroutines for sparse unsymmetric linear equations. Report AERE R8730, HMSO, London.

Duff, I. S. (1979). Practical comparisons of codes for the solution of sparse linear systems. In Duff and Stewart (1979), 107-134.

Duff, I. S. (ed.) (1981a). *Sparse matrices and their uses*. Academic Press, New York and London.

Duff, I. S. (1981b). A sparse future. In Duff (1981a), 1-29.

Duff, I. S. (1981c). On algorithms for obtaining a maximum transversal. *ACM Trans. Math. Softw.* **7**, 315-330.

Duff, I. S. (1981d). Algorithm 575. Permutations for a zero-free diagonal. *ACM Trans. Math. Softw.* **7**, 387-390.

Duff, I. S. (1981e). MA32 – A package for solving sparse unsymmetric systems using the frontal method. Report AERE R10079, HMSO, London.

Duff, I. S. (1983). Enhancements to the MA32 package for solving sparse unsymmetric equations. Report AERE R11009, HMSO, London.

Duff, I. S. (1984a). The solution of sparse linear systems on the CRAY-1. In Kowalik (1984), 293-309.

Duff, I. S. (1984b). A survey of sparse matrix software. In Cowell (1984), 165-199.

Duff, I. S. (1984c). Design features of a frontal code for solving sparse unsymmetric linear systems out-of-core. *SIAM J. Sci. Stat. Comput.* **5**, 270-280.

Duff, I. S. (1984d). Direct methods for solving sparse systems of linear equations. *SIAM J. Sci. Stat. Comput.* **5**, 605-619.

Duff, I. S. (1984e). Comments on the vectorization of a frontal code. In Engquist and Smedsaas (1984), 343-350.

Duff, I. S. (1984f). The solution of nearly symmetric sparse linear systems. In Glowinski and Lions (1984), 57-64.

Duff, I. S. (1985). Data structures, algorithms and software for sparse matrices. In Evans (1985), 1-29.

Duff, I. S. (1986). Parallel implementation of multifrontal schemes. *Parallel Computing* **3**, 193-204.

Duff, I. S. and Nowak, U. (1987). On sparse solvers in a stiff integrator of extrapolation type. *IMA J. Numer. Anal.* **7**, 391-405.

Duff, I. S. and Reid, J. K. (1974). A comparison of sparsity orderings for obtaining a pivotal sequence in Gaussian elimination. *J. Inst. Maths. Applics.* **14**, 281-291.

Duff, I. S. and Reid, J. K. (1976). A comparison of some methods for the solution of sparse overdetermined systems of linear equations. *J. Inst. Maths. Applics.* **17**, 267-280.

Duff, I. S. and Reid, J. K. (1978a). An implementation of Tarjan's algorithm for the block triangularization of a matrix. *ACM Trans. Math. Softw.* **4**, 137-147.

Duff, I. S. and Reid, J. K. (1978b). Algorithm 529. Permutations to block triangular form. *ACM Trans. Math. Softw.* **4**, 189-192.

Duff, I. S. and Reid, J. K. (1979b). Performance evaluation of codes for sparse matrix problems. In Fosdick (1979), 121-135.

Duff, I. S. and Reid, J. K. (1982). MA27 – A set of Fortran subroutines for solving sparse symmetric sets of linear equations. Report AERE R10533, HMSO, London.

Duff, I. S. and Reid, J. K. (1983). The multifrontal solution of indefinite sparse symmetric linear systems. *ACM Trans. Math. Softw.* **9**, 302-325.

Duff, I. S. and Reid, J. K. (1984). The multifrontal solution of unsymmetric sets of linear systems. *SIAM J. Sci. Stat. Comput.* **5**, 633-641.

Duff, I. S. and Stewart, G. W. (eds.) (1979). *Sparse matrix proceedings 1978.* SIAM, Philadelphia.

Duff, I. S., Erisman, A. M., Gear, C. W., and Reid, J. K. (1988). Sparsity structure and Gaussian elimination. *SIGNUM Newsletter,* Association for Computing Machinery, New York **23**, No. 2, 2-8.

Duff, I. S., Erisman, A. M., and Reid, J. K. (1976). On George's nested dissection method. *SIAM J. Numer. Anal.* **13**, 686-695.

Duff, I. S., Grimes, R. G., and Lewis, J. G. (1987). Sparse matrix test problems. Report CSS 191, Computer Science and Systems Division, Harwell Laboratory. *ACM Trans. Math. Softw.* (To appear).

Duff, I. S., Reid, J. K., Munksgaard, N., and Neilsen, H. B. (1979). Direct solution of sets of linear equations whose matrix is sparse, symmetric and indefinite. *J. Inst. Maths. Applics.* **23**, 235-250.

Eisenstat, S. C., Gursky, M. C., Schultz, M. H., and Sherman, A. H. (1982). Yale sparse matrix package. 1: The symmetric codes. *Int. J. Numer. Meth. Engng.* **18**, 1145-1151.

Eisenstat, S. C., Schultz, M. H., and Sherman, A. H. (1979). Software for sparse Gaussian elimination with limited core storage. In Duff and Stewart (1979), 135-153.

Engquist, B. and Smedsaas, T. (eds.) (1984). *PDE software : modules, interfaces and systems.* North-Holland, Amsterdam, New York, and London.

Eriksson, J. (1976). A note on the decomposition of systems of sparse nonlinear equations. *BIT* **16**, 462-465.

Erisman, A. M. (1972). Sparse matrix approach to the frequency domain analysis of linear passive electrical networks. In Rose and Willoughby (1972), 31-40.

Erisman, A. M. (1973). Decomposition methods using sparse matrix techniques with application to certain electrical network problems. In Himmelblau (1973), 69-80.

Erisman, A. M. (1981). Sparse matrix problems in electric power system analysis. In Duff (1981a), 31-56.

Erisman, A. M. and Reid, J. K. (1974). Monitoring the stability of the triangular factorization of a sparse matrix. *Numerische Math.* **22**, 183-186.

Erisman, A. M. and Spies, G. E. (1972). Exploiting problem characteristics in the sparse matrix approach to frequency domain analysis. *IEEE Trans. Circuit Theory* CT–**19**, 260-269.

Erisman, A. M. and Tinney, W. F. (1975). On computing certain elements of the inverse of a sparse matrix. *Communications ACM* **18**, 177-179.

Erisman, A. M., Grimes, R. G., Lewis, J. G., and Poole, W. G. Jr. (1985). A structurally stable modification of Hellerman-Rarick's P⁴ algorithm for reordering unsymmetric sparse matrices. *SIAM J. Numer. Anal.* **22**, 369-385.

Erisman, A. M., Grimes, R. G., Lewis, J. G., Poole, W. G. Jr., and Simon, H. D. (1987). Evaluation of orderings for unsymmetric sparse matrices. *SIAM J. Sci. Stat. Comput.* **7**, 600-624.

Erisman, A. M., Neves, K. W., and Dwarakanath, M. H. (eds.) (1980). *Electric power problems: the mathematical challenge.* SIAM, Philadelphia.

Evans, D. J. (ed.) (1985). *Sparsity and its applications.* Cambridge University Press, Cambridge.

Everstine, G. C. (1979). A comparison of three resequencing algorithms for the reduction of matrix profile and wavefront. *Int. J. Numer. Meth. Engng.* **14**, 837-853.

Felippa, C. A. (1975). Solution of linear equations with skyline-stored symmetric matrix. *Computers and Structures* **5**, 13-29.

Fletcher, R. (1980). *Practical methods of optimization. Volume 1. Unconstrained optimization.* John Wiley & Sons, Ltd., Chichester, New York, Brisbane, and Toronto.

Fletcher, R. (1981). *Practical methods of optimization. Volume 2. Constrained optimization.* John Wiley & Sons, Ltd., Chichester, New York, Brisbane, and Toronto.

Forsythe, G. and Moler, C. B. (1967). *Computer solution of linear algebraic equations.* Prentice-Hall, Englewood Cliffs, New Jersey.

Fosdick, L. D. (ed.) (1979). *Performance evaluation of numerical software.* North-Holland, Amsterdam, New York, and London.

Fox, L. (1964). *An introduction to numerical linear algebra.* Oxford University Press, London.

Gear, C. W. (1975). Numerical errors in sparse linear equations. Report UIUCDCS-F-75-885, Department of Computer Science, University of Illinois at Urbana-Champaign, Illinois.

George, A. (1971). Computer implementation of the finite-element method. Report STAN CS-71-208, Ph.D Thesis, Department of Computer Science, Stanford University, Stanford, California.

George, A. (1973). Nested dissection of a regular finite-element mesh. *SIAM J. Numer. Anal.* **10**, 345-363.

George, A. (1974). On block elimination for sparse linear systems. *SIAM J. Numer. Anal.* **11**, 585-603.

George, A. (1977a). Numerical experiments using dissection methods to solve $n \times n$ grid problems. *SIAM J. Numer. Anal.* **14**, 161-179.

George, A. (1977b). Solution of linear systems of equations: direct methods for finite-element problems. In Barker (1977), 52-101.

George, A. (1980). An automatic one-way dissection algorithm for irregular finite-element problems. *SIAM J. Numer. Anal.* **17**, 740-751.

George, A. and Heath, M. T. (1980). Solution of sparse linear least squares problems using Givens rotations. *Linear Alg. and its Applics.* **34**, 69-83.

George, A. and Liu, J. W. H. (1975). Some results on fill for sparse matrices. *SIAM J. Numer. Anal.* **12**, 452-455.

George, A. and Liu, J. W. H. (1978a). Algorithms for matrix partitioning and the numerical solution of finite-element systems. *SIAM J. Numer. Anal.* **15**, 297-327.

George, A. and Liu, J. W. H. (1978b). An automatic nested dissection algorithm for irregular finite-element problems. *SIAM J. Numer. Anal.* **15**, 1053-1069.

George, A. and Liu, J. W. H. (1979a). An implementation of a pseudoperipheral node finder. *ACM Trans. Math. Softw.* **5**, 284-295.

George, A. and Liu, J. W. H. (1979b). The design of a user interface for a sparse matrix package. *ACM Trans. Math. Softw.* **5**, 139-162.

George, A. and Liu, J. W. H. (1980). An optimal algorithm for symbolic factorization of symmetric matrices. *SIAM J. Computing* **9**, 583-593.

George, A. and Liu, J. W. H. (1981). *Computer solution of large sparse positive-definite systems.* Prentice-Hall, Englewood Cliffs, New Jersey.

George, A. and Ng, E. (1983). On row and column orderings for sparse least squares problems. *SIAM J. Numer. Anal.* **20**, 326-344.

George, A. and Ng, E. (1984). SPARSPAK : Waterloo sparse matrix package user's guide for SPARSPAK-B. CS-84-37, Department of Computer Science, University of Waterloo, Ontario, Canada.

George, A. and Ng, E. (1985). An implementation of Gaussian elimination with partial pivoting for sparse systems. *SIAM J. Sci. Stat. Comput.* **6**, 390-409.

George, A. and Ng, E. (1987). Symbolic factorization for sparse Gaussian elimination with partial pivoting. *SIAM J. Sci. Stat. Comput.* **8**, 877-898.

George, A., Liu, J. W. H., and Ng, E. (1988). A data structure for sparse **QR** and **LU** factorizations. *SIAM J. Sci. Stat. Comput.* **9**, 100-121.

Gibbs, N. E., Poole, W. G., Jr., and Stockmeyer, P. K. (1976). An algorithm for reducing the bandwidth and profile of a sparse matrix. *SIAM J. Numer. Anal.* **13**, 236-250.

Gill, P. E., Murray, W., and Wright, M. H. (1981). *Practical optimization*. Academic Press, New York and London.

Glowinski, R. and Lions, J.-L. (eds.) (1984). *Computing methods in applied sciences and engineering, VI*. North-Holland, Amsterdam, New York, and London.

Golub, G. H. and Meurant, G. A. (1983). *Résolution numérique des grands systegravemes linéaires*. Eyrolles, Paris.

Golub, G. H. and Van Loan, C. F. (1983). *Matrix computations*. North Oxford Academic, Oxford, and John Hopkins Press, Baltimore.

Grimes, R. G. and Lewis, J. G. (1981). Condition number estimation for sparse matrices. *SIAM J. Sci. Stat. Comput.* **2**, 384-388.

Gustavson, F. G. (1972). Some basic techniques for solving sparse systems of linear equations. In Rose and Willoughby (1972), 41-52.

Gustavson, F. G. (1976). Finding the block lower-triangular form of a sparse matrix. In Bunch and Rose (1976), 275-289.

Gustavson, F. G. (1978). Two fast algorithms for sparse matrices: multiplication and permuted transposition. *ACM Trans. Math. Softw.* **4**, 250-269.

Gustavson, F. G., Liniger, W. M., and Willoughby, R. A. (1970). Symbolic generation of an optimal Crout algorithm for sparse systems of linear equations. *J. ACM* **17**, 87-109.

Hachtel, G. D. (1972). Vector and matrix variability type in sparse matrix algorithms. In Rose and Willoughby (1972), 53-64.

Hachtel, G. D. (1976). The sparse tableau approach to finite-element assembly. In Bunch and Rose (1976), 349-363.

Hachtel, G. D., Brayton, R. K., and Gustavson, F. G. (1971). The sparse tableau approach to network analysis and design. *IEEE Trans. Circuit Theory* CT–**18**, 101-113.

Hageman, L. A. and Young, D. M. (1981). *Applied iterative methods*. Academic Press, New York and London.

Hall, M., Jr (1956). An algorithm for distinct representatives. *Amer. Math. Monthly* **63**, 716-717.

Hamming, R. W. (1971). *Introduction to applied numerical analysis*. McGraw-Hill, New York.

Happ, H. H. (1980). The application of diakoptics to the solutions of power system problems. In Erisman, Neves, and Dwarakanath (1980), 69-103.

Harary, F. (1969). *Graph theory*. Addison-Wesley, Massachusetts, Palo Alto, and London.

Harary, F. (1971). Sparse matrices and graph theory. In Reid (1971a), 139-150.

Hellerman, E. and Rarick, D. C. (1971). Reinversion with the preassigned pivot procedure. *Math. Programming* **1**, 195-216.

Hellerman, E. and Rarick, D. C. (1972). The partitioned preassigned pivot procedure (P^4). In Rose and Willoughby (1972), 67-76.

Hennart, J. P. (ed.) (1982). *Numerical analysis.* Springer-Verlag, Berlin, Heidelberg, New York, and Tokyo.

Himmelblau, D. M. (ed.) (1973). *Decomposition of large-scale problems.* North-Holland, Amsterdam, New York, and London.

Hockney, R. W. and Jesshope, C. R. (1981). *Parallel computers.* Adam Hilger Ltd., Bristol.

Hoffman, A. J., Martin, M. S., and Rose, D. J. (1973). Complexity bounds for regular finite difference and finite element grids. *SIAM J. Numer. Anal.* **10**, 364-369.

Hood, P. (1976). Frontal solution program for unsymmetric matrices. *Int. J. Numer. Meth. Engng.* **10**, 379-400.

Hopcroft, J. E. and Karp, R. M. (1973). An $n^{\frac{5}{2}}$ algorithm for maximum matchings in bipartite graphs. *SIAM J. Computing* **2**, 225-231.

Hsieh, H. Y. and Ghausi, M. S. (1972). A probabilistic approach to optimal pivoting and prediction of fill-in for random sparse matrices. *IEEE Trans. Circuit Theory* CT-**19**, 329-336.

IBM (1976). *IBM system/360 and system/370 IBM 1130 and IBM 1800 subroutine library – mathematics. User's guide.* Program Product 5736-XM7. IBM catalogue ♯SH12-5300-1.

Irons, B. M. (1970). A frontal solution program for finite-element analysis. *Int. J. Numer. Meth. Engng.* **2**, 5-32.

Jennings, A. (1966). A compact storage scheme for the solution of symmetric linear simultaneous equations. *Computer J.* **9**, 351-361.

Jennings, A. (1977). *Matrix computation for engineers and scientists.* John Wiley & Sons, Ltd., Chichester, New York, Brisbane, and Toronto.

Jennings, A. and Malik, G. M. (1977). Partial elimination. *J. Inst. Maths. Applics.* **20**, 307-316.

Jordan, T. L. (1979). A performance evaluation of linear algebra software in parallel architectures. In Fosdick (1979), 59-76.

Karp, R. M. (1986). Combinatorics, complexity, and randomness. *Communications ACM* **29**, 98-109.

Kaufman, L. (1982). Usage of the sparse matrix programs in the PORT library. Report 105, AT&T Bell Laboratories, Murray Hill, New Jersey.

Knuth, D. E. (1969). *Fundamental algorithms, The art of computer programming I.* Addison-Wesley, Massachusetts, Palo Alto, and London.

Knuth, D. E. (1973). *Sorting and searching, The art of computer programming III.* Addison-Wesley, Massachusetts, Palo Alto, and London.

Kowalik, J. S. (ed.) (1984). *High-speed computation.* NATO ASI Series Volume F7, Springer-Verlag, Berlin, Heidelberg, New York, and Tokyo.

König, Ð. (1950). *Theorie der endlichen und unendlichen Graphen.* Chelsea, New York.

Kron, G. (1963). *Diakoptics.* Macdonald, London.

Kron, G. (1968). Special issue on G. Kron's works. *J. Franklin Inst.* **286**.

Kuhn, H. W. (1955). The Hungarian method for solving the assignment problem. *Naval Research Logistics Quarterly* **2**, 83-97.

Kung, H. T. and Leiserson, C. E. (1979). Systolic arrays (for VLSI). In Duff and Stewart (1979), 256-282.

Law, K. H. and Fenves, S. J. (1981). Sparse matrices, graph theory and reanalysis. Proceedings First International Conference on Computing in Civil Engineering, ASCE, 234-249.

Lewis, J. G. (1982). Implementation of the Gibbs-Poole-Stockmeyer and Gibbs-King algorithms. *ACM Trans. Math. Softw.* **8**, 180-189 and 190-194.

Lewis, J. G. (1983). Numerical experiments with SPARSPAK. *SIGNUM Newsletter,* Association for Computing Machinery, New York **18** (3), 12-22.

Lewis, J. G. and Poole, W. G. (1980). Ordering algorithms applied to sparse matrices in electrical power problems. In Erisman, Neves, and Dwarakanath (1980), 115-124.

Lin, T. D. and Mah, R. S. H. (1977). Hierarchical partition — a new optimal pivoting algorithm. *Math. Programming* **12**, 260-278.

Liu, J. W. H. (1986a). On the storage requirement in the out-of-core multifrontal method for sparse factorization. *ACM Trans. Math. Softw.* **12**, 249-264.

Liu, J. W. H. (1986b). A compact row storage scheme for Cholesky factors using elimination trees. *ACM Trans. Math. Softw.* **12**, 127-148.

Liu, J. W. H. (1986c). On general row merging schemes for sparse Givens transformations. *SIAM J. Sci. Stat. Comput.* **7**, 1190-1211.

Liu, J. W. H. and Sherman, A. H. (1976). Comparative analysis of the Cuthill-McKee and the reverse Cuthill-McKee ordering algorithms for sparse matrices. *SIAM J. Numer. Anal.* **13**, 198-213.

McCormick, S. T. (1983). Optimal approximation of sparse Hessians and its equivalence to a graph coloring problem. *Math. Programming* **26**, 153-171.

Manteuffel, T. A. (1980). An incomplete factorization technique for positive-definite linear systems. *Math. Comp.* **34**, 473-498.

Markowitz, H. M. (1957). The elimination form of the inverse and its application to linear programming. *Management Sci.* **3**, 255-269.

Meijerink, J. A. and van der Vorst, H. A. (1977). An iterative solution method for linear systems of which the coefficient matrix is a symmetric M-matrix. *Math. Comp.* **31**, 148-162.

Munksgaard, N. (1980). Solving sparse symmetric sets of linear equations by preconditioned conjugate gradients. *ACM Trans. Math. Softw.* **6**, 206-219.

Munro, I. (1971b). Some results in the study of algorithms. Report 32, Ph.D. thesis, Department of Computer Science, University of Toronto, Ontario, Canada.

Noor, A. K., Kamel, H. A., and Fulton, R. E. (1977). Substructuring techniques – status and projections. *Computers and Structures* **8**, 621-632.

Oettli, W. and Prager, W. (1964). Compatibility of approximate solution of linear equations with given error bounds for coefficients and right-hand sides. *Numerische Math.* **6**, 405-409.

O'Leary, D. P. (1980). Estimating condition numbers. *SIAM J. Sci. Stat. Comput.* **1**, 205-209.

Ogbuobiri, E. C., Tinney, W. F., and Walker, J. W. (1970). Sparsity directed decomposition for Gaussian elimination on matrices. *IEEE Trans. Power* PAS–**89**, 141-150.

Østerby, O. and Zlatev, Z. (1983). *Direct methods for sparse matrices. Lecture notes in computer science* **157**, Springer-Verlag, Berlin, Heidelberg, New York, and Tokyo.

Pagallo, G. and Maulino, C. (1983). A bipartite quotient graph model for unsymmetric matrices. In Pereyra and Reinoza (1983), 227-239.

Pan, V. (1984). How can we speed up matrix multiplication? *SIAM Review* **26**, 393-415.

Parter, S. V. (1961) The use of linear graphs in Gaussian elimination. *SIAM Review* **3**, 119-130.

Pereyra, V. and Reinoza, A. (eds.) (1983). *Numerical methods. Proceedings, Caracas 1982.* Lecture notes in mathematics **1005**. Springer-Verlag, Berlin, Heidelberg, New York, and Tokyo.

Powell, M. J. D. (1970). A new algorithm for unconstrained optimization. In Rosen, Mangasarian, and Ritter (1970), 31-65.

Powell, M. J. D. and Toint, Ph. L. (1979). On the estimation of sparse Hessian matrices. *SIAM J. Numer. Anal.* **16**, 1060-1074.

Read, R. (ed.) (1972). *Graph theory and computing.* Academic Press, New York and London.

Reid, J. K. (ed.) (1971a). *Large sparse sets of linear equations.* Academic Press, New York and London.

Reid, J. K. (1971b). A note on the stability of Gaussian elimination. *J. Inst. Maths. Applics.* **8**, 374-375.

Reid, J. K. (1972). Two Fortran subroutines for direct solution of linear equations whose matrix is sparse, symmetric and positive definite. Report AERE R7119, HMSO, London.

Reid, J. K. (1981). Frontal methods for solving finite-element systems of linear equations. In Duff (1981a), 265-281.

Reid, J. K. (1982). A sparsity exploiting variant of the Bartels-Golub decomposition for linear programming bases. *Math. Programming* **24**, 55-69.

Reid, J. K. (1984). TREESOLVE. A Fortran package for solving large sets of linear finite-element equations. Report CSS 155, Computer Science and Systems Division, Harwell Laboratory.

Rose, D. J. (1972). A graph-theoretic study of the numerical solution of sparse positive-definite systems of linear equations. In Read (1972), 183-217.

Rose, D. J. and Bunch, J. R. (1972). The role of partitioning in the numerical solution of sparse systems. In Rose and Willoughby (1972), 177-187.

Rose, D. J. and Tarjan, R. E. (1978). Algorithmic aspects of vertex elimination on directed graphs. *SIAM J. Appl. Math.* **34**, 176-197.

Rose, D. J. and Willoughby, R. A. (eds.) (1972). *Sparse matrices and their applications.* Plenum Press, New York.

Rosen, J. B., Mangasarian, O. L., and Ritter, K. (eds.) (1970). *Nonlinear Programming.* Academic Press, New York and London.

Sangiovanni-Vincentelli, A. (1976). An optimization problem arising from tearing methods. In Bunch and Rose (1976), 97-110.

Sargent, R. W. H. and Westerberg, A. W. (1964). Speed-up in chemical engineering design. *Trans. Inst. Chem. Engrgs.* **42**, 190-197.

Sato, N. and Tinney, W. F. (1963). Techniques for exploiting the sparsity of the network admittance matrix. *IEEE Trans. Power* PAS–**82**, 944-949.

Saunders, M. A. (1972). Product form of the Cholesky factorization for large-scale linear programming. Report STAN-CS-72-301, Department of Computer Science, Stanford University, Stanford, California.

Schneider, H. (1977). The concepts of irreducibility and full indecomposability of a matrix. *Linear Alg. and its Applics.* **18**, 139-162.

Schrem, E. (1971). Computer implementation of the finite-element procedure. ONR symposium on numerical and computer methods in structural mechanics, Urbana, Illinois.

Sherman, A. H. (1975). On the efficient solution of sparse linear and non-linear equations. Report 46, Ph. D. Thesis. Department of Computer Science, Yale University, Connecticut.

Sherman, A. H. (1978). Algorithm 533. NSPIV, a Fortran subroutine for sparse Gaussian elimination with partial pivoting. *ACM Trans. Math. Softw.* **4**, 391-398.

Sherman, J. and Morrison, W. J. (1949). Adjustment of an inverse matrix corresponding to changes in the elements of a given column or a given row of the original matrix. *Ann. Math. Stat.* **20**, 621.

Skeel, R. D. (1979). Scaling for numerical stability in Gaussian elimination. *J. ACM* **26**, 494-526.

Sloan, S. W. and Randolph, M. F. (1983). Automatic element reordering for finite-element analysis with frontal schemes. *Int. J. Numer. Meth. Engng.* **19**, 1153-1181.

Sorensen, D. C. (1981). An example concerning quasi-Newton estimation of a sparse Hessian. *SIGNUM Newsletter,* Association for Computing Machinery, New York **16** (2), 8-10.

Speelpenning, B. (1978). The generalized element method. Report UIUCDCS-R-78-946, Department of Computer Science, University of Illinois at Urbana-Champaign, Illinois.

Steward, D. V. (1965). Partitioning and tearing systems of equations. *SIAM J. Numer. Anal.* **2**, 345-365.

Steward, D. V. (1969). Tearing analysis of the structure of disorderly sparse matrices. In Willoughby (1969), 65-75.

Stewart, G. W. (1973). *Introduction to matrix computations.* Academic Press, New York and London.

Stewart, G. W. (1974). Modifying pivot elements in Gaussian elimination. *Math. Comp.* **28**, 537-542.

Strang, G. (1980). *Linear algebra and its applications (second edition).* Academic Press, New York and London.

Strassen, V. (1969) Gaussian elimination is not optimal. *Numerische Math.* **13**, 354-356.

Swarztrauber, P. N. (1977). The methods of cyclic reduction, Fourier analysis and the FACR algorithm for the discrete solution of Poisson's equation on a rectangle. *SIAM Review* **19**, 490-501.

Szyld, D. B. (1981). Using sparse matrix techniques to solve a model of the world economy. In Duff (1981a), 357-365.

Takahashi, K., Fagan, J., and Chin, M. (1973). Formation of a sparse bus impedance matrix and its application to short circuit study. Proceedings 8th PICA Conference, Minneapolis, Minnesota.

Tarjan, R. E. (1972). Depth-first search and linear graph algorithms. *SIAM J. Computing* **1**, 146-160.

Tarjan, R. E. (1975). Efficiency of a good but not linear set union algorithm. *J. ACM* **22**, 215-225.

Tarjan, R. E. (1976). Graph theory and Gaussian elimination. In Bunch and Rose (1976), 3-22.

Tewarson, R. P. (1973). *Sparse matrices.* Academic Press, New York and London.

Thapa, M. N. (1983). Optimization of unconstrained functions with sparse Hessian matrices — quasi-Newton methods. *Math. Programming* **25**, 158-182.

Tinney, W. F. and Walker, J. W. (1967). Direct solutions of sparse network equations by optimally ordered triangular factorization. *Proc. IEEE* **55**, 1801-1809.

Tinney, W. F., Powell, W. L., and Peterson, N. M. (1973). *Sparsity-oriented network reduction*. Proceedings 8th PICA Conference, Minneapolis, Minnesota.

Toint, Ph. L. (1977). On sparse and symmetric matrix updating subject to a linear equation. *Math. Comp.* **31**, 954-961.

Toint, Ph. L. (1981). A note about sparsity exploiting quasi-Newton updates. *Math. Programming* **21**, 172-181.

Tomlin, J. A. (1972). Pivoting for size and sparsity in linear programming inversion routines. *J. Inst. Maths. Applics.* **10**, 289-295.

Tosovic, L. B. (1973). Some experiments on sparse sets of linear equations. *SIAM J. Appl. Math.* **25**, 142-148.

Varga, R. S. (1962). *Matrix iterative analysis*. Prentice-Hall, Englewood Cliffs, New Jersey.

Wang, H. H. (1981). A parallel method for tridiagonal equations. *ACM Trans. Math. Softw.* **7**, 170-183.

Westerberg, A. W. and Berna, T. J. (1979). LASCALA – A Language for large scale linear algebra. In Duff and Stewart (1979), 90-106.

Wilkinson, J. H. (1961). Error analysis of direct methods of matrix inversion. *J. ACM* **8**, 281-330.

Wilkinson, J. H. (1965). *The algebraic eigenvalue problem*. Oxford University Press, London.

Willoughby, R. A. (ed.) (1969). Sparse matrix proceedings. Report RA1(#11707), IBM Thomas J. Watson Research Center, Yorktown Heights, New York.

Willoughby, R. A. (1971). Sparse matrix algorithms and their relation to problem classes and computer architecture. In Reid (1971a), 255-277.

Woodbury, M. (1950). Inverting modified matrices. Memorandum 42, Statistics Research Group, Princeton, New Jersey.

Yannakakis, M. (1981). Computing the minimum fill-in is NP-complete. *SIAM J. Alg. Disc. Meth.* **2**, 77-79.

Zlatev, Z. (1980). On some pivotal strategies in Gaussian elimination by sparse technique. *SIAM J. Numer. Anal.* **17**, 18-30.

Zlatev, Z., Barker, V. A., and Thomsen, P. G. (1978). SLEST: A Fortran IV subroutine for solving sparse systems of linear equations. User's guide. Report NI-78-01, Numerisk Institut, Lyngby, Denmark.

Zlatev, Z., Wasniewski, J., and Schaumburg, K. (1981). *Y12M. Solution of large and sparse systems of linear algebraic equations*. Lecture notes in computer science **121**, Springer-Verlag, Berlin, Heidelberg, New York, and Tokyo.

Author index

Subject index